LONDON MATHEMATICAL SOCIETY LECTURE NOTE SERIES

Managing Editor: Professor J.W.S. Cassels, Department of Pure Mathematics and Mathematical Statistics, University of Cambridge, 16 Mill Lane, Cambridge CB2 1SB, England

The titles below are available from booksellers, or, in case of difficulty, from Cambridge University Press.

London Mathematical Society Lecture Note Series. 243

Geometric Galois Actions

2. The Inverse Galois Problem, Moduli Spaces and Mapping Class Groups

Edited by

Leila Schneps
CNRS

and

Pierre Lochak
CNRS

CAMBRIDGE
UNIVERSITY PRESS

CAMBRIDGE UNIVERSITY PRESS
Cambridge, New York, Melbourne, Madrid, Cape Town, Singapore, São Paulo, Delhi

Cambridge University Press
The Edinburgh Building, Cambridge CB2 8RU, UK

Published in the United States of America by Cambridge University Press, New York

www.cambridge.org
Information on this title: www.cambridge.org/9780521596411

First published 1997

A catalogue record for this publication is available from the British Library

ISBN 978-0-521-59641-1 paperback

Transferred to digital printing 2009

Table of Contents

Table of Contents

Introduction vii

Part I: ...

Part I. ...

Introduction

This volume grew out of the conference which was held at Luminy in August 1995 on the theme "Geometry and Arithmetic of Moduli Spaces". In some sense, it was conceived as a sequel to the 1993 Luminy conference on "The Grothendieck Theory of Dessins d'Enfants", which gave rise to proceedings bearing the same title (this series, number 200). The second conference revolved mostly around some "multidimensional" versions of the themes considered in the first one. All these themes are developments of ideas expressed in Grothendieck's *Esquisse d'un programme*. This seminal text has now been published in the companion volume to this one (this series, number 242).

In the second section of the *Esquisse*, Grothendieck sketches out what he terms "Galois-Teichmüller theory". This theory is an approach to the study of the absolute Galois group $\mathrm{Gal}(\overline{\mathbb{Q}}/\mathbb{Q})$, via the action of this group on the –profinite– Teichmüller modular groups (alias mapping class groups), which are the fundamental groups of the moduli spaces of Riemann surfaces with marked points. By Grothendieck's theory of the fundamental group, if X is a scheme defined over $\mathbb{Q}$, then $\mathrm{Gal}(\overline{\mathbb{Q}}/\mathbb{Q})$ acts on the algebraic fundamental group $\hat{\pi}_1(X \otimes \overline{\mathbb{Q}})$. In the *Esquisse* Grothendieck singles out the case of the moduli spaces as being of particular interest. Note that the "fine" moduli spaces are not schemes but stacks, however a similar theory applies (one can consult the article by T. Oda in the companion volume to this one).

In the third section of the *Esquisse*, Grothendieck points out that "even" the simplest case is already extremely interesting. This case concerns the smallest non-trivial moduli space, that of spheres with four ordered marked points, which is isomorphic to the sphere with three points removed. The study of the Galois action on the fundamental group of this space (the profinite completion of the free group on two generators) is equivalent to the study of the Galois action on finite covers of the projective line ramified over at most three points. By Belyi's famous theorem, every algebraic curve defined over a number field can be realized as such a cover. The study of this action is known as the theory of "dessins d'enfants", and formed the subject of the 1993 conference. The 1995 conference dealt with the generalization of this theme to all moduli spaces, as discussed in the second section of the *Esquisse*.

During the conference, we learned that Jean Malgoire had obtained from Grothendieck permission to publish some of his mathematical manuscripts, among which was the *Esquisse*. We are very grateful to him for suggesting to viii Introduction

us that it appear as part of a volume of proceedings. It took some thought to decide on the best way of including the *Esquisse* in such a volume; moreover we were agreeably surprised by the quantity of papers submitted both by participants at the conference and also by certain non-participants who were unable to come but would have liked to. The result is the two-volume series *Geometric Galois Actions*, of which the first volume contains the *Esquisse* itself, a letter from Grothendieck to Faltings on the subject of anabelian geometry, and various contributions all of which play a role in clarifying some of the specific themes introduced by Grothendieck – or some of the themes raised by these clarifications! – whereas the second, present, volume contains the original research papers submitted at the conference.

Let us briefly survey the contents of this volume. To begin with we include the abstracts of the talks which were actually given at Luminy; they were divided into five short courses of two lectures each, a series of individual talks, and an evening seminar given by graduate students, aimed at understanding the basics of Teichmüller and moduli spaces. The rest of the volume is divided into four separate but related sections. The first part, *Dessins d'enfants*, contains five articles forming a bridge with the previous conference. The first one, by N. Adrianov and G. Shabat, considers a special subclass of genus zero dessins, with the aim of determining finer Galois invariants than the well-known valency lists. The paper by G. Jones and M. Streit contains some introductory sections to dessins which are basically self-contained, and also gives a flavour of what has been happening recently in the field. The next paper by T. Hsu illustrates how group theoretic and combinatorial techniques can be used to bear on arithmetic problems, thanks to the very visual nature of dessins. Lastly, the paper by L. Zapponi concentrates on the genus one case, exploiting some of the specific techniques pertaining to the theory of elliptic curves.

The projective line with three points removed can be viewed, as Grothendieck saw it, as the simplest moduli space, but of course it can also be generalised to the projective line minus n points for arbitrary n, $n \geq 3$. Then one can go on to study families of unramified coverings of these objects, and this leads to the theory of Hurwitz spaces, which plays an important role in the investigation of the *Inverse Galois problem*. A recent aspect of this story can be found in the article by P. Dèbes and B. Deschamps, which surveys the state of affairs of the regular inverse Galois problem over so-called "large" fields, a notion introduced by Florian Pop. The paper by K. Strambach and H. Völklein deals with the regular realization of finite groups as Galois groups via the existence of rational points on Hurwitz spaces; it introduces the new idea of considering certain subvarieties of these spaces, rather than just its irreducible components. The third paper in this section, by M. Fried and Y. Kopeliovich, uses Hurwitz spaces to study the ramification properties of realizations as Galois groups of the characteristic finite quotients of the universal p-Frattini cover of a finite group.

The third part, *Galois actions and mapping class groups*, contains two papers which exemplify the kind of techniques that are currently being used in the study of the Galois action on fundamental groups. In his contribution, M. Matsumoto determines the Galois action on the mapping class groups of genus 3, via a remarkable connection between the genus 3 moduli space and the deformation space of E_7-singularity. The paper by Z. Wojtkowiak studies the monodromy of iterated integrals, making ample use of Hodge theory for fundamental groups, which makes it possible to use Lie algebras and differential techniques.

In the fourth section, *Universal Teichmüller Theory*, we have gathered three closely related papers. R. Penner first gives a gentle survey (in his own words) of a recent and promising theory of his which grew out from his previous studies of the "classical" decorated Teichmüller and moduli spaces. His paper serves as an introduction to this theory, but it also contains new features. It investigates the applications and completions of his "universal Ptolemy group" G, a group which was already known to group theorists as Richard Thompson's group, and which is in fact isomorphic to the group $\text{PPSL}_2(\mathbb{Z})$ of piecewise $\text{PSL}_2(\mathbb{Z})$ transformations. The identification of G with the Thompson group is not apparent at first sight; first noted by M. Kontsevitch, it was proved by M. Imbert and forms the subject of his contribution. In the last paper, by P. Lochak and L. Schneps, it is shown that by suitably enlarging and completing the Ptolemy group into a profinite Ptolemy-Teichmüller groupoid, one can let the Galois group act on this new object. This action actually occurs naturally as the restriction to $\text{Gal}(\overline{\mathbb{Q}}/\mathbb{Q})$ of an action of the Grothendieck-Teichmüller group (cf. the survey on this group in the companion volume to this one).

Abstracts of the talks

Short Courses

Fields of Definition of Covers; Embedding Problems over Large Fields. *Pierre Dèbes.*

I. The first talk deals with joint work with Jean-Claude Douai. Let $f : X \to B$ be a finite cover defined over the separable closure K_s of a field K, with B an algebraic variety defined over K. Assume that f is *isomorphic* to each of its conjugates under $G(K_s/K)$. The field K is called the *field of moduli* of the cover. Does it follow that the given cover can be defined over K? The answer is "No" in general: there is an obstruction to the field of moduli being a field of definition. Still, how can the obstruction be measured? We present a general approach for this problem. The obstruction is entirely of a cohomological nature. This was known only in the case of G-covers, *i.e.*, Galois covers given together with their automorphisms. This special case happens to be the simplest one. In the situation of *mere* covers, the problem is shown to be controlled not by one, as for G-covers, but by several characteristic classes in $H^2(K, Z(G), L)$ (for a certain action L of $G(K_s/K)$ on the center $Z(G)$ the group of the cover). Furthermore our approach reveals a more hidden obstruction coming on top of the main one, called the first obstruction and which does not exist for G-covers.

Our Main Theorem yields quite concrete criteria for the field of moduli to be a field of definition. Such criteria were not available in the general situation of mere covers. Furthermore the base space B can be here an algebraic variety of any dimension and the ground field K a field of any characteristic. All classical results, for which the base space was the projective line $\mathbf{P}^1$ over $\overline{\mathbb{Q}}$, are contained as special cases.

Our Main Theorem also leads to some local-global type results. For example we prove this *local-to-global* principle: a G-cover $f : X \to B$ is defined over $\mathbb{Q}$ if and only if it is defined over $\mathbb{Q}_p$ for all primes p. This was conjectured by E. Dew and proved by the author in the special case of G-covers of $\mathbf{P}^1$. We will develop this local-to-global result and other related questions. We will prove in particular this *global-to-local* principle for covers (or G-covers): if a cover (or a G-cover) $f : X \to B$ has a number field K as field of moduli, then it may not be defined over K, but it is necessarily defined over all but finitely many completions K_v of K.

II. The Inverse Galois Problem — is each finite group a quotient of $G(\mathbb{Q})$? — and Safarevic's conjecture — $G(\mathbb{Q}^{ab})$ is a free profinite group — are two main questions generally asked about the absolute Galois group $G(\mathbb{Q})$ of $\mathbb{Q}$.

A fruitful approach consists in studying the action of $G(\mathbb{Q})$ on covers of $\mathbf{P}^1$ defined over $\overline{\mathbb{Q}}$, or equivalently, on extensions of $\overline{\mathbb{Q}}(T)$.

The goal of this talk is twofold. First we present a "Main Conjecture" that contains all basic conjectures of the area, including the Inverse Galois Problem and Safarevic's conjecture but also the Fried-Völklein conjecture — $G(K)$ projective $+$ K hilbertian $\Rightarrow G(K)$ pro-free — . A special case of the Main Conjecture is that, if K is any given field, then all split embedding problems for $G(K(T))$ have a strong solution. The general form of the Main Conjecture is that the same is true for split embedding problems with "some extra constraint on $\overline{K}$".

In the second part, we present a "Main Theorem" that unifies a whole series of results that have recently appeared about the absolute Galois group $G(K(T))$ when K is a algebraically closed, or Pseudo Algebraically Closed, or is a local field, or is Pseudo S Closed (*e.g.* the field of totally real (or p-adic) algebraic numbers). The Main Theorem is that the Main Conjecture is true if the field K is *large*. By definition, a field K is large if each smooth geometrically irreducible curve defined over K has infinitely many K-rational points provided that there is at least one. The examples above (K algebraically closed, etc.) are examples of large fields.

The main contributions to the Main Theorem are due to D. Harbater who introduced some very efficient "patching and glueing techniques" for covers of $\mathbf{P}^1$, F. Pop for the arithmetic ingredients of the proof, in particular, his work on the property "K large" and M. Fried, for the idea of working on *families* of covers. Q. Liu, H. Völklein, D. Haran and the author were other contributors to this result.

Coordinates on Teichmüller and moduli space. *Adrien Douady.*

I. In the first lecture, pants decompositions (also known as maximal muticurves) of a surface are described, together with some elementary topological and enumerative properties. We then show how, picking a fixed decomposition, one can define the corresponding Fenchel-Nielsen coordinates which provide a global *real* analytic coordinatization of Teichmüller space. These coordinates can also be used in the study of the moduli spaces and their compactifications.

II. The second lecture is devoted to an introduction to the *complex* analytic theory. Starting from the definition and existence statement of Strebel quadratic differentials, we show how to introduce deformation parameters which afford a local complex analytic coordinatization of Teimüller space. More specifically, to a quadratic differential are associated horizontal and vertical foliations; picking one of them (this is a matter of convention), one associates to a Strebel differential a decomposition of the underlying Riemann surface into cylinders. The deformation parameters which are used in the coordinatization may be intuitively described as the mutual sliding and twisting of these cylinders, or more visually "stove pipes". Once these holomorphic coordinates on Teichmüller space are obtained, they descend to the moduli space, thus providing one of the several ways to endow this last space with a complex structure.

Topological field theory and connections with CFT, and the topology of configuration spaces. *Ruth Lawrence.*

I. This short course aims to show some of the structures naturally arising from topological, geometric and combinatorial approaches to topological quantum field theory as well as relations between them. The first lecture concentrates on the definition of a topological field theory and shows how simple topology relating to decompositions of manifolds into 'elementary pieces' forces the existence of a tight algebraic structure for TQFTs in two and three dimensions.

II. In the second lecture, we concentrate on the connections with conformal field theory and the geometry of configuration spaces of points. In particular, we see that the combinatorics involved in the representation theory of quantum groups forms a bridge between the two approaches.

On universal monodromy representations of Galois-Teichmüller modular groups. *Hiroaki Nakamura.*

I. In this course, I explain the basic setting and recent advanced results in the study of pro-ℓ exterior Galois representations arising from algebraic curves. When the algebraic curve varies, the Galois representation is deformed, but it turns out that there is an invariant portion common to all algebraic curves. These towers are defined by the universal monodromy representations of Galois-Teichmüller modular groups together with the weight filtrations in the so-called pro-ℓ mapping class groups. The problem of showing stability properties of these fields of definition with respect to genera and marking points was posed by Takayuki Oda, and recently fairly established by cooperation of Y. Ihara, M. Matsumoto, N. Takao, R. Ueno and the author.

II. In the second part, we explain graphs of profinite groups associated with combinatorial data on maximally degenerate curves.

A survey of Grothendieck-Teichmüller theory. *Leila Schneps.*

I. The first part of this short course surveys recent work on the following topics: new properties of the Grothendieck-Teichmüller group GT analogous to known properties of the absolute Galois group known to be contained in it and conjecturally equal to it, such as torsion, behavior of complex conjugation, action on combinatorial structures such as dessins and braid groups, and others, particularly the Galois-compatible action of GT on various fundamental groupoids of geometric objects, particularly moduli spaces of Riemann surfaces with marked points.

II. In the second part, we consider universal Teichmüller theory. We introduce the isomorphism (proved by M. Imbert) between Richard u Thompson's group, the group $\mathrm{PPSL}_2(\mathbb{Z})$ of piecewise $\mathrm{PSL}_2(\mathbb{Z})$-transformations, and Penner's universal Ptolemy group G. Penner's presentation of this group emphasizes a remarkable pentagon equation which indicates a connection with the Grothendieck-Teichmüller group. This connection is made explicit via the result that $\widehat{GT}$ acts on a suitable *groupoid-profinite* completion of the universal Ptolemy group.

Individual talks

Characterizing curves by their dessins. *Jean-Marc Couveignes.* We recall that a Belyi function ϕ is a function from some curve $\mathcal{C}$ defined over $\overline{\mathbb{Q}}$ unramified outside $\{0, 1, \infty\}$. Then $(\mathcal{C}, \phi)$ is called a Belyi pair. A morphism between two such pairs is a map I from $\mathcal{C}_1$ to $\mathcal{C}_2$ such that $\phi_2 I = \phi_1$. An isomorphism class of Belyi pairs is called a dessin following Grothendieck. We first give a sketch of proof for some slight improvement of Belyi's theorem stating the existence of a Belyi function with no automorphisms on any curve defined over a number field $\mathbf{K}$. To each curve $\mathcal{C}$ defined over $\mathbf{K}$ we associate a dessin $(\mathcal{C}, \phi)$ with no automorphisms (or the set of all of them). This is enough to characterize this curve up to isomorphisms defined over $\mathbf{K}$. We then look for some explicit examples in order to test which kind of arithmetic information on $\mathcal{C}$ can be read on the topological structure of its characteristic dessins. We insist that a naive use of Belyi's theorem is not likely to provide any such non trivial example. Instead, we obtain families of dessins by an indirect way, using coverings ramified over four points and the corresponding moduli spaces (called Hurwitz spaces after Fried). For example, we present a family of dessins of genus zero with no automorphims, indexed by four parameters m, n, p, q, and associated to the curves $\mathcal{C}_{m,n,p,q}$ in $\mathbf{P}_3$ given by the equations $ma + nb + pc + qd = 0$ and $ma^2 + nb^2 + pc^2 + qd^2 = 0$. We give both the algebraic and topological description of these dessins and prove that any conic can be set in the form of some $\mathcal{C}_{m,n,p,q}$ for suitables values of the parameters. We finish by a brief study of the case $m = 1$, $n = 2$, $p = 3$, $q = 7$ where we prove that the conic has bad reduction at 7 while 7 is prime to the degree of and any ramification in our associated dessin.

Presentation of the Mapping Class Group. *Sylvain Gervais.* The Mapping Class Group $M(g, n)$ of a genus g surface S with n boundary components is the group of diffeomorphisms of S which leave fixed its boundary, modulo whose which are isotopic to the identity. $M(g, n)$ is generated by Dehn twists for all g and n; we will give a presentation of $M(g, n)$ considering all twists as generators. When g is greater than or equal to 2, $M(g, n)$ is generated by twists along non-separating curves. A presentation is also given with these generators. In both cases, all the relations live in surfaces of genus 0 with 4 boundary components or of genus 1 with 1 or 2 boundary components.

Arithmetic aspects of Teichmüller theory. *Bill Harvey.* We focus attention on two ways in which the theory of uniformisation for Riemann surfaces X which are hyperbolic (covered by the real hyperbolic disc D^2) has points of contact with fields of definition for X as an algebraic curve.

1) Following Belyi's theorem, we know that X can be defined over $\overline{\mathbb{Q}}$ if and only if a Fuchsian uniformising group for X is contained in a triangle group. For many (perhaps all) such surfaces, there exist holomorphic quadratic differential forms ω on X such that the associated Teichmüller deformation disc $\mathbf{D}([X], \omega) \subseteq T(X)$, (here $T(X)$ denotes the Teichmüller space of the closed surface X), has the property that the subgroup of the mapping class group $\mathrm{Mod}(X)$ which preserves $\mathbf{D}$ is a Fuchsian group (of the first kind) intermediate between a triangle group of type $\{p, q, \infty\}$ and a subgroup representing $\pi_1(X^*)$, the surface X with the zero set of ω removed.

2) The classical Schottky uniformisation of a (complex) Riemann surface has an analogue for non-Archimedean fields due to Mumford. This motivates the search for a choice of Schottky uniformisation of curves definable over a given number field. If we choose a class of stable degeneration for a genus g curve (via its dual graph), the work of Bers and Maskit provides complex coordinates for a natural uniformisation of such curves. I have shown

Theorem. A modification of Maskit's coordinates determines Schottky coordinates which define Mumford curves at a specific set of places of the field generated by the matrix entries of the (classical) Schottky group.

A complex interpretation of Fricke coordinates. *John Hubbard.* No abstract submitted.

On π_1 of curves over arithmetic ground fields. *Yasutaka Ihara.* Let X be a smooth irreducible curve over a field k (assumed to be finite over a prime field). The first topic is a simple but useful observation of A. Tamagawa. When k is finite and $g(X) > 0$, he characterized group-theoretically which section $s : \mathrm{Gal}(\overline{k}/k) \to \pi_1(X)$ of the projection $\pi_1(X) \to \mathrm{Gal}(\overline{k}/k) \simeq \hat{\mathbb{Z}}$ comes from a k-rational point $P \in X(k)$. The second subject is the quotient of $\pi_1(X)$ by the normal subgroup generated by all conjugates of $s(\mathrm{Gal}(\overline{k}/k))$, where s is a section corresponding to some P. This quotient, denoted by $\pi_1(X)_{(P)}$, is the Galois group of the tower of finite etale covers of X in which P splits completely. We first discuss what can be said in general about this group $\pi_1(X)_{(P)}$ (various "derived" quotient groups must be "small", etc.) using "abelian" mathematics, and then restrict to Shimura curves to discuss deeper phenomena.

A cohomological interpretation of the Grothendieck-Teichmüller group. *Pierre Lochak.* The result given in this talk, following from joint work with L. Schneps, is the following. The three defining relations of the Grothendieck-Teichmüller group are cocycle relations for certain noncommutative cohomology sets of cyclic groups with values in braid groups. These cohomology sets are very simple, and one can actually compute explicit coboundaries representing the elements of GT. The methods for calculating the cohomology sets are due to Serre, Brown and Scheiderer; the same methods also gives the result that complex conjugation is self-normalizing in GT.

Etale covers of a generic curve in characteristic $p > 0$ and ordinarity. *Michel Matignon.* Contrary to the affine case (Abhyankar's conjecture) there is no conjecture concerning finite quotients of the fundamental group of a smooth projective curve. The fundamental group codes the Hasse-Witt invariants of etale covers; this gives an infinite set of invariants whose study was begun by H. Katsurada and S. Nakajima. In this talk we concentrate on π_1 of a generic curve of genus g over an algebraically closed field k of characteristic $p > 0$; such a curve degenerates into proper stable curves over k of arithmetic genus g; then a theorem due to M. Saïdi gives the profinite fundamental group of the graph of groups, built on the intersection graph of the stable curve and with group data the tame fundamental group of irreducible components minus the intersection points and amalgamation along tame inertia at these points, as a quotient of π_1. The case of maximal degenerations is related to $\pi_1^t(\mathbb{P}^1 - \{0, 1, \infty\})$. An optimistic conjecture is that every etale cover of a generic curve is ordinary; this is known for abelian covers. A first attempt towards this conjecture is the study of covers of the Legendre elliptic curve which are étale outside ∞ (such covers appear by degenerating into a chain of generic elliptic curves); one expects that those which don't factorise through an isogeny are still ordinary; the case of hyperelliptic and (in particular) genus two curves is examined. In the last part we look at the geometry of the "Deligne-Mumford" boundary of the Teichmüller tower of M_g; in this way one hopes to give a group theoretic description of quotients of π_1 which give rise to ordinary etale covers.

Galois actions on braid-like groups. *Makoto Matsumoto.* Let V be a geometrically connected variety defined over a number field k. Then, by Grothendieck's theory, we have an outer action of the absolute Galois group G_k of k on the profinite completion of the topological fundamental group of $V(\mathbf{C})$. We denote this representation by ρ_V. By Belyi's theorem, this action is faithful if V is the projective line minus three points, and each element σ of G_k has its unique "coordinate" $(\chi(\sigma), f_\sigma)$ in $\hat{\mathbb{Z}} \times [F_2, F_2]$. The basic question is: can we use this coordinate in order to describe ρ_V for V other than projective line minus three points? This occured in Ihara's geometric proof of the embedding of $G_\mathbb{Q}$ into $\hat{GT}$ (done by Drinfeld's) using two-dimensional tangential base point.

We show two examples where same kinds of arguments can be applied: (1) V configuration space of some points on an open curve, and (2) moduli space of curves of genus g, with $g = 2, 3$. The former example can be applied to show a generalization of Belyi's Injectivity theorem for nonabelian affine curves. The latter example for $g = 3$ uses a close relationship between the deformation space of E_7-singularity and the moduli stack of genus 3 curves.

On Grothendieck's anabelian conjecture. *Florian Pop.* The idea of Grothendieck's anabelian geometry is that under certain "anabelian" hypotheses the isomorphy type, hence the geometry and the arithmetic of schemes, is encoded in their étale fundamental groups. Relatively speaking, if S is some base scheme, and $\mathcal{A}$ is an anabelian category of S-schemes, then for every object X and Y of $\mathcal{A}$ one should have the following

$$\mathrm{Isom}(X, Y) \cong \mathrm{Out}_{\pi_1(S)}(\pi_1(X), \pi_1(Y))$$

in a functorial way. Here π_1 denotes the étale fundamental group, Isom denotes the space of all $\mathcal{A}$-isomorphisms, and Out denotes the space of all outer isomorphisms.

Richard Thompson's chameleon groups: old and new. *Vlad Sergiescu.* Around 1965, Richard J. Thompson (unpublished) constructed two finitely presented infinite simple groups. They can be defined as groups of affine dyadic homeomorphisms (resp. interval exchanges) of S^1. Recently, it became clear that these groups relate to the universal Ptolemy groupoid constructed by R. J. Penner. The talk surveys these and other developments leading to the recent Lochak-Schneps construction of an action of the Grothendieck-Teichmüller group on an enlarged braids version of Ptolemy; suitably completed.

Singularities of moduli spaces of curves, new Galois invariants of Grothendieck dessins and finite projective geometry. *George Shabat.* In this talk, reflecting joint work with N. Adrianov, we discuss the connections between the Grothendieck *dessins d'enfant* theory and the moduli families of curves. An explicit formula for a Strebel differential on any curve corresponding to a given dessin is suggested. The arithmetical nature of curves with many automorphisms is mentioned.

The equivalence between the classical and the *cartographical* Galois theories is established. From this point of view for a particular case of genus 0 dessins with one cell a new invariant, the *edge rotation group*, is introduced. The examples of its non-trivial behaviour, including the occurrence of two Mathieu groups, are presented. The list of examples when this group is finite projective is discussed.

On groups acting on dessin-labeled objects. *Vasily Shabat.* Grothendieck's theory of *dessins d'enfant* offers a possibility to visualize the action of Galois group $Gal(\overline{\mathbb{Q}}/\mathbb{Q})$ on Belyi pairs.

We use a purely combinatorial approach and introduce the *edge group* that acts on dessins with one marked oriented edge. To do this we define an operation *semiflip* and use it together with known operations of cartographic group. The action of the edge group on the set of dessins of fixed genus and fixed number of edges with one marked flag is transitive. We also discuss an alternative definition of semiflip that preserves the number of vertices (and cells). The well-known group generated by elementary moves of ideal triangulations turns out to be a subgroup of the edge group. Since triangulations parametrize cells of cellular decomposition of Teichmuller space and moduli space, we may speculate that the moves between cells can be realized in terms of (semi)flips. We also discuss relations within the edge group.

Non-abelian unipotent periods. *Zdislaw Wojtkowiak.* We study the monodromy of the universal unipotent connection on $V = \mathbf{C} \setminus \{0, 1\}$. To the monodromy homomorphism we associate a certain torsor and we calculate this torsor partially. The group corresponding to this torsor is closely related to the image of $\mathrm{Gal}(\overline{\mathbb{Q}}/\mathbb{Q})$ in $\mathrm{Aut}(\pi_1(V)_{\mathrm{et}})$.

Hypergeometric functions and moduli of abelian varieties. *Jürgen Wolfart.*

In this talk it was shown how monodromy groups of classical hypergeometric functions and Appell Lauricella functions, i.e. in particular Fuchsian triangle groups and Picard - Terada - Mostow - Deligne groups can be seen as modular groups of suitable families of complex abelian varieties, even in the case where the groups in question are not arithmetic. One obtains so-called modular embeddings of the groups into modular groups acting on higher dimensional complex symmetric domains compatible with analytic embeddings of the domains themselves. Singular points of the differential equations play a particular role for complex multiplication of the parametrized abelian varieties. (Joint work with Paula Beazley Cohen, Ann. ENS 1993)

Evening seminar on Teichmüller and moduli space.

Decompositions of surfaces and compactification of moduli space. *Xavier Buff*

Cellulation de l'espace des modules. *Jean Nicolas Dénarié.*

Fuchsian Model of Teichmüller Space according to Imayoshi. *Ivan Faucheux.*

Penner's cells are cells. *Jérôme Fehrenbach.*

Part I. Dessins d'enfants

Unicellular cartography and Galois orbits of plane trees

Nicolai Adrianov and George Shabat

Galois groups, monodromy groups and cartographic groups

Gareth Jones and Manfred Streit

Permutation techniques for coset representations of modular subgroups

Tim Hsu

Dessins d'enfants en genre 1

Leonardo Zapponi

Unicellular Cartography and Galois Orbits of Plane Trees

Nikolai Adrianov* and George Shabat**

§0. Introduction

After several years of extensive research in the directions outlined by Alexander Grothendieck in his Esquisse d'un Programme [Gr], there still exists a considerable gap between what we are able to *calculate* (Belyi pairs, fields of moduli, etc.) and what we are able to *see* (the combinatorial topology of dessins d'enfants). In this situation it is important to find the *visualizable* Galois invariants of dessins.

Several versions of finite groups of substitutions associated to dessins deliver some of such invariants. They are known as cartographic groups, monodromy groups, etc.; we use the term *group of edge rotations*. Their Galois invariance is established in [Mat], [JS].

Edge rotation groups do not completely solve the problem of determining the Galois orbits of dessins d'enfants. For example, they do not separate the trees from the orbit of *Leila's flower* [Sch2]. However, the non-trivial behavior of edge rotations points out certain interesting phenomena.

One of the goals of the present paper is to draw attention to a class of dessins which seems to occupy a reasonable intermediate position between general dessins and plane trees. This is the class of *unicellular* dessins (the maps with only one face). Theorem 1.1 shows that this case is not restrictive from the point of view of moduli of curves.

Theorem 2.1 describes the composition factors of unicellular dessins. In section 2.2 we present a complete list of primitive edge rotation groups in the genus zero case. In part 3 we show the techniques of calculations of edge rotation groups with "bare hands". We conclude the paper by applying these groups to some exceptional cases of Galois orbits of plane trees; they help us to compute the corresponding generalized Chebyshev polynomials defined over imaginary quadratic fields.

We are indebted to the participants of G.Shabat's seminar "Graphs on surfaces and curves over number fields", and to T.Ekedahl and A.Zvonkin for useful discussions; starting from section 2.2 we use some joint results with Yu.Kochetkov and A.Suvorov.

§1. Belyi theorem for unicellular dessins

We use the concepts and terminology from [Gr], [Sch1], [ShVo]. However, throughout the paper we mean by a *dessin* a *bicolored dessin*. Thus our dessins are equipped with "coloring" functions on the sets of their vertices, having values in the two-element set, such that any two vertices adjacent to the same edge are colored differently. The colors will be called *black* and *white*.

Theorem 1.1. *Suppose $X \ni P$ are a curve and a point on it, both defined over $\overline{\mathbb{Q}}$. Then there exists a Belyi function on X having a single pole at P.*

Proof. If X is of genus g, then by Riemann-Roch [GrHa] there exists a non-constant function $f \in \overline{\mathbb{Q}}(X) \backslash \overline{\mathbb{Q}}$ (say, of degree $g+1$) with the only pole at P. Applying the usual procedure to the critical values of f (see [Be]) we obtain a polynomial $F \in \mathbb{Q}[x]$ such that the composition $F \circ f$ is a Belyi function. Obviously it has no poles apart from P.   ◇

By the Grothendieck correspondence between dessins d'enfants and Belyi pairs we see that the above theorem implies the existence of unicellular dessins on all curves over $\overline{\mathbb{Q}}$.

§2. Edge rotation groups of unicellular dessins

Given a dessin, denote by $r_\bullet$ (resp. $r_\circ$) the transformation of the set of its edges defined by the counterclockwise "rotation" of any edge around its black (resp. white) end.

Definition. For a dessin D the group of permutations of its edges $\mathrm{ER}(D)$ generated by $r_\bullet$ and $r_\circ$ is called the *edge rotation group* of D.

The complete description of edge rotation groups for general dessins is basically equivalent to the description of all the transitive groups of substitutions generated by two elements and hence is out of reach. However, these groups are relatively well understood in the case of unicellular dessins. Theorem 2.3 below gives their complete description for plane trees in the *irreducible case*.

2.1. Composition factors

Theorem 2.1. *Let $G = \mathrm{Mon}(f)$ be the monodromy group of a meromorphic function f having a single pole on a compact Riemann surface X. Then the following list contains all the possible composition factors of G:*

  (a) cyclic groups $\mathbf{C}_n$;
  (b) alternating groups $\mathbf{A}_n$;
  (c) Mathieu groups $\mathbf{M}_{11}$, $\mathbf{M}_{23}$;
  (d) projective groups $PSL_n(\mathbb{F}_q)$.

Proof. Consider a function $f : X \longrightarrow \mathbb{P}^1(\mathbb{C})$ with a single pole and represent it as a maximal chain of morphisms of compact Riemann surfaces

$$X = X_0 \xrightarrow{f_1} X_1 \xrightarrow{f_2} X_2 \xrightarrow{f_3} \ldots X_{m-1} \xrightarrow{f_m} X_m = \mathbb{P}^1(\mathbb{C}).$$

Maximality of the chain implies that all the monodromy groups $\mathrm{Mon}(f_i)$ $(i = 1, \ldots, m)$ are primitive.

Let $f : X \longrightarrow Y$ and $g : Y \longrightarrow Z$ be morphisms of algebraic curves. Then the set of composition factors of the monodromy group $\mathrm{Mon}(g \circ f)$ is contained in the union of the sets of composition factors of $\mathrm{Mon}(f)$ and $\mathrm{Mon}(g)$ (see [GTh], Prop. 2.1). Hence it suffices to prove that the composition factors of monodromy groups of any morphism f_i are contained in the above list.

Since f has only one pole, for any morphism f_i there exists a point $P_i \in X_i$ such that the ramification degree of f_i over P_i equals the total degree of f_i. The monodromy group of f_i is a factor of the fundamental group $\pi(X_i \backslash \text{ramification points})$ and hence it contains a cyclic permutation which corresponds to a loop around P_i.

Therefore we have reduced the problem to the case of primitive permutation group which contains a cyclic permutation. At this point we use two classical results. Burnside's theorem (1911) says that every non-solvable transitive group of prime degree is doubly transitive (see [Wie], Th.11.7). Schur's theorem (1933) guarantees that every primitive transitive group of composite degree which contains a cyclic permutation is doubly transitive (see [Wie, Th.25.3]).

A more recent result was proved by Feit (1980) using the classification of finite simple groups (see [F], Th.1.49): Let G be non-solvable doubly transitive group of degree n which contains a cyclic permutation: then one of the following statements holds

(i) $G \simeq \mathbf{A}_n$ or $\mathbf{S}_n$;

(ii) $n = 11$ and $G \simeq \mathrm{PSL}_2(\mathbb{F}_{11})$ or $G \simeq \mathbf{M}_{11}$;

(iii) $n = 23$ and $G \simeq \mathbf{M}_{23}$;

(iv) $n = (q^m - 1)/(q - 1)$ and G is isomorphic to a subgroup of $\mathrm{P\Gamma L}_m(\mathbb{F}_q)$ containing $\mathrm{PSL}_m(\mathbb{F}_q)$.

Here $\mathrm{P\Gamma L}_m(\mathbb{F}_q)$ is the so-called *semilinear* group, i.e. the extension of the group $\mathrm{PGL}_m(\mathbb{F}_q)$ by the Frobenius automorphism. For $m \geq 3$ it can be also considered as the group of collineations of the projective space $\mathbb{P}^{m-1}(\mathbb{F}_q)$.

If G is solvable then its composition factors are cyclic. Otherwise applying the Burnside, Schur and Feit theorems completes the proof. $\diamond$

Since the edge rotation group of a dessin is isomorphic to the monodromy group of the corresponding Belyi function we get the following corollary.

Corollary 2.2. *The composition factors of the edge rotation groups of unicellular dessins are contained in the list of theorem 2.1.*

Remark. We suppose that for any fixed genus of X there exists only a finite number of possible pairs (n, q) such that the group $\mathrm{PSL}_n(\mathbb{F}_q)$ can occur as a composition factor of $\mathrm{Mon}(f)$ for a meromorphic function f with a single pole on X.

This is a particular case of the Guralnick-Thompson conjecture (see [GTh]) according to which the analogous statement holds for all simple non-abelian groups except $\mathbf{A}_n$ and an arbitrary function f.

2.2. Edge rotation groups of irreducible plane trees

We call a plane tree T *irreducible* if it has no non-trivial morphisms $T \longrightarrow T'$. The irreducibility of a plane tree T is equivalent to the primitivity of the group $\mathrm{ER}(T)$.

Theorem 2.3. *Let T be an irreducible plane tree with n edges and $G = \mathrm{ER}(T)$. Then one of the following statements holds:*

(1) $G \simeq \mathbf{A}_n$ or $\mathbf{S}_n$;

(2) $G \simeq \mathbf{C}_n$ or $\mathbf{D}_n$ and n is prime;

(3) $G \simeq PSL_2(\mathbb{F}_{11})$ and $n = 11$;

(4) $G \simeq \mathbf{M}_{11}$ or $\mathbf{M}_{23}$ and $n = 11$ or 23;

(5) $G \simeq P\Gamma L_m(\mathbb{F}_q)$, $n = (q^m - 1)/(q - 1)$ and

    (a) $m = 2$ and $q = 5, 7, 8, 9$;

    (b) $m = 3$ and $q = 2, 3, 4$ or

    (c) $m = 4, 5$ and $q = 2$.

See the proof in [Adr]; an equivalent result for primitive monodromy groups of polynomials was obtained by Müller [Mül]. The complete list of irreducible plane trees with $\mathrm{ER}(T)$ isomorphic to one of the groups (3)-(5) will be published in [AKS]. Below we present all such plane trees with 9, 10 and 11 edges; for the trees with $\mathrm{ER}(T) \simeq \mathbf{M}_{23}$ see also [AKSS].

§3. Combinatorial structures help to see the edge rotation groups

The common way of calculating the edge rotation groups is to switch on the computer and use some software based on Sims' algorithm. (This is how most of ER's of the present paper were calculated.) However, in this section we give two examples of calculations by hand.

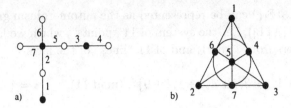

Figure 1. A plane tree with 7 edges and the Fano plane.

3.1. Fano plane

Consider the 7-edged plane tree in fig. 1a. In the specified numeration of the edges of the tree the generating substitutions have the following form:

$$a := r_\bullet = (2567)(34), \qquad b := r_o = (12)(35)$$

(the edges are numbered in such a way that $ab = (1234567)$).

We assign the edges of the tree to the points of the Fano plane as shown on fig. 1b. Under such a correspondence the group $\mathrm{ER}(T)$ preserves the collinearity of the points of the Fano plane and therefore it is a subgroup of the collineation group $\mathrm{PSL}_3(\mathbb{F}_2) \simeq \mathrm{PSL}_2(\mathbb{F}_7)$. We show that the order of the group $\mathrm{ER}(T)$ is 168.

Indeed, the group $\mathrm{ER}(T)$ is transitive on the seven lines of the Fano plane (e.g., the operator ab permutes them cyclically), hence the stabilizer of the line $(4, 6, 7)$ is a subgroup in $\mathrm{ER}(T)$ of index 7. In particular, this stabilizer contains the substitutions $(ab)^{-3}a(ab)^3 = (1235)(67)$ and $(ab)a^2(ab)^{-1} = (15)(46)$. The last substitutions, acting on the set $\{1, 2, 3, 5\}$, generate the full symmetric group $\mathbf{S}_4$. Therefore, the order of the edge rotation group is $\#\mathrm{ER}(T) \geq 7 \cdot 4! = 168$, and hence $\mathrm{ER}(T) \simeq \mathrm{PSL}_3(\mathbb{F}_2)$.

3.2. Biplane

The second example is related to the exceptional action (known already to Evariste Galois, see [Con]) of $\mathrm{PSL}_2(\mathbb{F}_{11})$ on 11 points.

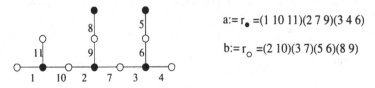

$$a := r_\bullet = (1\ 10\ 11)(2\ 7\ 9)(3\ 4\ 6)$$
$$b := r_o = (2\ 10)(3\ 7)(5\ 6)(8\ 9)$$

Figure 2. A plane tree with 11 edges.

Consider the plane tree in fig. 2. We are going to show that its edge rotation group is isomorphic to the group $\mathrm{PSL}_2(\mathbb{F}_{11})$. It is known that the group $PSL_2(\mathbb{F}_{11})$ can be represented as the automorphism group of the "biplane" (see [ATL]), i.e. the system of 11 "points", which we label by the natural numbers from 1 to 11, and of 11 "lines" of the form

$$L_i = \{i+1, i+3, i+4, i+5, i+9\} \pmod{11}, \qquad i = 1, \ldots 11.$$

There are exactly two lines passing through any two different points and any two different lines meet in exactly two points.

We identify the points of biplane with the edges of 11-edged tree (fig. 2) in such a way that the edge rotation group coincides with the automorphism group of the biplane. Similarly to the previous example we label the edges of the tree in such a way that the product ab has the form $(1, 2, \ldots, 11)$.

It is easy to see that the permutations a and b preserve the lines; hence the group $ER(T)$ is a subgroup of $PSL_2(\mathbb{F}_{11})$. Since 11 is prime our tree is irreducible, hence by theorem 2.3 we have $ER(T) = PSL_2(\mathbb{F}_{11})$.

§4. Generalized Chebyshev polynomials of cartographically special trees

We call an irreducible plane tree T with n edges *cartographically special* if $ER(T)$ is not isomorphic to one of the groups $\mathbf{C}_n$, $\mathbf{D}_n$, $\mathbf{A}_n$ and $\mathbf{S}_n$. We present all the cartographically special trees with 9, 10 and 11 edges and their generalized Chebyshev polynomials (see e.g. [ShZv]).

Some of these polynomials defined over $\mathbb{Q}$; they were computed earlier by Matzat (see [Mül]). Computation of the generalized Chebyshev polynomial for the plane tree with $ER(T) \simeq \mathbf{M}_{11}$ was also performed by Yu.Matiyasevich with the help of the Couveignes-Granboulan algorithm (cf. [CoGr]).

4.1. Results

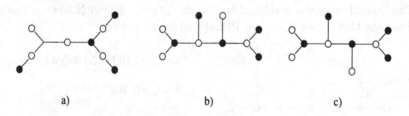

Figure 3. $e = 9$ and $ER(T) \simeq P\Gamma L_2(\mathbb{F}_8)$.

(1) The generalized Chebyshev polynomial for the 9-edged tree in fig. 3a is given by

$$P = \left(x^2 - x + \frac{-13 + 5\sqrt{-3}}{2}\right)^3 \times$$
$$\left(x^3 + 3x^2 + 3 \cdot 7\frac{-1 + \sqrt{-3}}{2}x - 2^2(2 - \sqrt{-3})^2\frac{7 + \sqrt{-3}}{2}\right).$$

Its critical values are 0 and $c = -2^5(1 - \sqrt{-3})(2 - \sqrt{-3})^7$, and we have

$$P - c = (x - 4)\left(x^4 + 2x^3 + 3(-4 + 3\sqrt{-3})x^2 + 2(2 + \sqrt{-3})\frac{7 + 13\sqrt{-3}}{2}x\right.$$
$$\left. + (2 - \sqrt{-3})\frac{31 - 13\sqrt{-3}}{2}\right)^2.$$

(2) The generalized Chebyshev polynomial for the 9-edged tree in fig. 3b is given by

$$P = x^9 + 2^2 3^3 7^2 x^7 + 2^3 3^3 7^3 x^6 + 2 \cdot 3^7 7^4 x^5 + 2^3 3^5 7^6 x^4 +$$
$$2^2 3^4 7^6 307 x^3 + 2^3 3^6 7^7 37 x^2 + 3^6 7^8 37^2 x + 2^3 3^6 7^9 139.$$

Its critical values are $\pm c$, where $c = 2^{13} 3^4 7^9 \sqrt{-3}$.

(3) The generalized Chebyshev polynomial for the 9-edged tree in fig. 3c is given by

$$P = x^9 - 2 \cdot 3^2 7(2 + \sqrt{-3})x^7 + 3^2 7^4 \sqrt{-3}x^5 +$$
$$3 \cdot 7^6(1 - \sqrt{-3})(2 + \sqrt{-3})x^3 - 3^2 7^8\frac{1 + \sqrt{-3}}{2}x.$$

Its critical values are $\pm c$, where $c = 2^2 7^8(1 + \sqrt{-3})(2 + \sqrt{-3})$.

$$P + c = \left(x^2 + 7x - 7^2\frac{1 + \sqrt{-3}}{2}\right)^3 \times$$
$$\left(x^3 - 3 \cdot 7x^2 - 3 \cdot 7\frac{1 + \sqrt{-3}}{2}(2 + \sqrt{-3})^2 x + 2^2 7^2(1 + \sqrt{-3})(2 + \sqrt{-3})\right)$$

$$P - c = \left(x^2 - 7x - 7^2\frac{1 + \sqrt{-3}}{2}\right)^3 \times$$
$$\left(x^3 + 3 \cdot 7x^2 - 3 \cdot 7\frac{1 + \sqrt{-3}}{2}(2 + \sqrt{-3})^2 x - 2^2 7^2(1 + \sqrt{-3})(2 + \sqrt{-3})\right)$$

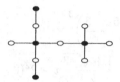

Figure 4. $e = 10$ and $ER(T) \simeq P\Gamma L_2(\mathbb{F}_9)$.

(4) The generalized Chebyshev polynomial for the 10-edged tree in fig. 4 is given by

$$P = (x^2 - 20x + 180)(x^2 + 5x - 95)^4.$$

Its critical values are 0 and $c = 2^4 3^{12} 5^5$.

$$P - c = (x^4 + 30x^3 + 75x^2 - 4850x - 39375)(x^3 - 15x^2 + 75x + 550)^2.$$

(5) The generalized Chebyshev polynomial for the 11-edged tree in fig. 2 is given by

$$P = \left(x + \frac{-3 + \sqrt{-11}}{2}\right)^3 \left(x^2 + \frac{5 - \sqrt{-11}}{2}x - \frac{1 + \sqrt{-11}}{2}\right)^3$$
$$\times (x^2 - 3x + (5 + \sqrt{-11})).$$

Its critical values are 0 and $c = -1728$.

$$P - c = \left(x^3 + 4x^2 + \frac{7 + 5\sqrt{-11}}{2}x + 2(2 + 3\sqrt{-11})\right) \times \left(x - \frac{1 + \sqrt{-11}}{2}\right)^2$$
$$\times \left(x^3 + \frac{-3 + \sqrt{-11}}{2}x^2 + (-2 + \sqrt{-11})x - \frac{3(1 + \sqrt{-11})}{2}\right)^2.$$

Figure 5. $e = 11$ and $ER(T) \simeq M_{11}$.

(6) The generalized Chebyshev polynomial for the 11-edged tree in fig. 5 is given by

$$P = (x^2 + 2(4 - \sqrt{-11})x - 9(19 + 8\sqrt{-11}))(x - 18) \times$$
$$\left(x^2 + \frac{5 + \sqrt{-11}}{2}x - \frac{221 + 19\sqrt{-11}}{2}\right)^4$$

Its critical values are 0 and

$$c = -2^6 3^6 (4 - \sqrt{-11})^5 (4 + \sqrt{-11})^2 (3 + \sqrt{-11})^2 (9 - 2\sqrt{-11}).$$

We have

$$P - c = \left(x^3 + 3(5 + \sqrt{-11})x^2 + 3\frac{-71 + 29\sqrt{-11}}{2}x + (-2119 + 37\sqrt{-11}) \right)$$
$$\times \left(x^4 - 3\frac{5 + \sqrt{-11}}{2}x^3 - \frac{591 + 75\sqrt{-11}}{2}x^2 + \frac{4267 + 1775\sqrt{-11}}{2}x \right.$$
$$\left. + \frac{6057 + 4743\sqrt{-11}}{2} \right)^2.$$

4.2. Techniques of computation

In this section we show by an example that a priori knowledge of the field of definition of a dessin can essentially simplify calculations of corresponding Belyi function.

Consider the plane tree in fig. 5. This tree and its mirror reflection are the only 11-edged plane trees with edge rotation group $\mathbf{M}_{11}$. Hence they constitute the Galois orbit over some imaginary quadratic field. Using techniques similar to those in [Mat], one determines that this field is $\mathbb{Q}(\sqrt{-11})$.

We can write the desired generalized Chebyshev polynomial in the form

$$P = (x^3 - 4x^2 + ax + b)(x^2 + x + c)^4.$$

with the coefficients in the field of definition of the trees. Then

$$P' = (x^2 + x + c)^3 Q,$$

where Q is a polynomial of degree 4. The roots of Q are the coordinates of the white vertices of the tree T.

The polynomial P assumes the same values at the roots of the polynomial Q. Dividing P by Q, write

$$P = QX + c_3 x^3 + c_2 x^2 + c_1 x + c_0.$$

The above condition implies that $c_1 = c_2 = c_3 = 0$, or:

$$144239ac^2b - 100362a^2cb + 2373872acb + 5253120c^2 - 13152000a$$
$$- 23961600b + 21043200c + 523008c^3 + 15648c^4 - 11126ac^4$$
$$- 81920bc^3 - 2049056ac^2 - 12453120ac - 123440ac^3 - 8239040bc$$
$$- 1022648bc^2 + 10831160ba + 2108528a^2c + 2509740b^2 + 5275200a^2$$
$$- 1183439ba^2 + 179456a^2c^2 - 87614a^3c + 499152b^2c - 371048b^2a$$
$$- 563130a^3 + 24183ba^3 + 1814a^2c^3 - 850a^3c^2 + 162a^4c$$
$$- 16352acb^2 - 26048b^3 + 21768b^2c^2 + 4248b^2a^2 + 13050a^4 = 0$$

$$34585ac^2b - 23779a^2cb + 899443acb + 1085184c^2 - 52642560a$$
$$- 13375040b + 13704960c - 5472c^3 + 24576c^4 - 22281bc^3$$
$$- 783020ac^2 - 10463168ac - 32258ac^3 - 2670632bc - 209351bc^2 \qquad (*)$$
$$+ 6056367ba + 2040186a^2c + 669724b^2 + 13434684a^2 - 574642ba^2$$
$$+ 103128a^2c^2 - 86746a^3c + 91960b^2c - 77176b^2a - 1169544a^3$$
$$+ 6075ba^3 - 2112b^3 + 63129600 + 24300a^4 = 0$$

$$120864c^3 + 240306acb + 25557760c + 12144b^2c - 126966640b$$
$$- 2465672bc - 14854ac^3 - 270334a^3 - 608732ac^2 - 177419bc^2$$
$$+ 2681547ba + 740302a^2c + 243584b^2 + 14854a^2c^2 - 7362a^3c$$
$$- 6160b^2a - 92147ba^2 - 9057728ac - 48328960a + 1458a^4$$
$$- 4096c^4 + 104313600 + 2448384c^2 + 6669644a^2 = 0$$

The direct solving of this system leads to an equation of large degree (there are 10 trees with the same valency sets). Straightforward computations using Gröbner basis techniques applied to the system demand considerable computer resources.

However, we are not looking for all the solutions of the system $(*)$ but only for those from $\mathbb{Q}(\sqrt{-11})$. We eliminate variables using resultants and obtain the polynomial

$$R(a) = Res_c(Res_b(c_1, c_2), Res_b(c_2, c_3))$$

of degree 156. We find a prime (say, $p = 29$) such that
1) p does not divide the leading coefficient of $R(a)$;
2) $a^2 + 11$ is irreducible modulo p;
3) there exists only one quadratic irreducible factor of $R(a)$ modulo p.

Using standard techniques (see e.g. [Akr]) we lift this quadratic factor to a divisor $d(a)$ of $R(a)$ in $\mathbb{Z}[a]$. Adding $d(a)$ to the system $(*)$ we easily compute the Gröbner basis:

$$\left\{ \begin{aligned} 49b + 256a - 18 &= 0 \\ 49c - 18a - 73 &= 0 \\ 3a^2 + 155a + 4203 &= 0 \end{aligned} \right\}$$

and the desired polynomial P is found. The corresponding polynomial presented in 4.1 (which has integer algebraic coefficients) is obtained from P by a suitable linear substitution.

References

[Adr]   N.M.Adrianov, Classification of primitive edge rotation groups of plane trees (in Russian), to appear in *Fundamentalnaya i prikladnaya matematika*.

[AKS]   N.M.Adrianov, Yu.Yu.Kochetkov, A.D.Suvorov, Plane trees with special primitive edge rotation groups (in Russian), to appear in *Fundamentalnaya i prikladnaya matematika*.

[Akr]   A.G.Akritas, *Elements of computer algebra with applications*, John Wiley & Sons, 1989.

[AKSS] N.M.Adrianov, Yu.Yu.Kochetkov, A.D.Suvorov, G.B.Shabat, Mathieu groups and plane trees (in Russian), *Fundamentalnaya i prikladnaya matematika*, **v.1**, no. 2 (1995), 377-384.

[ATL]   J.H.Conway, R.T.Curtis, S.P.Norton, R.A.Parker, R.A.Wilson, *An ATLAS of finite groups*. Oxford University Press, London, 1985.

[Be]   G.V.Belyi, On Galois extensions of a maximal cyclotomic field (in Russian), *Izv. Akad. Nauk SSSR* **43**, (1979), 269-276.

[Con]   J.Conway, Three lectures on exceptional groups, in *Finite simple groups* (eds. M.P.Pawel, G.Higman), AC Press, New York, 1971, 215-247.

[CoGr] J.-M.Couveignes, L.Granboulan, Dessins from a geometric point of view, in [GTDE], 79-114.

[F]   W.Feit, Some consequences of classification of finite simple groups. Santa Cruz Conference, Proc. Sympos. Pure Math. **37**, AMS, Providence, R.I. (1980), 175-183.

[GrHa] Ph.Griffiths, J.Harris, *Principles of algebraic geometry*, Pure Appl. Math., John Wiley & Sons, N.Y., 1978.

[Gr]   A.Grothendieck, *Esquisse d'un programme*, in *Geometric Galois Actions*, volume I.

[GTh]   R.M.Guralnick, J.G.Thompson, Finite groups of genus zero. *J. Algebra* **131** no.1 (1990), 303-341.

[GTDE] *The Grothendieck Theory of Dessins d'Enfants*, ed. L. Schneps, London Math. Soc. Lecture Notes Series **200**, Cambridge University Press, 1994.

[JS]   G.A.Jones, M.Streit, Galois groups, monodromy groups and cartographic groups, this volume.

[Mat]   B.H.Matzat, *Konstruktive Galoistheorie*, Springer Lecture Notes 1284, Berlin-Heidelberg-New York, 1987.

[Mül]   P. Müller, Primitive monodromy groups of polynomials, in *Recent developments in the inverse Galois problem* (M. Fried, ed.), Contemporary Mathematics, vol. **186**, 1995, 385-401.

[Sch1]   L.Schneps, Introduction to [GTDE], 1-15.

[Sch2]   L.Schneps, Dessins d'enfants on the Riemann sphere, in [GTDE], 47-78.

[ShVo]   G.B.Shabat, V.A.Voevodsky, Drawing curves over number fields, in *The Grothendieck Festschrift III*, Progress in Math. **88**, Birkhäuser, Basel, 1990, 199-227.

[ShZv]   G.Shabat, A.Zvonkin, Plane trees and algebraic numbers, in *Jerusalem Combinatorics 93* (H. Barcelo, G. Kalai, eds.), Contemporary Mathematics, vol. 178, 1994, 233-275.

[Wie]   H.Wielandt, *Finite permutation groups*. Academic Press, New York-London, 1964.

* Moscow State University
adrianov@nw.math.msu.su

** Institute of New Technologies
shabat@int.glas.apc.org

Galois Groups, Monodromy Groups and Cartographic Groups

Gareth A. Jones and Manfred Streit

Abstract. The Riemann surfaces defined over the algebraic numbers are those admitting Belyĭ functions; such functions can be represented combinatorially by maps called *dessins d'enfants*, and these are permuted faithfully by the Galois group of the algebraic numbers. We define the monodromy group and the cartographic group of a *dessin*, and show that these permutation groups (and hence many other properties of *dessins*) are invariant under this action. We give examples to show how these groups can be computed and how they can be used to distinguish Galois orbits on *dessins*.

§0. Introduction

One of the most interesting but intractable groups in mathematics is the *absolute Galois group*, the automorphism group $\mathbf{G} = \mathrm{Gal}(\overline{\mathbf{Q}}/\mathbf{Q})$ of the field $\overline{\mathbf{Q}}$ of algebraic numbers. Since $\overline{\mathbf{Q}}$ is the union of all the algebraic number fields $K \subset \mathbf{C}$, it follows that $\mathbf{G}$ is the inverse limit of the finite Galois groups $\mathrm{Gal}\,(K/\mathbf{Q})$, where K ranges over the Galois extensions of $\mathbf{Q}$. Thus $\mathbf{G}$ is a profinite group, which one can regard as embodying the whole of classical Galois theory over $\mathbf{Q}$.

In recent years there has developed a geometric approach to this important group, through *dessins d'enfants*, which are essentially maps drawn on compact Riemann surfaces. It is well-known that a Riemann surface X is compact if and only if it corresponds to an equivalence class of algebraic curves. We may choose some polynomial $f \in \mathbf{C}[x,y]$ so that the curve $f(x,y) = 0$ is a plane model (possibly singular) representing this class. Belyĭ [Bel] showed that X is defined over $\overline{\mathbf{Q}}$ (in the sense that we can take $f \in \overline{\mathbf{Q}}[x,y]$) if and only if there is a Belyĭ function $\beta : X \to \Sigma = \mathbf{P}^1(\mathbf{C}) = \mathbf{C} \cup \{\infty\}$, that is, a meromorphic function which is unbranched outside $\{0, 1, \infty\}$. One can form a 'picture' of such a Belyĭ pair (X, β) by taking a simple *dessin* $\mathcal{D}_1$ on the Riemann sphere Σ, such as the bipartite map with two vertices at 0 and 1 joined by a single edge $[0, 1]$, and using β to lift $\mathcal{D}_1$ to a *dessin* $\mathcal{D} = \beta^{-1}(\mathcal{D}_1)$ on X. The maps on Riemann surfaces obtained in this way are called *dessins d'enfants* in view of the rather childish appearance of some of the simplest examples.

Grothendieck [Gro] observed that the natural action of $\mathbf{G}$ on Belyĭ pairs induces an action on these *dessins*. What is remarkable is that this action of $\mathbf{G}$ is faithful, in the sense that each non-identity element $\sigma \in \mathbf{G}$ sends some *dessin* $\mathcal{D}$ to a non-isomorphic *dessin* $\mathcal{D}^\sigma$; indeed, a result of Lenstra and Schneps [Sch1] shows that $\mathbf{G}$ acts faithfully on plane trees (maps on Σ with a single face). These apparently simple and explicit combinatorial objects therefore provide a direct insight into the group $\mathbf{G}$. Many examples of this connection between Galois theory and plane trees have been studied by Shabat and Voevodsky [SV] and by Bétréma, Péré and Zvonkin [BPZ], and a general theory is emerging in the papers of Couveignes [Cou], Schneps [Sch1] and Shabat and Zvonkin [SZ]; Wolfart [Wol1] gives examples of the action of $\mathbf{G}$ on torus *dessins*, and further examples of positive genus are considered in [Jon]. For general surveys on different aspects of *dessins d'enfants* see [CIW, Jon, JS3], and for recent research in this and related areas, see the proceedings of the 1993 Luminy conference [Sch2].

In its action on *dessins*, $\mathbf{G}$ preserves all the obvious numerical parameters such as the genus, the numbers of vertices, edges and faces, the valencies of the vertices, and so on. It also preserves the group of orientation-preserving automorphisms, though not necessarily those which reverse orientation. (These facts seem to be widely-known, but explicit proofs are not so easy to find.) This makes it a non-trivial task to determine whether two *dessins* are conjugate under $\mathbf{G}$: it is necessary but not sufficient that these invariants should be equal. A rather finer invariant is provided by the monodromy group of a *dessin*, which we define to be the monodromy group of the branched covering $\beta : X \to \Sigma$. This is a 2-generator transitive subgroup $G = \langle g_0, g_1 \rangle$ of the symmetric group S_N, where N is the degree of β; the permutations g_0, g_1 and $g_\infty = (g_0 g_1)^{-1}$ describe the branching-pattern of β above $0, 1$ and ∞. Two *dessins* $\mathcal{D}$ and $\mathcal{D}'$ are isomorphic if and only if their monodromy generators g_i and g'_i are simultaneously conjugate, that is, some $x \in S_N$ satisfies $g'_i = g_i^x$ for $i = 0, 1, \infty$. Our main result (proved in the Appendix) is the following:

Theorem. *Two dessins $\mathcal{D}$ and $\mathcal{D}^\sigma$ which are conjugate under* **G** *have monodromy groups G and G^σ which are conjugate in S_N, that is, which are isomorphic as permutation groups.*

Since the orientation-preserving automorphism group of a *dessin* is the centraliser of its monodromy group in S_N, it follows that this is also invariant under **G**. We shall show that for each $i = 0, 1, \infty$, the generators g_i and g_i^σ of G and G^σ are conjugate in S_N; they therefore have the same cycle-structure, which implies that $\mathcal{D}$ and $\mathcal{D}^\sigma$ share the same numerical parameters listed above. (Note that these generators need not be *simultaneously* conjugate, so that $\mathcal{D}$ and $\mathcal{D}^\sigma$ can be non-isomorphic.) Unfortunately, although these parameters are invariant under **G**, they are not always sufficient to distinguish its different orbits: we shall give examples where two *dessins* share the same parameters, but the monodromy groups (and hence the *dessins*) are not conjugate.

Similarly, the monodromy group does not always distinguish orbits of **G**; thus non-conjugate *dessins* can have conjugate monodromy groups, so the converse of our theorem is false. Under these circumstances, a finer invariant is needed, and for this one can use the cartographic group C of a *dessin*; this transitive subgroup of S_{2N} is the monodromy group of the Belyĭ function $4\beta(1-\beta) : X \to \Sigma$, and our main theorem also shows that conjugate *dessins* have conjugate cartographic groups. Now conjugacy of cartographic groups implies conjugacy of monodromy groups, but the converse is false: we shall give examples of *dessins* which have conjugate monodromy groups but non-conjugate cartographic groups. This shows that C is more effective than G in distinguishing orbits of $\mathbf{G}$, but there is a price to be payed for this: doubling the degree of a permutation group can have a disproportionate effect on its size, so it can be much harder to compute C than G for a given *dessin*. Moreover, we will show by means of examples that in some cases even C can fail to distinguish orbits of $\mathbf{G}$.

This paper is organised as follows. After briefly describing $\mathbf{G}$ in §1, we state Belyĭ's Theorem with a few simple examples in §2, and then in §3 we show how this theorem is related to various types of *dessins*, such as triangulations, hypermaps, and bipartite maps. These are represented algebraically in §4 in terms of certain permutation groups, namely the monodromy group and the cartographic group, and geometrically in §5 as quotients of certain universal structures on the hyperbolic plane. We describe how $\mathbf{G}$ acts on these *dessins*, and in the Appendix we prove that various parameters are invariant under $\mathbf{G}$. In §6 we concentrate our attention on the simplest *dessins* which admit a faithful action of $\mathbf{G}$, namely the plane trees, showing how these are derived from a class of polynomials which generalise the Chebyshev polynomials. In §7 we give a number of examples of this action of $\mathbf{G}$, showing how the monodromy and cartographic groups can be used to distinguish the orbits of $\mathbf{G}$ on plane trees of a given combinatorial type.

We are very grateful to the European Union Human Capital and Mobility Program on Computational Conformal Geometry for enabling this collaboration to take place, to David Singerman for many valuable comments about Riemann surfaces and related matters, and to Sasha Zvonkin for his constructive criticisms of an earlier draft of this paper.

§1. The absolute Galois group

The algebraic numbers are the elements $a \in \mathbf{C}$ which generate a finite extension $\mathbf{Q}(a) \supseteq \mathbf{Q}$ of the rational field $\mathbf{Q}$; they form a field $\overline{\mathbf{Q}}$, the algebraic closure of $\mathbf{Q}$ in $\mathbf{C}$. For each $a \in \overline{\mathbf{Q}}$, the minimal polynomial of a over $\mathbf{Q}$ has a splitting field $K_a \subset \overline{\mathbf{Q}}$ which is a Galois (finite normal) extension of $\mathbf{Q}$, and conversely the primitive element theorem implies that every Galois extension $K \supseteq \mathbf{Q}$ arises in this way; thus

$$\overline{\mathbf{Q}} = \bigcup_{K \in \mathcal{K}} K,$$

where $\mathcal{K}$ is the set of all Galois extensions K of $\mathbf{Q}$ in $\mathbf{C}$. For each $K \in \mathcal{K}$, the Galois group $G(K) = \mathrm{Gal}\,(K/\mathbf{Q})$ of K over $\mathbf{Q}$ is a finite group of order equal to the degree $|K : \mathbf{Q}|$. If $K \supseteq L$ where $K, L \in \mathcal{K}$, then every automorphism of K leaves L invariant, so the restriction mapping is a homomorphism $\rho_{K,L} : G(K) \to G(L)$; in fact, $\rho_{K,L}$ is an epimorphism, since every automorphism of L can be extended (in $|K : L|$ ways) to an automorphism of K.

The groups $G(K)$ $(K \in \mathcal{K})$ and the epimorphisms $\rho_{K,L}$ form an inverse (or projective) system, and $\mathbf{G}$ can be identified with its inverse limit:

$$\mathbf{G} = \varprojlim G(K).$$

This is the subgroup of the cartesian product $\prod_{K \in \mathcal{K}} G(K)$ consisting of all elements (g_K) such that $g_K \rho_{K,L} = g_L$ whenever $K \supseteq L$, and one identifies each $g \in \mathbf{G}$ with the element (g_K) where g_K is the restriction of g to K. (In simple language, each $g \in \mathbf{G}$ is a consistent choice of automorphisms of the fields $K \in \mathcal{K}$.) The discrete topologies on the finite groups $G(K)$ impose a topology on $\prod_{K \in \mathcal{K}} G(K)$, which is compact by Tychonoff's Theorem. The subgroup $\mathbf{G}$ inherits a topology, called the *Krull topology*, which is compact since $\mathbf{G}$ is closed. Under the Galois correspondence, the subfields of $\overline{\mathbf{Q}}$ correspond to the closed subgroups of $\mathbf{G}$. (See [Jac, Ch.8] for further details.)

It follows from the above identification that $\mathbf{G}$ has cardinality $2^{\aleph_0}$, so one should not expect the structure of $\mathbf{G}$ to be particularly straightforward; for example, being uncountable $\mathbf{G}$ cannot be finitely generated. Neukirch [Neu] has shown that $\mathbf{G}$ is complete (that is, its centre $Z(\mathbf{G})$ and its outer automorphism group

Out($\mathbf{G}$) are both trivial), so $\mathbf{G}$ is naturally isomorphic to its automorphism group Aut($\mathbf{G}$); this means that one cannot imbed $\mathbf{G}$ as a normal subgroup in any larger group, other than as a direct factor, so one cannot obtain more information about $\mathbf{G}$ from such imbeddings. Ideally, one would like to have a fairly explicit faithful representation of $\mathbf{G}$ on some easily-studied structure $\mathcal{S}$, so that $\mathbf{G} \leq \text{Aut}(\mathcal{S})$ and properties of $\mathbf{G}$ can be deduced from its action on $\mathcal{S}$. We shall show how Grothendieck's ideas lead us to examples of this situation.

§2. Belyĭ's Theorem

A Riemann surface X is compact if and only if it is isomorphic (that is, conformally equivalent) to the Riemann surface X_f of an algebraic curve $f(x,y) = 0$ for some polynomial $f(x,y) \in \mathbf{C}[x,y]$. We say that X is *defined over* K for some subfield K of $\mathbf{C}$ if $X \cong X_f$ where $f(x,y) \in K[x,y]$. We will be particularly interested in Riemann surfaces defined over the field $\overline{\mathbf{Q}}$ and over algebraic number fields $K \subset \overline{\mathbf{Q}}$.

Let Σ denote the Riemann sphere (or complex projective line) $\mathbf{C} \cup \{\infty\} = \mathrm{P}^1(\mathbf{C})$; this is, up to isomorphism, the unique compact Riemann surface of genus 0. A *Belyĭ function* on a compact Riemann surface X is a meromorphic function $\beta : X \to \Sigma$ whose critical values are contained in $\{0, 1, \infty\}$; topologically, this is a finite covering which is unbranched outside these three points. In these circumstances we will call (X, β) a *Belyĭ pair*.

Example 1. Let $X = \Sigma$ and let $\beta(x) = x^n$ for some fixed integer $n > 1$. This is an n-sheeted covering, branched over 0 and ∞ (where the n sheets come together in cycles of length n), so it is a Belyĭ function on Σ.

Example 2. Let X be the n-th degree Fermat curve X_n, given by the algebraic function $x^n + y^n = 1$; this is a compact Riemann surface of genus $(n-1)(n-2)/2$, defined over $\mathbf{Q}$. The projection $\pi : X_n \to \Sigma$, $(x,y) \mapsto x$ is not a Belyĭ function, since its critical values are the n-th roots of unity, but if we compose π with the function $x \mapsto x^n$ (which maps all these points to 1) we obtain a Belyĭ function $\beta : X_n \to \Sigma$, $(x,y) \mapsto x^n$ of degree n^2. Above the critical values $0, 1$ and ∞, the n^2 sheets come together in n cycles of length n.

Example 3. An *elliptic curve* is a Riemann surface X of genus 1; algebraically, it can be put into Legendre normal form

$$y^2 = x(x-1)(x-\lambda) \qquad (\lambda \in \mathbf{C} \setminus \{0,1\}).$$

Such a surface $X = X_\lambda$ is defined over $\overline{\mathbf{Q}}$ if and only if $\lambda \in \overline{\mathbf{Q}}$. In these cases it is possible to obtain a Belyĭ function β on X by composing the projection $\pi : (x,y) \mapsto x$ (which has critical values $0, 1, \infty$ and λ) with suitable rational functions $\Sigma \to \Sigma$. A good example is the case $\lambda = 1 + \sqrt{2}$ considered by Wolfart [Wol1]: this is an algebraic number, with minimal polynomial $p(x) = x^2 - 2x - 1$ over $\mathbf{Q}$, and the function $x \mapsto w = -p(x)$ sends the four critical values of π to $1, 2, \infty$ and 0 respectively; the function $w \mapsto w^2/4(w-1)$ now sends these points to $\infty, 1, \infty$ and 0, and one can verify that the composition

$$\beta : (x,y) \mapsto \frac{w^2}{4(w-1)} = \frac{(x^2 - 2x - 1)^2}{4x(2-x)}$$

of these functions with π is a Belyĭ function of degree 8 on X.

In 1979, Belyĭ [Bel] proved the following theorem:

Theorem. *A compact Riemann surface X is defined over $\overline{\mathbf{Q}}$ if and only if there is a Belyĭ function $\beta : X \to \Sigma$.*

The fact that this condition is sufficient follows from Weil's rigidity theorem [Wei]; see [Wol2] for the details. Belyĭ proved that this condition is necessary by composing a projection $\pi : X \to \Sigma$, with critical values only in $\overline{\mathbf{Q}} \cup \{\infty\}$, with suitably chosen rational functions $\Sigma \to \Sigma$, these eventually force the critical values into $\{0, 1, \infty\}$, thus giving rise to a Belyĭ function $\beta : X \to \Sigma$ as in Examples 2 and 3.

There is a natural action of the absolute Galois group $\mathbf{G}$ on Belyĭ pairs (X, β): both X and β are defined over $\overline{\mathbf{Q}}$, so one can let each automorphism $\sigma \in \mathbf{G}$ act naturally on the defining coefficients to give a Belyĭ pair (X^σ, β^σ). Let us define two Belyĭ pairs (X, β) and (X', β') to be *equivalent* if there is an isomorphism $i : X \to X'$ such that $\beta' \circ i = \beta$; then there is an induced action of $\mathbf{G}$ on equivalence classes of Belyĭ pairs. In Examples 1 and 2, for instance, everything is defined over $\mathbf{Q}$ and is therefore fixed by $\mathbf{G}$, but in Example 3 we see a non-trivial Galois action. Wolfart's curve $X_\lambda = X_{1+\sqrt{2}}$ is defined over $\mathbf{Q}(\sqrt{2})$, which has a Galois group of order 2 generated by $\sqrt{2} \mapsto -\sqrt{2}$. This automorphism (which extends to an element $\sigma \in \mathbf{G}$) transforms X_λ to the conjugate elliptic curve $X_{\lambda^\sigma} = X_{1-\sqrt{2}}$, and leaves the formula for the Belyĭ function β unchanged since this is defined over $\mathbf{Q}$. Now in general, two elliptic curves $X = X_\lambda$ are isomorphic (as Riemann surfaces) if and only if they have the same J-invariant $J(X) = 4(1 - \lambda + \lambda^2)^3/27\lambda^2(1 - \lambda)^2$; in our case, $X_{1+\sqrt{2}}$ and $X_{1-\sqrt{2}}$ have J-invariants $(19 - 3\sqrt{2})/27$ and $(19 + 3\sqrt{2})/27$ respectively, so they are non-isomorphic and hence the two Belyĭ pairs are inequivalent.

This action of $\mathbf{G}$ on equivalence classes of Belyĭ pairs is faithful. In fact, it is easy to see that it remains faithful when restricted to Belyĭ functions on elliptic curves: if σ is any non-identity element of $\mathbf{G}$ then some $a \in \overline{\mathbf{Q}}$ is moved by σ, and by solving the equation $J(X) = 4(1 - \lambda + \lambda^2)^3/27\lambda^2(1 - \lambda)^2 = a$ we obtain an elliptic curve $X = X_\lambda$ which is not isomorphic to its conjugate curve X^σ; since $\lambda \in \overline{\mathbf{Q}}$, there is a Belyĭ function β on X, and the Belyĭ pair (X, β) is transformed by σ to an inequivalent pair. As we shall see later, there is a more useful (but less obviously faithful) action on Belyĭ pairs of genus 0.

§3. Belyĭ functions and dessins

A good way of visualising a Belyĭ pair (X, β) (and hence of seeing how $\mathbf{G}$ acts) is to take some simple combinatorial structure on Σ, such as a map, a triangulation or a hypermap, and to use β to lift it to X, thus obtaining a similar structure on X which covers the original structure on Σ. The action of $\mathbf{G}$ on Belyĭ pairs then induces actions of $\mathbf{G}$ on these structures.

3.1. Triangulations. Let us form a triangulation $\mathcal{T}_1$ of Σ

by choosing three vertices $0, 1$ and ∞, and three edges along the line-segments in $\mathbf{R}$ joining these vertices, so that there are two triangular faces, corresponding to the upper and lower half-planes of $\mathbf{C}$. If $\beta : X \to \Sigma$ is a Belyĭ function, then $\beta^{-1}(\mathcal{T}_1)$ is a triangulation $\mathcal{T}$ of X. Since there is no branching away from the vertices, each of the two faces of $\mathcal{T}_1$ lifts to N triangular faces on X, where $N = \deg(\beta)$, and similarly each of the three edges lifts to N edges on X, so $\mathcal{T}$ has $2N$ faces and $3N$ edges. Now suppose that over a vertex $v = 0, 1$ or ∞ of $\mathcal{T}_1$, the N sheets come together in cycles of lengths $n_{v,1}, \ldots, n_{v,k_v}$ (where $n_{v,1} + \cdots + n_{v,k_v} = N$); since v has valency 2, it follows that $\beta^{-1}(v)$ consists of k_v vertices of valencies $2n_{v,1}, \ldots, 2n_{v,k_v}$. Using this triangulation $\mathcal{T}$ of X, one can compute the Euler characteristic χ and the genus g of X:

$$2 - 2g = \chi = (k_0 + k_1 + k_\infty) - 3N + 2N = k_0 + k_1 + k_\infty - N . \quad (1)$$

There is a natural 2-colouring of the faces of $\mathcal{T}$: one can label the faces $+$ or $-$ as they cover the upper or lower half-plane of $\mathbf{C}$, so each edge separates faces with different labels. Similarly the vertices can be 3-coloured, or labelled $0, 1$ or ∞, depending on which vertex v of $\mathcal{T}_1$ they cover, so that each edge joins vertices with different labels. The three vertices on each face of $\mathcal{T}$ have different labels, and the face is called *positive* or *negative* as its orientation (induced by β from the orientation of Σ) corresponds to the cyclic order (01∞) or $(\infty10)$ of these labels.

For instance, the Belyĭ function $\beta : \Sigma \to \Sigma, x \mapsto x^n$ in Example 1 of §2 gives rise to a triangulation $\mathcal{T}$ of Σ with two vertices of valency $2n$ at 0 and ∞, and n vertices of valency 2 at the n-th roots $\exp(2\pi i j / n)$ of 1 (where there is no branching, so $n_{1,j} = 1$ for each j). There are $3n$ edges, one joining each $\exp(2\pi i j / n)$ to 0 and one joining it to ∞, and n joining 0 to ∞ (in this case, though not in general, the edges are all euclidean line-segments in $\mathbf{C}$); these edges enclose $2n$ triangular faces, n labelled $+$ and n labelled $-$. In Example 2, the Belyĭ function $\beta : (x, y) \mapsto x^n$ of degree n^2 gives a triangulation of the Fermat curve X_n with $3n$ vertices (all of valency $2n$), $3n^2$ edges, and $2n^2$ triangular faces.

This arises in topological graph theory as a minimum-genus orientable embedding of the complete tripartite graph $K_{n,n,n}$ [RY]: there are three sets of n vertices, labelled $0, 1$ and ∞, and each pair with distinct labels are joined by a single edge.

3.2. Bipartite maps. If we delete the vertex at ∞ and its two incident edges in $\mathcal{T}_1$, we are left with a map $\mathcal{B}_1$ consisting of a graph $\mathcal{G}_1$ embedded in Σ. This has one edge $I = [0, 1] \subset \mathbf{R}$ joining two vertices 0 and 1, and a single face $\Sigma \setminus I$. Our Belyĭ function β lifts $\mathcal{B}_1$ to a map $\mathcal{B} = \beta^{-1}(\mathcal{B}_1)$ on X, which can be obtained from $\mathcal{T}$ by deleting its vertices labelled ∞ and their incident edges. The embedded graph $\mathcal{G} = \beta^{-1}(I)$ is bipartite: its vertices can be coloured black or white as they cover 0 or 1, so that each edge of $\mathcal{G}$ joins vertices of different colours. There are k_0 black vertices and k_1 white vertices, of valencies $n_{0,1}, \ldots, n_{0,k_0}$ and $n_{1,1}, \ldots, n_{1,k_1}$ respectively. There are N edges in $\mathcal{G}$ (each covering I), and $\mathcal{B}$ has k_∞ faces with $2n_{\infty,1}, \ldots, 2n_{\infty,k_\infty}$ sides. As in §3.1, the genus and characteristic of X are given by (1). Just as one can obtain $\mathcal{B}$ from $\mathcal{T}$ by deleting vertices and edges, one can obtain $\mathcal{T}$ from $\mathcal{B}$ by reversing this process: an extra vertex (labelled ∞) is placed in each face of $\mathcal{B}$, and this is joined by non-intersecting edges to the vertices of $\mathcal{B}$ incident with that face. The resulting triangulation $\mathcal{T}$ is called the *stellation* of $\mathcal{B}$.

For instance, the Belyĭ function $\beta : \Sigma \to \Sigma, x \mapsto x^n$ in Example 1 of §2 yields a bipartite map $\mathcal{B}$ with one black vertex of valency n at 0, joined by a single edge to each of the n-th roots of 1 (which are coloured white). This is an example of a plane tree (an embedding of a tree $\mathcal{G}$ in $\mathbf{C}$), a situation which always arises if $X = \Sigma$ and the Belyĭ function β is a polynomial; we will examine this connection with trees more generally in §6. Example 2 of §2 gives an embedding of the complete bipartite graph $K_{n,n}$ in the Fermat curve X_n: there are n black and n white vertices, each black and white pair joined by a single edge, giving a total of n^2 edges; the map has n faces, each of them a $2n$-gon formed from $2n$ faces of the Fermat triangulation described in §3.1. For more examples of bipartite maps arising from Belyĭ pairs, see [Jon].

3.3. Hypermaps. The dual of the triangulation $\mathcal{T}$ in §3.1 is a trivalent map on X, together with a 3-colouring of its faces with the labels $0, 1$ and ∞. These three types of faces form the hypervertices, hyperedges and hyperfaces of a hypermap $\mathcal{H}$ on X, which can equivalently be obtained by using β to lift the trivial hypermap $\mathcal{H}_1$ (with one hypervertex, one hyperedge and one hyperface) from Σ to X. The Walsh map $W(\mathcal{H})$ associated with $\mathcal{H}$ (see [Wal]) is the bipartite map $\mathcal{B}$ we have just described, with black vertices, white vertices and faces of $\mathcal{B}$ corresponding to hypervertices, hyperedges and hyperfaces of $\mathcal{H}$. For a good survey of hypermaps, see [CM], and for their relevance to Belyĭ functions see [JS2, JS3].

3.4. Maps. For our next example we need the idea of a *clean* or *pure* Belyĭ function: this is a Belyĭ function with the property that (in the notation of §3.1) $n_{1,j} = 2$ for all $j = 1, \ldots, k_1$, so that the sheets come together in pairs over 1 (and hence the degree $N = 2k_1$ must be even). It is easily seen that if $\beta : X \to \Sigma$ is any Belyĭ function then $\gamma = 4\beta(1 - \beta) : X \to \Sigma$ is a clean Belyĭ function of degree $2\deg(\beta)$; this is formed by composing β with the quadratic polynomial $q : \Sigma \to \Sigma$, $z \mapsto 4z(1 - z)$ (which is itself a clean Belyĭ function).

Given any clean Belyĭ function $\gamma : X \to \Sigma$, we can construct a map on X. Let us first draw a map $\mathcal{M}_1$ on Σ consisting of a single vertex at 0, and a single "free edge" (or "half-edge ") along the real line-segment $I = [0, 1]$ from 0 to the point 1 (which is not itself a vertex); the rest of Σ forms the single face of $\mathcal{M}_1$. We then define $\mathcal{M}$ to be the map $\gamma^{-1}(\mathcal{M}_1)$ on X. Since the vertex 0 has valency 1, $\gamma^{-1}(0)$ consists of k_0 vertices of valencies $n_{0,j}$ ($j = 1, \ldots, k_0$). Since $n_{1,j} = 2$ for all j, the half-edge in $\mathcal{M}_1$ lifts to $k_1 = N/2$ edges on X, each containing an element of $\gamma^{-1}(1)$. Since the only branching over the unique face of $\mathcal{M}_1$ is at ∞, there are k_∞ faces in $\mathcal{M}$ with $n_{\infty,j}$ sides ($j = 1, \ldots, k_\infty$), each containing an element of $\gamma^{-1}(\infty)$. In particular, if $\gamma = 4\beta(1-\beta)$ for some Belyĭ function β on X, then $\mathcal{M}$ is the bipartite map $\mathcal{B}$ formed from β in §3.2: this is because $q^{-1}(\mathcal{M}_1) = \mathcal{B}_1$, so q lifts the single vertex 0 of $\mathcal{M}_1$ to the two vertices 0 and 1 of $\mathcal{B}_1$, and then β lifts these to the black and white vertices of $\mathcal{M} = \mathcal{B}$.

The combinatorial structures obtained from Belyĭ functions as above are sometimes known as *dessins d'enfants*. The natural action of **G** on Belyĭ pairs induces actions of **G** on the various types of *dessins* representing these pairs, and explicit examples of these actions are given in §7. As shown in the Appendix, one can give algebraic definitions of the degree N and the partitions $n_{v,1} + \cdots + n_{v,k_v}$ of N associated with the critical values $v = 0, 1, \infty$, in such a way that these are invariant under this action of **G**. It follows that **G** preserves such numerical parameters as the genus, the numbers of vertices, edges and faces, the valencies of the vertices, etc. For example, **G** leaves invariant the set of clean Belyĭ functions and hence acts on the set of maps $\mathcal{M}$ in §3.4. Similarly, since **G** preserves the genus of X and the branching-pattern over ∞, it acts on the set of plane trees associated with Belyĭ polynomials $\beta : \Sigma \to \Sigma$ in §3.2. Other **G**-invariant properties include coverings and orientation-preserving automorphism groups.

Although **G** preserves all the above properties, it nevertheless acts faithfully in the sense that each non-identity element of **G** transforms some *dessin* into a non-isomorphic *dessin* of the same type. This remains true even when one restricts the action of **G** to very simple *dessins*: for example, Schneps [Sch1] has used results of Lenstra on polynomials to show that **G** acts faithfully on plane trees. This allows one to study **G** in a very explicit way, both visually and computationally, and it motivates the choice of plane trees for our examples in §7. Before this, we will show how to use permutations to represent Belyĭ pairs; it is possible to do this directly in terms of branched coverings, without reference to combinatorial structures, but conceptually it is perhaps easiest to understand the connection in terms of bipartite maps.

§4. Belyĭ pairs and permutations

Let (X, β) be a Belyĭ pair, and let $\mathcal{B}$ be the bipartite map $\beta^{-1}(\mathcal{B}_1)$ on X described in §3.2. The N sheets of the covering $\beta : X \to \Sigma$ can be identified with the set E of edges of the bipartite graph $\mathcal{G} = \beta^{-1}(I)$, one edge lying on each sheet. The positive orientation of Σ lifts, *via* β, to an orientation of the surface X, and this induces a cyclic ordering of the edges around each vertex of $\mathcal{G}$. Each edge $e \in E$ is incident with a unique black vertex in $\beta^{-1}(0)$ and a unique white vertex in $\beta^{-1}(1)$, and so the cyclic orderings around the black and white vertices of $\mathcal{G}$ form the disjoint cycles of a pair of permutations g_0 and g_1 of E. These permutations describe how the sheets are permuted by using β to lift rotations in Σ around 0 and 1, or equivalently how the edges of $\mathcal{G}$ are permuted by rotations around their incident black and white vertices; the cycle-lengths of g_0 and g_1 are therefore the valencies of these vertices, forming the partitions $n_{v,1} + \cdots + n_{v,k_v}$ of N associated with the critical values $v = 0$ and 1 in §3.1. Let G denote the subgroup $\langle g_0, g_1 \rangle$ generated by g_0 and g_1 in the symmetric group $S^E \cong S_N$ of all permutations of E; since $\mathcal{G}$ is connected, G is transitive. We call G the *monodromy group* of (X, β), or of $\mathcal{B}$, since it is the monodromy group of the branched covering $\beta : X \to \Sigma$. We will show in §5 that every two-generator transitive permutation group arises in this way from some Belyĭ pair.

The following result is well-known, and is easily proved:

Lemma. *Let (X, β) and (X', β') be Belyĭ pairs, with bipartite maps $\mathcal{B}$ and $\mathcal{B}'$, and with monodromy groups $G = \langle g_0, g_1 \rangle$ and $G' = \langle g_0', g_1' \rangle$ in S_N. Then the following are equivalent:*

a) *the Belyĭ pairs (X, β) and (X', β') are equivalent;*

b) *the bipartite maps $\mathcal{B}$ and $\mathcal{B}'$ are isomorphic;*

c) *the pairs (g_0, g_1) and (g_0', g_1') are conjugate in S_N, that is, some $g \in S_N$ satisfies $g^{-1} g_i g = g_i'$ for $i = 0, 1$.*

Notice that (c) implies that G and G' are conjugate in S_N, so in particular they are isomorphic. Our definition of G may seem to give undue prominence to the critical values 0 and 1 of β, whereas Belyĭ's Theorem gives equal status to ∞. In fact, the element $g_\infty := (g_0 g_1)^{-1} \in G$ describes the permutation of sheets induced by a rotation around ∞, each cycle of length n corresponding to a $2n$-gonal face of $\mathcal{B}$. One can obtain a more symmetric representation of Belyĭ pairs in terms of permutations by using the triangulations $\mathcal{T}$ in place of the bipartite maps $\mathcal{B}$. The set to be permuted now consists of the positive triangles of $\mathcal{T}$, and the permutations g_v are obtained by using the orientation of X to rotate these triangles around their vertices labelled $v = 0, 1$ and ∞, so that $g_0 g_1 g_\infty = 1$. If we identify each positive triangle with its unique edge labelled 01, then g_0 and g_1 are identified with the permutations of E defined earlier, so $\langle g_0, g_1, g_\infty \rangle$ is the monodromy group G.

By representing G as the Galois group of an extention of function-fields, we will show in the Appendix that the permutation group (G, E) is invariant under $\mathbf{G}$, up to permutation-isomorphism. By this we mean that if (G, E) and (G^σ, E^σ) correspond to conjugate Belyĭ pairs (X, β) and (X^σ, β^σ), where $\sigma \in \mathbf{G}$, then there is an isomorphism $G \to G^\sigma, g \mapsto g'$ and a bijection $E \to E^\sigma, e \mapsto e'$ such that $(eg)' = e'g'$ for all $e \in E$ and $g \in G$; equivalently, one can identify E and E^σ with $\{1, \ldots, N\}$ so that G and G^σ are conjugate subgroups of S_N. We will also show that the partitions of N associated with the critical values $v = 0, 1$ and ∞ are invariant under $\mathbf{G}$, so the corresponding generators g_v and g_v^σ of G and G^σ have the same cycle-structures and are therefore conjugate in S_N for each v. However, they need not be *simultaneously* conjugate, by the same conjugating permutation, so that although $G \cong G^\sigma$, there need not necessarily be an isomorphism $G \to G^\sigma$ carrying each g_v to g_v^σ. Thus (X, β) and (X^σ, β^σ) need not be equivalent, and it is this possibility which allows $\mathbf{G}$ to act non-trivially (and indeed faithfully) on the equivalence classes of Belyĭ pairs.

There is another closely-related permutation group associated with a Belyĭ pair (X, β), namely its *cartographic group* C. This depends on a method for representing maps by permutations which was developed in the 1970s by Malgoire and Voisin [MV] and by Jones and Singerman [JS1], though the basic idea goes back at least as far as Hamilton [Ham]. The cartographic group is rather more general than the monodromy group, since it can be defined for any oriented map $\mathcal{M}$, whether bipartite or not. If the embedded graph $\mathcal{G}$ has N edges, then the set Ω permuted by

C consists of the $2N$ *darts* (or directed edges) of $\mathcal{M}$; we define $C = \langle r_0, r_1 \rangle \leq S^\Omega$, where r_0 uses the orientation of X to rotate darts around the vertex to which they point, while r_1 is the involution which reverses the direction of each dart (so $r_2 := (r_0 r_1)^{-1}$ rotates darts around faces). Since $\mathcal{G}$ is connected, C acts transitively on Ω. Indeed, C can be identified with the monodromy group of the bipartite map $W(\mathcal{M})$ formed by inserting a new vertex of valency 2 in each edge of $\mathcal{G}$, so that darts of $\mathcal{M}$ correspond to edges of $W(\mathcal{M})$, with r_0 and r_1 corresponding to rotations of edges around the old and new vertices. (If we regard $\mathcal{M}$ as a hypermap then $W(\mathcal{M})$ is its Walsh map, described in §3.3.)

Any Belyĭ pair (X, β) determines a bipartite map $\mathcal{B} = \beta^{-1}(\mathcal{B}_1)$; as an oriented map, $\mathcal{B}$ has a cartographic group $C = \langle r_0, r_1 \rangle \leq S^\Omega$, which we also define to be the cartographic group of (X, β). Our main theorem shows that (C, Ω), like the monodromy group (G, E), is invariant under **G**. Unlike G, which may be primitive, C is always imprimitive, preserving a non-trivial equivalence relation on Ω: two darts of $\mathcal{B}$ are equivalent if they point to vertices of the same colour, so that r_0 preserves the two equivalence classes Ω_0 and Ω_1 of N black and N white darts while r_1 transposes them. Thus C acts imprimitively on Ω, with two blocks Ω_0 and Ω_1 of N black and N white darts, so C has a normal subgroup $D = \langle r_0, r_0^{r_1} \rangle$ of index 2 which preserves the colours, while the elements of $C \setminus D$ transpose them. One easily checks that r_0 and $r_0^{r_1}$ permute the black darts in Ω_0 as g_0 and g_1 permute their underlying edges, whereas they act on Ω_1 as g_1 and g_0 respectively act on E; thus D acts as G on both blocks, these two actions differing only by a transposition of generators. This gives an embedding $D \leq G \times G$, and likewise C is embedded in the wreath product $G \wr S_2$ (an extension of $G \times G$ by a complement $\langle r_1 \rangle \cong S_2$ which transposes the two factors by conjugation). In particular, $2|G| \leq |C| \leq 2|G|^2$ (and we shall see that both bounds are attained infinitely often), so that although C carries a little more information than G, it can be significantly larger and hence harder to compute and to work with.

§5. Belyĭ's Theorem and uniformisation

Uniformisation theory allows one to formulate equivalent versions of Belyĭ's Theorem, which characterise the Riemann surfaces defined over $\overline{\mathbf{Q}}$ in terms of certain Fuchsian groups. Let $\mathcal{U}$ be the upper half-plane $\{ z \in \mathbf{C} \mid \text{Im}(z) > 0 \}$. This is a non-compact, simply-connected Riemann surface, with automorphism group

$$\text{Aut}(\mathcal{U}) = PSL_2(\mathbf{R})$$

consisting of the Möbius transformations

$$z \mapsto \frac{az + b}{cz + d}$$

where $a, b, c, d \in \mathbf{R}$ and $ad - bc = 1$. The *modular group* is the subgroup $\Gamma = PSL_2(\mathbf{Z})$ consisting of those transformations with $a, b, c, d \in \mathbf{Z}$ and $ad - bc = 1$, and the *principal congruence subgroup* $\Gamma(2)$ of level 2 is the normal subgroup of index 6 in Γ consisting of those elements for which a and d are odd while b and c are even (this is the kernel of the reduction of Γ mod (2)). A *triangle group* $\Delta = \Delta(l, m, n)$ is the subgroup of $\text{Aut}(\mathcal{U})$ generated by rotations through angles $2\pi/l, 2\pi/m$ and $2\pi/n$ about the vertices of a hyperbolic triangle with internal angles $\pi/l, \pi/m$ and π/n, where l, m and n are integers greater than 1 (such triangles exist in $\mathcal{U}$ if and only if $l^{-1} + m^{-1} + n^{-1} < 1$).

Belyĭ's Theorem can now be restated as follows (see [Bel, CIW, Wol2] for the proof):

Theorem. *If X is a compact Riemann surface then the following are equivalent:*

a) X is defined over $\overline{\mathbf{Q}}$;

b) $X \cong \mathcal{U}/K$ for some subgroup K of finite index in a triangle group $\Delta < PSL_2(\mathbf{R})$;

c) $X \cong \overline{\mathcal{U}/L}$ for some subgroup L of finite index in the modular group Γ;

d) $X \cong \overline{\mathcal{U}/M}$ for some subgroup M of finite index in the principal congruence subgroup $\Gamma(2)$.

In (c), $\overline{\mathcal{U}/L}$ denotes the compactification of $\mathcal{U}/L$, formed by adding finitely many points (called the *cusps*) corresponding to the orbits of L on the rational projective line $\mathbf{P}^1(\mathbf{Q}) = \hat{\mathbf{Q}} = \mathbf{Q} \cup \{\infty\}$; the same applies to $\overline{\mathcal{U}/M}$ and $\Gamma(2)$ in (d). When X satisfies the conditions of the Theorem, one obtains Belyĭ functions β of the form

$$X \cong \mathcal{U}/K \to \mathcal{U}/\Delta \cong \Sigma$$

in (b),

$$X \cong \overline{\mathcal{U}/L} \to \overline{\mathcal{U}/\Gamma} \cong \Sigma$$

in (c), or

$$X \cong \overline{\mathcal{U}/M} \to \overline{\mathcal{U}/\Gamma(2)} \cong \Sigma$$

in (d). Here the arrows represent the natural projections, induced by the inclusions $K \leq \Delta$, $L \leq \Gamma$ and $M \leq \Gamma(2)$, while the isomorphisms $\mathcal{U}/\Delta \cong \Sigma$, $\overline{\mathcal{U}/\Gamma} \cong \Sigma$ and $\overline{\mathcal{U}/\Gamma(2)} \cong \Sigma$ are induced by the Δ-invariant Schwarz triangle function, by the Γ-invariant modular function $J : \mathcal{U} \to \Sigma$, and by the $\Gamma(2)$-invariant lambda-function $\lambda : \mathcal{U} \to \Sigma$. In each case, the degree of β is equal to the index of the relevant subgroup.

The natural action of $\mathbf{G}$ on Belyĭ pairs now induces actions of $\mathbf{G}$ on the various sets of subgroups in (b), (c) and (d), or more precisely, on conjugacy classes of such subgroups, since conjugacy corresponds to equivalence of Belyĭ pairs.

One can use this version of Belyĭ's Theorem to show how pairs of permutations give rise to Belyĭ pairs. The universal covering space of $\Sigma_0 = \Sigma \setminus \{0, 1, \infty\}$ is the upper half-plane $\mathcal{U}$, with the projection $\mathcal{U} \to \Sigma_0$ given by the lambda-function λ. The group of covering transformations of λ is the principal congruence subgroup $\Gamma(2)$ of Γ, a free group of rank 2 generated by the Möbius transformations

$$T_0 : z \mapsto \frac{z}{-2z+1} \qquad \text{and} \qquad T_1 : z \mapsto \frac{z-2}{2z-3}.$$

(One can therefore identify $\Gamma(2)$ with the fundamental group of Σ_0, which is freely generated by the homotopy classes of loops in

C which wind once around 0 and 1 respectively in the positive direction.) If permutations g_0, g_1 generate a transitive subgroup G of S_N, then there is an epimorphism $\Gamma(2) \to G$, $T_i \mapsto g_i$, and the inverse image M of a point-stabiliser in G is a subgroup of index N in $\Gamma(2)$. If we take $X = \overline{\mathcal{U}/M}$ as in part (d) of the Theorem, and if β is the projection $X \to \Sigma$ induced by λ, then (X, β) is a Belyǐ pair with monodromy group G.

Conversely, given a Belyǐ pair (X, β) one can obtain its monodromy group $G = \langle g_0, g_1 \rangle$ by taking g_0 and g_1 to be the permutations of the cosets of M in $\Gamma(2)$ induced by the generators T_0 and T_1. Similarly, the cartographic group C can be identified with the action of the congruence subgroup $\Gamma_0(2)$ on the cosets of M. This group $\Gamma_0(2)$, sometimes called *Grothendieck's cartographic group*, consists of those elements of Γ with c even; it has index 3 in Γ, it contains $\Gamma(2)$ as a normal subgroup of index 2, and it is a free product of two cyclic groups of orders ∞ and 2, whose generators

$$U_0 = T_0 : z \mapsto \frac{z}{-2z + 1} \quad \text{and} \quad U_1 : z \mapsto \frac{z - 1}{2z - 1},$$

fixing 0 and 1, induce the permutations r_0 and r_1 of the darts. In this action, $\Gamma(2)$ acts as the colour-preserving subgroup D of index 2 in C, its generators T_0 and T_1 (which are conjugate under U_1) inducing the permutations r_0 and $r_0^{r_1}$.

The various combinatorial structures associated with a Belyǐ pair (X, β) in §3 can now be obtained as quotients of similar structures on $\mathcal{U}$ or $\overline{\mathcal{U}}$. For example, the triangulation $\mathcal{T}$ of X has the form $\hat{\mathcal{T}}/L$, where L is the subgroup of Γ in (c), and $\hat{\mathcal{T}}$ is the *universal triangulation* of $\overline{\mathcal{U}}$: this has vertex-set $\hat{\mathbf{Q}} = \mathbf{Q} \cup \{\infty\}$, with vertices a/b and c/d (in reduced form) joined by a hyperbolic geodesic (euclidean semicircle or half-line) if and only if $ad - bc = \pm 1$, so that $\text{Aut} \, \hat{\mathcal{T}} = \Gamma$. Similarly, the bipartite map $\mathcal{B}$ has the form $\hat{\mathcal{B}}/M$ where M is as in (d) and $\hat{\mathcal{B}}$ is the *universal bipartite map*: again this lies on $\overline{\mathcal{U}}$, but in this case the vertices are the elements $a/b \in \hat{\mathbf{Q}}$ with b odd, coloured black or white as a is even or odd; as before, the condition for an edge between

a/b and c/d is that $ad - bc = \pm 1$, so the automorphism group of $\hat{\mathcal{B}}$ (preserving vertex-colours) is $\Gamma(2)$.

§6. Plane trees and Shabat polynomials

In this section, we will consider the simplest situation in which one can find a faithful action of $\mathbf{G}$. The simplest Riemann surface X is the Riemann sphere Σ, the unique Riemann surface of genus $g = 0$. This has automorphism group

$$
\begin{aligned}
\mathrm{Aut}(\Sigma) \; &= \; PSL_2(\mathbf{C}) \\
&= \; \{\, T : z \mapsto \frac{az + b}{cz + d} \mid a, b, c, d \in \mathbf{C}, \; ad - bc = 1 \,\}.
\end{aligned}
$$

The meromorphic functions on Σ are the rational functions $P(z)/Q(z)$, where P and Q are polynomials with complex coefficients. Perhaps the simplest rational functions are the polynomials themselves, which can be characterised as the meromorphic functions with a unique pole, located at ∞. A polynomial P has ∞ as a critical value (unless $\deg(P) \le 1$), and it will be a Belyĭ function on Σ (called a *Belyĭ polynomial*) provided its only finite critical values lie in $\{0, 1\}$. More generally, we call P a *Shabat polynomial*, or a *generalised Chebyshev polynomial*, if it has at most two critical values in $\mathbf{C}$; by choosing an affine transformation of $\mathbf{C}$ which sends these into $\{0, 1\}$, we obtain a Belyĭ polynomial $aP + b$ of the same degree, so one can regard Shabat polynomials and Belyĭ polynomials as essentially equivalent.

Example. The n-th degree Chebyshev polynomial $T_n(z) = \cos(n \cos^{-1} z)$ has its finite critical values all equal to ± 1 [BSMM, §16.6.3.1], so T_n is a Shabat polynomial. It follows that the functions $(T_n + 1)/2$, T_n^2 and $1 - T_n^2$ are all Belyĭ polynomials, of degrees $N = n, 2n$ and $2n$ respectively. Since T_n has only simple zeros, $1 - T_n^2$ is, in fact, a clean Belyĭ polynomial, formed as in §3.4 by composing $(T_n + 1)/2$ with the quadratic polynomial q. The recurrence relation $T_n = 2zT_{n-1} - T_{n-2}, T_0 = 1, T_1 = z$ shows that T_n has integer coefficients, so T_n is invariant under $\mathbf{G}$, as are these other Belyĭ polynomials obtained from T_n.

In §3.2 we showed how to construct a bipartite map from a Belyĭ pair; we will now extend this construction to the case of a Shabat polynomial P. Let P have finite critical values c_0 and c_1, let L denote the euclidean line-segment from c_0 to c_1 in $\mathbf{C}$, and let $\mathcal{G} = \mathcal{G}_P = P^{-1}(L) \subset \mathbf{C}$. This is a graph embedded in Σ, with $N = \deg(P)$ edges mapped homeomorphically onto L by P. Its vertices are the elements of $P^{-1}(c_0)$ and $P^{-1}(c_1)$, which we will colour black and white respectively; since each edge joins a black vertex to a white vertex, $\mathcal{G}$ is bipartite. Let $\mathcal{P} = \mathcal{P}_P$ be the bipartite map formed from this embedding of $\mathcal{G}_P$ in Σ. Since the N sheets of the covering $P : \Sigma \to \Sigma$ all come together in a single cycle at ∞, this map has a single face, so $\mathcal{G}$ can contain no circuits. Since this face is simply-connected, its complement $\mathcal{G}$ is connected and is therefore a tree. We call this the *plane tree* associated with P, since it is embedded in the complex plane. It is uniquely determined by P, apart from the choice of vertex-colours, which can be transposed by renumbering c_0 and c_1. For example, if P is the Chebyshev polynomial T_n, with critical values ± 1, then $L = [-1, 1] \subset \mathbf{R}$ and $\mathcal{G} = T_n^{-1}(L)$ is a path along the real axis from 1 to -1, with vertices at the points $z = \cos(j\pi/n)$ for $j = 0, 1, \ldots, n$, alternately coloured black or white as they project onto -1 or 1 (see Figure 1, which shows the restriction of $T_6(z) = 32z^6 - 48z^4 + 18z^2 - 1$ to $\mathbf{R}$).

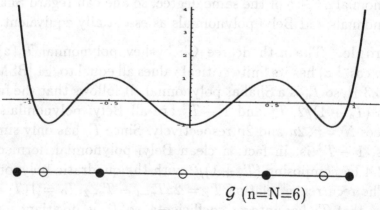

$\mathcal{G}$ (n=N=6)

Figure 1

In $\mathcal{P}_P$ there are k_0 black vertices and k_1 white vertices, of valencies $n_{0,1}, \ldots, n_{0,k_0}$ and $n_{1,1}, \ldots, n_{1,k_1}$ respectively, where $n_{0,1} + \cdots + n_{0,k_0} = n_{1,1} + \cdots + n_{1,k_1} = N$. To simplify the notation, let us write

$$k_0 = p, \quad k_1 = q, \quad n_{0,j} = \alpha_j \quad \text{and} \quad n_{1,j} = \beta_j ,$$

so that α_j and β_j are the orders of the zeros of P and $P - 1$, satisfying

$$\alpha_1 + \cdots + \alpha_p = \beta_1 + \cdots + \beta_q = N . \qquad (2)$$

Since $k_\infty = 1$ and $g = 0$, equation (1) gives $k_0 + k_1 = N + 1$, that is,

$$p + q = N + 1 . \qquad (3)$$

If α and β denote the partitions of N in (2), we will call the ordered pair $\langle \alpha; \beta \rangle$ the *type* of $\mathcal{P}_P$; by (3), these two partitions contain a total of $N + 1$ parts. Every finite tree is bipartite, and therefore has a type, which is unique up to transposition of α and β. Conversely, by a simple induction on N one can show that every pair of partitions of N, with $N + 1$ parts, is the type of some tree with N edges. In general, there may be several such trees, and each tree may have several plane embeddings, but the number of non-isomorphic plane trees of a given type must be finite. (We regard two plane trees as isomorphic if there is an orientation-preserving self-homeomorphism of $\mathbf{C}$ taking one to the other; in the case of bicoloured plane trees, we also require the vertex-colours to be preserved.)

As in the case of a Belyĭ function, one can represent $\mathcal{P}_P$ by a transitive pair g_0, g_1 of permutations of the edges, with cycle-lengths α_j and β_j. The condition that the map should have a single face is equivalent to the permutation $g_\infty = (g_0 g_1)^{-1}$ being a single cycle of length N, in which case the condition that the map should be planar is equivalent to g_0 and g_1 having a total of $N + 1$ cycles. Isomorphism classes of bicoloured plane trees are in one-to-one correspondence with conjugacy classes of such pairs in S_N. The subgroup $G = \langle g_0, g_1 \rangle$ of S_N generated by g_0 and g_1

is the monodromy group of the branched covering $P : \Sigma \to \Sigma$, permuting the edges transitively.

If the coefficients of P are all in $\overline{\mathbf{Q}}$, then each automorphism $\sigma \in \mathbf{G}$ transforms P into a Shabat polynomial P^σ, which has an associated plane tree $\mathcal{P}_P^\sigma$. This action of $\mathbf{G}$ preserves the type $\langle \alpha; \beta \rangle$ and the monodromy group G (up to conjugacy in S_N), though $\mathbf{G}$ does not generally preserve the isomorphism class of a bicoloured plane tree.

§7. Examples of plane trees and their groups

Example 1. The monodromy group G of a Shabat polynomial P of degree N must contain the cyclic group $\langle g_\infty \rangle \cong C_N$ of order N generated by g_∞. Let us consider the simplest case, in which $G = \langle g_\infty \rangle \cong C_N$. In this case, both g_0 and g_1 must be powers of the N-cycle g_∞, so they consist of p and q cycles of length N/p and N/q respectively, where p and q divide N. The genus formula (3) states that $p + q = N + 1$, so the only possibilities are that $p = 1$ and $q = N$, or vice versa. In the first case, $g_1 = 1$ and so $g_0 = g_\infty^{-1}$; the plane tree $\mathcal{P}_P$ has a single black vertex of valency N, and there is a white vertex of valency 1 on each of its N edges (see Figure 2). In the second case, the roles of g_0 and g_1 are interchanged, and the corresponding tree is obtained from Figure 2 by transposing the colours. The first tree can be obtained from the Shabat polynomial $P(z) = z^N$ (as we saw in §3.2), and the second arises from the polynomial $1 - z^N$.

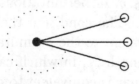

Figure 2

The cartographic groups of these trees are transitive subgroups $C \leq S_{2N}$. In each case, r_0 is a cycle of length N, while r_1

consists of N transpositions. The two disjoint N-cycles r_0 and $r_0^{r_1}$ commute, generating the colour-preserving subgroup $C_0 \cong C_N \times C_N$ of index 2 in C, and r_1 acts by conjugation on C_0, transposing the two direct factors, so C is the wreath product $C_N \wr S_2$, of order $2N^2$.

Example 2. The second simplest situation is that in which the monodromy group G has order $2N$. It contains a cyclic subgroup $\langle g_\infty \rangle \cong C_N$, which has index 2 and is therefore normal in G. Equation (3) implies that at least one of g_0 and g_1 must have a fixed-point; by transposing colours, if necessary, we can assume that it is g_0. We can then label the edges with the elements of $\mathbf{Z}_N$ so that g_∞ acts as the translation $i \mapsto i + 1$, and g_0 fixes the edge 0. Now the stabiliser in G of this edge has index N and hence has order 2, so g_0 must be an involution, generating this stabiliser. Since g_0 normalises $\langle g_\infty \rangle$, it permutes the edges as an automorphism of the additive group $\mathbf{Z}_N$, acting as $i \mapsto ui$ for some involution u in the multiplicative group U_N of units in $\mathbf{Z}_N$. If $u = -1$ (with $N > 2$), for example, then G is the dihedral group D_N of transformations $i \mapsto \pm i + b$ ($b \in \mathbf{Z}_N$) of $\mathbf{Z}_N$; one easily checks that $p = (N + 2)/2$ or $(N + 1)/2$ and $q = N/2$ or $(N + 1)/2$ as N is even or odd, so (3) is satisfied. The corresponding tree is the path of N edges shown in Figure 3, obtained from the Shabat polynomial $(T_N + 1)/2$ of degree N (see §6). The cartographic group is the dihedral group D_{2N} of order $4N$.

N odd   N even

Figure 3

For certain values of N, there may be other involutions $u \in U_N$: if $N = 2^e$ with $e \geq 3$, for example, there are two extra involutions $u = (N \pm 2)/2$, in which cases the permutations $g_0 : i \mapsto ui$ and $g_\infty : i \mapsto i + 1$ generate the semiabelian and semidihedral groups of order $2N$. For any involution $u \neq -1$, a simple but tedious argument, counting cycles of g_0 and g_1, shows that $p+q < N+1$, so $G = \langle g_0, g_\infty \rangle$ is the monodromy group of a single-faced bipartite map of genus greater than 0. Thus among the plane trees with N edges, only the paths of length $N > 2$ have monodromy groups of order $2N$.

Example 3.   There are two plane trees of type $\langle \alpha; \beta \rangle = \langle 4, 1, 1; 2, 2, 1, 1 \rangle$, shown in Figure 4, and in each case, the monodromy group G is a transitive subgroup of S_6.

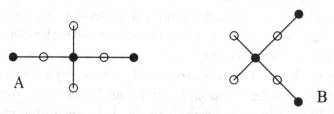

Figure 4

In case A, one can use the obvious rotational symmetry to divide the six edges into three pairs which are permuted by g_0 and g_1; thus G is imprimitive, and is contained in the wreath product $S_2 \wr S_3$, the largest subgroup of S_6 preserving this set of three pairs. One can verify by hand, or by means of a program such as GAP, that $|G| = 48 = |S_2 \wr S_3|$, so $G = S_2 \wr S_3$. (Alternatively, one can identify the three pairs of edges with antipodal vertices of an octahedron $\mathcal{O}$, so that g_0 acts as a rotation of order 4 and g_1 as a reflection, generating the full isometry group of $\mathcal{O}$.) The rotational symmetry of the tree corresponds to the fact that g_0 and g_1 (and hence G) commute with an element of order 2 in S_6: this is the central involution of $S_2 \wr S_3$, which transposes the edges in each pair, or equivalently the antipodal isometry of $\mathcal{O}$.

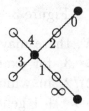

Figure 5

In case B, Figure 5 shows how one can identify the six edges with the points $0, 1, 2, 3, 4, \infty$ of the projective line $PG_1(5)$ over the field $GF(5)$ of five elements, so that g_0 and g_1 induce the projective transformations

$$t \mapsto 2t \quad \text{and} \quad t \mapsto \frac{t-2}{t-1}.$$

These generate the group $PGL_2(5)$ of all projective transformations of $PG_1(5)$, so $G \cong PGL_2(5)$ (a group of order 120, isomorphic as an abstract group to S_5). The reflectional symmetry in case B corresponds to the fact that the pair (g_0, g_1) is conjugate in S_6 to $(g_0^{-1}, g_1^{-1}) = (g_0^{-1}, g_1)$, so that the tree is invariant under orientation-reversal.

Since cases A and B have non-isomorphic monodromy groups, they cannot be conjugate under $\mathbf{G}$. Having no other conjugates, they must therefore be defined over $\mathbf{Q}$, and indeed it is shown in [BPZ] that they can be obtained from the Shabat polynomials

$$P(z) = z^4(z^2 - 1) \quad \text{and} \quad P(z) = z^4(z^2 - 2z + \frac{25}{9}).$$

(The imprimitivity in case A is reflected in the fact that this first function is a polynomial in z^2, so that $z \mapsto -z$ is an automorphism of the tree.)

Example 4. Simple trial and error shows that there are, up to isomorphism and colour-reversal, exactly four plane trees of type $\langle 4, 2, 1; 2, 2, 1, 1, 1 \rangle$; these are shown in Figure 6 (which is also Figure 18 of [SZ]).

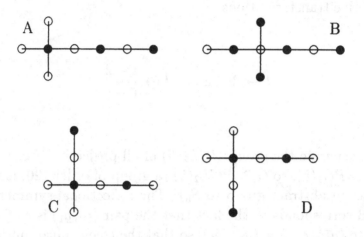

Figure 6

From their cycle-structures we see that g_0 and g_1 are always even permutations, so in each case the monodromy group G is contained in the alternating group A_7. Being a transitive group of prime degree, G must be primitive. In case A we find that $g_0^2 g_1$ has cycle-structure $3, 2, 2$, so $(g_0^2 g_1)^2$ is a 3-cycle; now a primitive group containing a 3-cycle must contain the alternating group [Wie, Theorem 13.3], so $G = A_7$. In case B we find that the commutator $[g_0, g_1] = g_0^{-1} g_1^{-1} g_0 g_1$ has cycle-structure $3, 2, 2$, so the same argument shows that $G = A_7$; in this case, however, $g_0^2 g_1$ has cycle-structure $4, 2, 1$, so although the two monodromy groups are the same, the two triples (G, g_0, g_1) are not isomorphic. This is an algebraic confirmation of what is visually obvious, that the two plane trees are not isomorphic.

In cases C and D, the groups G are again primitive subgroups of A_7, but now they are proper subgroups, isomorphic to $GL_3(2)$, the simple group of order 168. This group (which is also isomorphic to $PSL_2(7)$) is the automorphism group of the Fano plane

$\mathcal{F}$, the 7-point projective plane over $GF(2)$ shown in Figure 7.

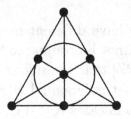

Figure 7

$GL_3(2)$ has two inequivalent faithful representations of degree 7, on the points and lines of $\mathcal{F}$, but these are transposed by the outer automorphism $g \mapsto g^* = (g^{-1})^T$ (the point-line duality of $\mathcal{F}$), so there is a unique conjugacy class of subgroups isomorphic to $GL_3(2)$ in S_7; there are 30 groups in this class (corresponding to the 30 Fano structures one can impose on a 7-element set), and within A_7, they split into two classes of size 15.

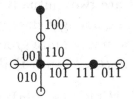

Figure 8

In Figure 8 the edges in C are labelled with triples ijk to identify them with the points (i, j, k) of $\mathcal{F}$, so that g_0 and g_1 act as the projective transformations of $\mathcal{F}$ induced by the matrices

$$\begin{pmatrix} 1 & & \\ 1 & 1 & \\ & & 1 \end{pmatrix} \quad \text{and} \quad \begin{pmatrix} 1 & 1 & \\ & 1 & \\ & & 1 \end{pmatrix}.$$

Thus $G \leq GL_3(2)$, and since one can easily show that these two matrices generate $GL_3(2)$, we have equality. Case D is the mirror-image of case C, and if we use this reflectional symmetry to transfer edge-labels from C to D we find that the permutations

g_0 and g_1 in case D are the inverses of those in case C, so again $G \cong GL_3(2)$.

Since cases C and D have different monodromy groups from cases A and B, they cannot be conjugate to A or B under **G**. In fact, it is shown in [BPZ] that these four plane trees can all be obtained from the Shabat polynomial $P(z) = z^4(z-1)^2(z-a)$, where $a = (3\pm2\sqrt{21})/9$ in cases A and B, and $a = (-1\pm i\sqrt{7})/4$ in cases C and D; in each case, the non-trivial automorphism of the field of definition extends (in uncountably many ways) to an element of **G**, so A and B are conjugate to each other, as are C and D. Thus **G** has two orbits of length 2 on these four trees.

One can confirm algebraically that there are no other plane trees of this type. The elements $g_0 \in S_7$ with cycle-structure $\alpha = 4, 2, 1$ are all even, and form a single conjugacy class in A_7 (consisting of its elements of order 4); similarly, the elements g_1 with cycle-structure $\beta = 2, 2, 1, 1$ form a single class in A_7, consisting of the involutions; the element $g_\infty = (g_0 g_1)^{-1}$ must be a 7-cycle, and there are two mutually inverse conjugacy classes of these in A_7, forming a single class in S_7. Now if A, B and C are conjugacy classes in any finite group H, then the number of triples $(a, b, c) \in A \times B \times C$ such that $abc = 1$ is given by the formula

$$\frac{|A| \cdot |B| \cdot |C|}{|H|} \sum_\chi \frac{\chi(a)\chi(b)\chi(c)}{\chi(1)},$$

where χ ranges over the irreducible complex characters of H (see [Ser, Theorem 7.2.1], for example). If we take $H = A_7$, with A and B its unique classes of elements of orders 4 and 2, and C either of its two classes of elements of order 7, then by using the character table of A_7 (in [CCNPW], for example) we find that there are 2.7! triples $(g_0, g_1, g_\infty) \in A \times B \times C$ satisfying $g_0 g_1 g_\infty = 1$, so S_7 contains 4.7! such triples in all. Any such triple must generate a subgroup G of A_7 of order divisible by $4.7 = 28$, and from knowledge of the groups of degree 7 we find that the only possibilities are that $G = A_7$ or $G \cong GL_3(2)$. By applying the same counting formula to $H = GL_3(2)$ (which has unique classes of elements of orders 4 and 2, and two mutually inverse classes of elements of order 7), we find that $GL_3(2)$ contains two classes of 168 triples; there are 30 subgroups $G \cong GL_3(2)$ in S_7, so the number of triples in S_7 generating such subgroups is $30.2.168 = 2.7!$, and the remaining $4.7! - 2.7! = 2.7!$ triples generate A_7. Now S_7 permutes its triples by conjugation, and since both A_7 and $GL_3(2)$ have trivial centralisers in S_7, each orbit has length 7!, so the triples form four conjugacy classes in S_7, two generating A_7 and two generating subgroups isomorphic to $GL_3(2)$. These correspond to the cases A, B, C and D.

Example 5. For a given type $\langle \alpha; \beta \rangle$, conjugate trees always have isomorphic monodromy groups, and in Examples 3 and 4, the converse was also true, but this is not always the case. Consider, for example, trees of type $\langle 3, 2, 1, 1; 3, 2, 1, 1 \rangle$. In any such example, G is a transitive (and hence primitive) group of degree 7 containing a transposition g_0^3, so $G = S_7$ by [Wie, Theorem 13.3]. However, it is shown in [BPZ] that there are nine trees of this type, forming two orbits (of lengths 3 and 6) under **G** (see Figure 9), so here we have non-conjugate trees of the same type with the same monodromy group.

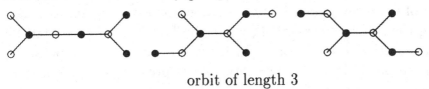

orbit of length 3

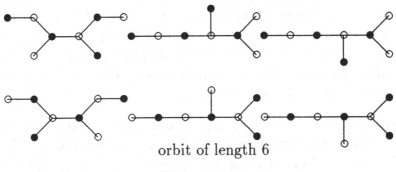

orbit of length 6

Figure 9

Thus, although the monodromy group will in many cases distinguish non-conjugate trees of the same type, it does not always do so. Nevertheless, one can use the monodromy permutations g_i to distinguish the two orbits of **G** in this example. Each of the three trees on the top of Figure 9 is isomorphic (by a rotation of order 2) to the tree obtained by transposing its vertex-colours, so that the pair (g_0, g_1) is conjugate in S_7 to (g_1, g_0); equivalently, if we place the centre of rotation at the origin, then $\beta(-z) = 1 - \beta(z)$, so $\beta - \frac{1}{2}$ is an odd polynomial. This last property is invariant under **G**, and it is not shared by the remaining six trees since they have no such colour-reversing rotation, so they cannot be conjugate to the first three trees.

Although their monodromy groups fail to distinguish the two orbits in this example, their cartographic groups C are different. In each case, C is a transitive but imprimitive group of degree 14, with two monochrome blocks of size 7, so $C \le S_7 \wr S_2$. In the case of the first orbit, the pairing of darts induced by the rotational symmetry gives a second system of imprimitivity, consisting of seven blocks of size 2, and C is the group $S_7 \times S_2$ of order 2.7! preserving both of these systems: the factor S_7 is the subgroup C_0 of C preserving each of the monochrome blocks, and permuting them both 'in parallel', while the factor S_2 is generated by the rotational symmetry, which transposes the two blocks. In the second orbit, this symmetry is not present, so there is only one system of imprimitivity. In this case C is a group of order $(7!)^2$, having index 2 in $S_7 \wr S_2$: the colour-preserving subgroup C_0 is the subgroup of index 2 in $S_7 \times S_7$ which induces permutations of the same parity on each of the two blocks, and C is a split extension of C_0 by a complement S_2 which transposes the blocks. This illustrates the principle that although the cartographic group can be harder to determine than the monodromy group, it is a rather finer invariant.

Example 6. Our final example shows that in some cases, even the cartographic group is not a fine enough invariant to distinguish orbits of **G**. A plane tree of type $\langle 5, 1^{(22)}; 6, 6, 5, 5, 5 \rangle$ has a 'central' black vertex v_0 of valency 5; this is adjacent to white vertices of valencies 6, 6, 5, 5 and 5, and these are in turn adja-

cent to $5, 5, 4, 4$ and 4 additional black vertices, all of valency
1. Up to isomorphism, there are just two possibilities, shown in
Figure 10: in the cyclic order of the white vertices around v_0,
the two 6-valent vertices could be consecutive or not.

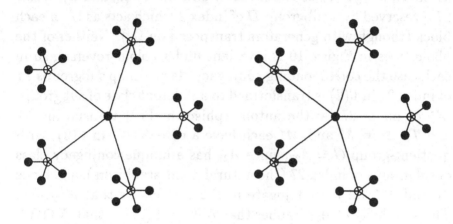

Figure 10

In [SZ, §5.1, Example 2] it is shown that these two plane trees
are defined over $\mathbf{Q}$, so they lie in separate orbits of $\mathbf{G}$ (this is
the special case $s = 4, t = 3, d = 1$ of an infinite class of such
inequivalent pairs). In each case, the monodromy group G is a
transitive group of degree 27, contained in A_{27} since g_0 and g_1 are
even permutations. If G were imprimitive, it would have three
blocks of size 9 or else nine blocks of size 3, permuted transitively
by G. In the first case, the elements g_0 and g_1^6 of order 5 must
permute the blocks trivially, so each of their 5-cycles would be
contained completely in some block; by inspection of Figure 10,
the three 5-cycles in g_1^6 (around the 5-valent white vertices) all
intersect the unique 5-cycle in g_0 (around v_0), so all 17 edges
in these 5-cycles would have to lie in a single block, which is
impossible. In the second case, each element of order 5 in G
must fix point-wise any block which it leaves invariant; it would
therefore fix $9 - 5 = 4$ blocks and hence have cycle-structure
$5^{(3)}, 1^{(12)}$ on the edges, whereas g_0 has cycle-structure $5, 1^{(22)}$.
Thus G acts primitively on the 27 edges, and since it contains
the 5-cycle g_0, where 5 is a prime not exceeding $27 - 3 = 24$, a

theorem of Jordan [Wie, Theorem 13.9] implies that G contains A_{27}, so $G = A_{27}$.

For each of these two plane trees, the cartographic group C acts as a transitive but imprimitive group on the 54 darts: as shown at the end of §4, it has two monochrome blocks Ω_0 and Ω_1 of size 27, preserved by a subgroup D of index 2 which acts as G on each block (though with generators transposed on Ω_1). Neither of the plane trees in Figure 10 is invariant under colour-reversal, so in each case the corresponding conjugacy class of map subgroups M of index 27 in $\Gamma(2)$ is transformed to a different class of subgroups M^* of index 27 by the automorphism of $\Gamma(2)$ transposing T_0 and T_1. Now M and M^* each have a core N, N^* in $\Gamma(2)$, with quotient-group $G = A_{27}$; since A_{27} has a unique conjugacy class of subgroups of index 27 (the natural point-stabilisers), and since M and M^* are not conjugate in $\Gamma(2)$, it follows that $N \neq N^*$. The simplicity of A_{27} implies that $NN^* = \Gamma(2)$, so that $N \cap N^*$, which is the core of M in $\Gamma_0(2)$, has quotient-group $\Gamma(2)/(N \cap N^*) \cong A_{27} \times A_{27}$. Thus $D \cong A_{27} \times A_{27}$, and since the two direct factors are conjugate in C we have $C \cong A_{27} \wr S_2$, the largest subgroup of S_{54} with two blocks of imprimitivity, each permuted evenly. There is a unique conjugacy class of such subgroups C in S_{54}, so the plane trees in Figure 10 have permutation-isomorphic cartographic groups, even though they are not conjugate under **G**.

Appendix

Invariance of monodromy and cartographic groups

Theorem. *Let (X, β) be a Belyĭ pair, with monodromy group (G, E) and cartographic group (C, Ω). If $\sigma \in \mathbf{G}$, then the monodromy group (G^σ, E^σ) and the cartographic group $(C^\sigma, \Omega^\sigma)$ of (X^σ, β^σ) are permutation-isomorphic to (G, E) and to (C, Ω).*

(This means that G and G^σ are conjugate in S_N, and C and C^σ are conjugate in S_{2N}, where $N = \deg(\beta)$.)

Proof. It is sufficient to prove that monodromy groups are invariant, since the cartographic group of β is permutation-isomorphic to the monodromy group of the Belyĭ function $q \circ \beta = 4\beta(1 - \beta)$. A complete proof for the monodromy group is given by Matzat in [Mat], in great detail and in considerably more generality than we require here, so we will simply outline the main steps in a rather simplified version of the argument. The basic idea is to realise the monodromy group as the action of a certain Galois group, which is preserved by $\mathbf{G}$. For background information on the connections between coverings, monodromy groups and Galois groups, see [Dou], and for the connections between Riemann surfaces and algebraic curves, see [Rey].

We have defined the monodromy group of β to be a permutation group (G, E), where E is the set of N edges of the bipartite map $\mathcal{B} = \beta^{-1}(\mathcal{B}_1)$ on X associated with β, and G is the group of permutations of E generated by the rotations g_0 and g_1 around the black and white vertices in $\beta^{-1}(0)$ and $\beta^{-1}(1)$. Let us choose base-points $\frac{1}{2} \in \Sigma_0$ and $p_0 \in \beta^{-1}(\frac{1}{2}) \subset X_0$, and define $\Pi = \pi_1(\Sigma_0, \frac{1}{2})$, the fundamental group of Σ_0. The restriction of β to X_0 induces a monomorphism $\beta_* : \pi_1(X_0, p_0) \to \Pi$ with image $\beta_*(\pi_1(X_0, p_0))$ a subgroup Φ of index N in Π. There is a natural bijection between E and the fibre $\Omega = \beta^{-1}(\frac{1}{2})$ of β above $\frac{1}{2}$, each edge containing a unique point in this fibre, so we can regard G as permuting Ω rather than E. In fact, this is the more usual definition of a monodromy group: there is a natural action of Π on Ω, each element $g \in \Pi$ sending a point $b \in \Omega$ to the common end-point $b^g \in \Omega$ of the paths in X_0 which start at b and project (under β) to the closed paths in Σ_0 represented by g; since Π is generated (freely) by the two homotopy classes of paths winding once round 0 and 1 respectively, and since these induce the permutations g_0 and g_1 on E, one can use the bijection $E \to \Omega$ to verify that (G, E) is permutation-isomorphic to the group of permutations induced by Π on Ω, or equivalently on the N cosets of Φ (the subgroup of Π fixing the point $p_0 \in \Omega$).

The kernel of this action of Π on Ω is the core $\tilde{\Phi}$ of Φ; this is the intersection of the conjugates of Φ in Π, a normal subgroup with $\Pi/\tilde{\Phi} \cong G$. Since $\tilde{\Phi}$ is the largest normal subgroup of Π contained in Φ, it corresponds to the minimal regular lifting

$\tilde{\beta} : \tilde{X} \to \Sigma$ of β; this is a regular covering of Σ which factors through β, and which has the property that every other regular lifting of β factors through $\tilde{\beta}$. If $\tilde{X}_0 = \tilde{\beta}^{-1}(\Sigma_0)$ and if $\tilde{p}_0 \in \tilde{X}_0$ projects onto p_0, then $\tilde{\Phi}$ is the image of $\pi_1(\tilde{X}_0, \tilde{p}_0)$ under the monomorphism $\tilde{\beta}_*$ of fundamental groups induced by $\tilde{\beta}$. One can use $\tilde{\beta}$ to lift the complex structure from Σ to $\tilde{X}$ so that $\tilde{X}$ is a compact Riemann surface and $\tilde{\beta}$ is meromorphic; being unbranched outside $\{0, 1, \infty\}$, $\tilde{\beta}$ is a Belyĭ function on $\tilde{X}$. The group of covering transformations of $\tilde{\beta}$ is isomorphic to G; in fact, this group acts regularly on the fibre $\tilde{\Omega} = \tilde{\beta}^{-1}(\frac{1}{2})$, and by projecting from $\tilde{\Omega}$ to Ω we obtain the monodromy action of G on Ω.

Let $\mathcal{M}(\ )$ denote the field of meromorphic functions on a Riemann surface. If we use coverings to lift meromorphic functions from one surface to another, then the chain of coverings $\tilde{X} \to X \to \Sigma$ induces a chain of field extensions

$$\mathcal{M}(\tilde{X}) \supseteq \mathcal{M}(X) \supseteq \mathcal{M}(\Sigma) = \mathbf{C}(\tilde{\beta}) \cong \mathbf{C}(t) \qquad (1)$$

(of degrees equal to those of the corresponding coverings). Since $\tilde{\beta}$ is a regular covering, $\mathcal{M}(\tilde{X})$ is a normal extension of $\mathcal{M}(\Sigma)$, with Galois group $G_1 = \text{Gal}(\mathcal{M}(\tilde{X})/\mathcal{M}(\Sigma)) \cong G$: the action of G on $\tilde{X}$ as the group of covering transformations of $\tilde{\beta}$ induces an action on meromorphic functions, with fixed-field $\mathcal{M}(\Sigma)$. Under the Galois correspondence, the intermediate field $\mathcal{M}(X)$ corresponds to the subgroup $H_1 = \text{Gal}(\mathcal{M}(\tilde{X})/\mathcal{M}(X))$ of index N in G_1 which fixes it point-wise. There are N coverings $X \to \Sigma$, all conjugate under G to β, which have minimal regular lifting $\tilde{\beta}$, and these correspond to N distinct monomorphisms from $\mathcal{M}(X)$ into $\mathcal{M}(\tilde{X})$, all conjugate under G_1. In this action of G_1 the stabiliser of the monomorphism induced by β is H_1, so there is a permutation-isomorphism $(G, \Omega) \cong (G_1, \Omega_1)$, where Ω_1 is the set of cosets of H_1 in G_1.

Belyĭ's Theorem implies that the surfaces X and $\tilde{X}$, and the coverings β and $\tilde{\beta}$, are all defined over $\overline{\mathbf{Q}}$, so the field-extensions in (1) can be obtained from a chain of field-extensions

$$\mathcal{M}_{\overline{\mathbf{Q}}}(\tilde{X}) \supseteq \mathcal{M}_{\overline{\mathbf{Q}}}(X) \supseteq \mathcal{M}_{\overline{\mathbf{Q}}}(\Sigma) = \overline{\mathbf{Q}}(\tilde{\beta}) \cong \overline{\mathbf{Q}}(t) \qquad (2)$$

by extension of constant-fields from $\overline{\mathbf{Q}}$ to $\mathbf{C}$. (In simple terms, this means that the surfaces and coverings in (1) can be defined by polynomials and rational functions with coefficients in $\overline{\mathbf{Q}}$, and the fields in (2) are the rational function fields on these surfaces, which we regard as varieties over $\overline{\mathbf{Q}}$.) The passage from (2) to (1) by extension of constant-fields preserves intermediate fields, their degrees, and their Galois groups, so $(G_2, \Omega_2) \cong (G_1, \Omega_1)$ where $G_2 = \mathrm{Gal}\,(\mathcal{M}_{\overline{\mathbf{Q}}}(\tilde{X})/\mathcal{M}_{\overline{\mathbf{Q}}}(\Sigma))$ and Ω_2 is the set of cosets of $H_2 = \mathrm{Gal}\,(\mathcal{M}_{\overline{\mathbf{Q}}}(\tilde{X})/\mathcal{M}_{\overline{\mathbf{Q}}}(X))$ in G_2.

The finite extension

$$\mathcal{M}_{\overline{\mathbf{Q}}}(\tilde{X}) \supseteq \mathcal{M}_{\overline{\mathbf{Q}}}(\Sigma) = \overline{\mathbf{Q}}(\tilde{\beta})$$

corresponds to a non–singular projective curve $C \subset P^m(\overline{\mathbf{Q}})$ together with a covering $\tilde{\beta} : C \to P^1(\overline{\mathbf{Q}})$ (strictly speaking, this is the restriction of $\tilde{\beta}$ to the subset C of $\overline{\mathbf{Q}}$-rational points of $\tilde{X}$, where X is embedded in $P^m(\mathbf{C})$). By abuse of language we may identify C with its plane model $C \subset P^2(\overline{\mathbf{Q}})$, which always exists, and which (like $\tilde{X}$) is defined by an equation $f(x, y, z) = 0$. Here f is a homogeneous polynomial in $\overline{\mathbf{Q}}[x, y, z]$, and $\tilde{\beta}$ is given by $[x, y, z] \mapsto x/z$ where y/z is a primitive element of $\overline{\mathbf{Q}}(x/z, y/z) = \mathcal{M}_{\overline{\mathbf{Q}}}(\tilde{X})/\overline{\mathbf{Q}}(\tilde{\beta}) = \overline{\mathbf{Q}}(x/z)$.

Any automorphism $\sigma \in \mathbf{G}$ sends f to a homogeneous polynomial $f^\sigma \in \overline{\mathbf{Q}}[x, y, z]$, by acting on its coefficients. It thus sends C to a curve $C^\sigma \subset P^2(\overline{\mathbf{Q}})$ defined by $f^\sigma(x, y, z) = 0$, together with a covering $\tilde{\beta}^\sigma : C^\sigma \to P^1(\overline{\mathbf{Q}})$, $[x, y, z] \mapsto x/z$. This corresponds to an extension-field

$$\mathcal{M}_{\overline{\mathbf{Q}}}(\tilde{X})^\sigma \supseteq \mathcal{M}_{\overline{\mathbf{Q}}}(\Sigma) = \overline{\mathbf{Q}}(\tilde{\beta}^\sigma)\,,$$

and since this action of $\mathbf{G}$ preserves intermediate fields and their Galois groups we get a chain of extensions

$$\mathcal{M}_{\overline{\mathbf{Q}}}(\tilde{X})^\sigma \supseteq \mathcal{M}_{\overline{\mathbf{Q}}}(X)^\sigma \supseteq \mathcal{M}_{\overline{\mathbf{Q}}}(\Sigma) = \overline{\mathbf{Q}}(\tilde{\beta}^\sigma) \cong \overline{\mathbf{Q}}(t)\,, \qquad (3)$$

where the outer extension is normal, with Galois group $G_3 \cong G_2$; in fact, since $\mathbf{G}$ commutes with the actions of these two

Galois groups, σ induces a permutation-isomorphism $(G_3, \Omega_3) \cong (G_2, \Omega_2)$ where Ω_3 is the set of cosets of $H_3 = \mathrm{Gal}\,(\mathcal{M}_{\overline{\mathbf{Q}}}(\tilde{X})^\sigma / \mathcal{M}_{\overline{\mathbf{Q}}}(X)^\sigma)$ in G_3.

By extending the constant-fields in (3) from $\overline{\mathbf{Q}}$ to $\mathbf{C}$ (again preserving intermediate fields and their Galois groups), we get a chain of field-extensions

$$\mathcal{M}(\tilde{X})^\sigma \supseteq \mathcal{M}(X)^\sigma \supseteq \mathcal{M}(\Sigma) = \mathbf{C}(\tilde{\beta}^\sigma) \cong \mathbf{C}(t). \qquad (4)$$

As before, the outer extension is normal with Galois group $G_4 \cong G_3$, and if Ω_4 is the set of cosets of $H_4 = \mathrm{Gal}\,(\mathcal{M}(\tilde{X})^\sigma / \mathcal{M}(X)^\sigma)$ in G_4 then $(G_4, \Omega_4) \cong (G_3, \Omega_3)$. The finite extensions of $\mathbf{C}(\tilde{\beta}^\sigma) \cong \mathbf{C}(t)$ in (4) correspond to a chain of coverings

$$\tilde{X}^\sigma \to X^\sigma \to \Sigma$$

of the same degrees as the original chain $\tilde{X} \to X \to \Sigma$; the outer covering $\tilde{\beta}^\sigma : \tilde{X}^\sigma \to \Sigma$ is the minimal regular lifting of the second covering $\beta^\sigma : X^\sigma \to \Sigma$, and β^σ is unbranched outside $\{0, 1, \infty\}^\sigma = \{0, 1, \infty\}$, so it is a Belyĭ function. By imitating the arguments we used for β, one can show that the monodromy group of β^σ is permutation-isomorphic to the action of the Galois group G_4 on Ω_4, so the isomorphisms $(G_4, \Omega_4) \cong (G_3, \Omega_3) \cong (G_2, \Omega_2) \cong (G_1, \Omega_1) \cong (G, \Omega) \cong (G, E)$ now show that it is permutation-isomorphic to the monodromy group of β.

Invariance of automorphism groups

A Belyĭ function $\beta : X \to \Sigma$ determines a bipartite map $\mathcal{B} = \beta^{-1}(\mathcal{B}_1)$ on X. The (orientation-preserving) automorphism group $\mathrm{Aut}(\mathcal{B})$ of $\mathcal{B}$ is the group of permutations of the darts of $\mathcal{B}$ which commute with the permutations r_0, r_1 and r_2. Group-theoretically, this is the centraliser of the cartographic group $C = \langle r_0, r_1, r_2 \rangle$ in S_{2N}, and topologically, it is the group of covering transformations of $q \circ \beta : X \to \Sigma$. Similarly, the subgroup $\mathrm{Aut}_0(\mathcal{B})$ of $\mathrm{Aut}(\mathcal{B})$ preserving the two vertex-colours is the centraliser of the monodromy group G in S_N, or equivalently the group of covering transformations of β. By the Theorem, $\mathbf{G}$

preserves the conjugacy classes of C in S_{2N} and of G in S_N; since conjugate subgroups have conjugate centralisers, it follows that both $\mathrm{Aut}(\mathcal{B})$ and $\mathrm{Aut}_0(\mathcal{B})$ are invariant under $\mathbf{G}$ (up to permutation-isomorphism).

By contrast, the *full* automorphism group (including orientation-reversing elements) and its colour-preserving subgroup need not be invariant under $\mathbf{G}$. In Example 5 of §7, for instance, Figure 9 shows that in the $\mathbf{G}$-orbit of length 3, one tree has a Klein four-group of automorphisms (generated by two reflections), while the other two trees have only rotations. Similarly in the orbit of length 6, two of the trees admit reflections, while the other four do not. This is analogous to the situation in classical Galois theory, where one cannot distinguish algebraically between real and imaginary roots of an irreducible polynomial.

Invariance of branching-partitions

At each of the possible critical values $v = 0, 1$ and ∞, a Belyĭ function $\beta : X \to \Sigma$ determines a partition

$$n_{v,1} + \cdots + n_{v,k_v} = N \tag{5}$$

of its degree N, which describes the branching of β at v; here k_v is the number of points in the fibre above v, and $n_{v,j}$ is the number of sheets of the covering which meet at the j-th point $v_j \in \beta^{-1}(v)$. (There is no canonical ordering for these points, so we regard (5) as an unordered partition.) We need to show that the action of $\mathbf{G}$ on Belyĭ pairs preserves this partition. For notational simplicity, we will restrict attention to the critical value $v = 0$, writing (5) in the form

$$n_1 + \cdots + n_k = N, \tag{6}$$

where $k = k_0$ and $n_j = n_{0,j}$; the arguments at $v = 1$ and $v = \infty$ are identical, except for an obvious change of local coordinates. Since full details can be found in [Mat], we will just sketch the proof.

Within the meromorphic function field $\mathcal{M}(\Sigma)$, there is a subring $\mathcal{R} = \mathcal{R}_0(\Sigma)$ (the valuation ring at 0), consisting of those meromorphic functions f which are finite (have no pole) at 0. Within this ring, there is a maximal ideal $\mathcal{I} = \mathcal{I}_0(\Sigma)$ (the valuation ideal, or place, at 0), consisting of those meromorphic functions f with a zero at 0; this is the kernel of the evaluation epimorphism $\mathcal{R} \to \mathbf{C}$, $f \mapsto f(0)$, and each r-th power $\mathcal{I}^r$ of $\mathcal{I}$ consists of those f with a zero of order at least r at 0. As we saw in the proof of the Theorem, β induces an embedding of $\mathcal{M}(\Sigma)$ in the field $\mathcal{M}(X)$, where we identify each $f \in \mathcal{M}(\Sigma)$ with $\beta \circ f \in \mathcal{M}(X)$. Under this embedding, the ideal $\mathcal{I}$ decomposes as a product

$$\mathcal{I} = \prod_{j=1}^{k} \mathcal{I}_j^{n_j} ,$$

where $\mathcal{I}_j$ is the valuation ideal in $\mathcal{M}(X)$ corresponding to v_j. (This simply says that a zero of $f \in \mathcal{M}(\Sigma)$ at 0 gives rise to a zero of $\beta \circ f$ at each v_j, the order of this zero being multiplied by n_j; in other words, we find local coordinates around v_j by taking n_j-th roots.) Just as in the proof of the Theorem, this decomposition is obtained by extension of constant-fields from a similar decomposition

$$\mathcal{I}_{\overline{\mathbf{Q}}} = \prod_{j=1}^{k} (\mathcal{I}_j)_{\overline{\mathbf{Q}}}^{n_j} ,$$

where $\mathcal{I}_{\overline{\mathbf{Q}}}$ and $(\mathcal{I}_j)_{\overline{\mathbf{Q}}}$ are the valuation ideals in $\mathcal{M}_{\overline{\mathbf{Q}}}(\Sigma)$ and $\mathcal{M}_{\overline{\mathbf{Q}}}(X)$ at 0 and v_j. (Notice that v_j is a $\overline{\mathbf{Q}}$-rational point of X since $v_j \in \beta^{-1}(0)$, β is a Belyĭ function and $0 \in \mathbf{Q}$.) The action of $\mathbf{G}$ preserves these decompositions of valuation ideals, and in particular preserves the multiset of exponents n_j, so the unordered partition (6) is invariant under $\mathbf{G}$, as required.

This result means that $\mathbf{G}$ preserves the types of the various combinatorial structures on which it acts; for instance, valencies of vertices are preserved. It also means that, as we have assumed throughout this paper, $\mathbf{G}$ preserves the genus of a surface X, since this can be expressed in terms of the invariant quantities k_0, k_1, k_∞ and N by equation (1) of §3.

References

[Bel] G. V. BELYĬ, 'On Galois extensions of a maximal cyclotomic field', *Izv. Akad. Nauk SSSR* 43 (1979) 269–276 (Russian); *Math. USSR Izvestiya* 14 (1980), 247–256 (English transl.).

[BPZ] J. BÉTRÉMA, D. PÉRÉ and A. ZVONKIN, 'Plane trees and their Shabat polynomials. Catalogue', preprint, Bordeaux, 1992.

[BSMM] I. N. BRONSTEIN, K. A. SEMENDJAJEW, G. MUSIOL and H. MÜHLIG, *Taschenbuch der Mathematik*, Verlag Harri Deutsch, Thun / Frankfurt am Main, 1993.

[CIW] P. B. COHEN, C. ITZYKSON and J. WOLFART, 'Fuchsian triangle groups and Grothendieck dessins. Variations on a theme of Belyi', *Comm. Math. Phys.* 163 (1994), 605–627.

[CCNPW] J. H. CONWAY, R. T. CURTIS, S. P. NORTON, R. A. PARKER and R. A. WILSON, *ATLAS of Finite Groups*, Clarendon Press, Oxford, 1985.

[CM] R. CORI and A. MACHÌ, 'Maps, hypermaps and their automorphisms: a survey I, II, III', *Expositiones Math.* 10 (1992), 403–427, 429–447, 449–467.

[Cou] J.-M. COUVEIGNES, 'Calcul et rationalité de fonctions de Belyi en genre 0', *Annales de l'Institut Fourier* 44 (1994).

[Dou] R. and A. DOUADY, *Algèbre et Théories Galoisiennes II*, CEDIC / Fernand Nathan, Paris, 1979.

[Gro] A. GROTHENDIECK, 'Esquisse d'un programme', preprint, Montpellier, 1984.

[Ham] W. R. HAMILTON, Letter to John T. Graves 'On the Icosian' (17th October 1856), *Mathematical papers, Vol. III, Algebra*, eds. H. Halberstam and R. E. Ingram, Cambridge University Press, Cambridge, 1967, pp. 612–625.

[Jac] N. JACOBSON, *Basic Algebra II*, Freeman, San Fransisco, 1980.

[Jon] G. A. JONES, 'Maps on surfaces and Galois groups', submitted.

[JS1] G. A. JONES and D. SINGERMAN, 'Theory of maps on orientable surfaces', *Proc. London Math. Soc.* (3) 37 (1978), 273–307.

[JS2] G. A. JONES and D. SINGERMAN, 'Maps, hypermaps and triangle groups', in *The Grothendieck Theory of Dessins d'Enfants* (ed. L. Schneps), London Math. Soc. Lecture Note Ser. 200 (1994), pp. 115–145.

[JS3] G. A. JONES and D. SINGERMAN, 'Belyĭ functions, hypermaps and Galois groups', *Bull. London Math. Soc.*, to appear.

[MV] J. MALGOIRE and C. VOISIN, 'Cartes cellulaires', *Cahiers Mathématiques de Montpellier* 12 (1977).

[Mat] B. H. MATZAT, *Konstruktive Galoistheorie*, Lecture Notes in Math. 1284, Springer-Verlag, Berlin, 1987.

[Neu] J. NEUKIRCH, 'Über die absoluten Galoisgruppen algebraischer Zahlkörper', *Astérisque* 41–42 (1977), 67–79.

[Rey] E. REYSSAT, *Quelques Aspects des Surfaces de Riemann*, Birkhäuser, Boston / Basel / Berlin, 1989.

[RY] G. RINGEL and J. W. T. YOUNGS, 'Das Geschlecht des vollständigen dreifärbbaren Graphen', *Comm. Math. Helv.* 45 (1970), 152–158.

[Sch1] L. SCHNEPS, 'Dessins d'enfants on the Riemann sphere', in *The Grothendieck Theory of Dessins d'Enfants* (ed Schneps), London Math. Soc. Lecture Note Ser. 200 (1994), pp. 47–77.

[Sch2] L. SCHNEPS (ed.), *The Grothendieck Theory of Dessins d'Enfants*, London Math. Soc. Lecture Note Ser. 200 (1994).

[Ser] J-P. SERRE, *Topics in Galois Theory*, Jones and Bartlett, Boston / London, 1992.

[SV] G. B. SHABAT and V. A. VOEVODSKY, 'Drawing curves over number fields', in *Grothendieck Festschrift III* (eds.P. Cartier *et al.*), Progress in Math. 88, Birkhäuser, Basel, 1990, pp. 199–227.

[SZ] G. SHABAT and A. ZVONKIN, 'Plane trees and algebraic numbers', in *Jerusalem Combinatorics '93* (eds. H. Barcelo and

G. Kalai), Amer. Math. Soc. Contemporary Math. Ser. 178, pp. 233–275.

[Wal] T. R. S. WALSH, 'Hypermaps versus bipartite maps', *J. Combinatorial Theory (B)* 18 (1975), 155–163.

[Wei] A. WEIL, 'The field of definition of a variety', *Amer. J. Math.* 78 (1956), 509–524.

[Wie] H. WIELANDT, *Finite Permutation Groups*, Academic Press, New York, 1964.

[Wol1] J. WOLFART, 'Mirror-invariant triangulations of Riemann surfaces, triangle groups and Grothendieck dessins: variations on a thema of Belyi', preprint, Frankfurt, 1992.

[Wol2] J. WOLFART, 'The 'obvious' part of Belyi's theorem and Riemann surfaces with many automorphisms', preprint, Frankfurt, 1995.

Department of Mathematics
University of Southampton
Southampton SO17 1BJ
United Kingdom

Fachbereich Mathematik
J.-W. Goethe Universität
Robert–Mayer–Str. 6–10
60054 – Frankfurt/Main
Germany

Permutation techniques for coset representations
of modular subgroups

Tim Hsu

§1. Introduction

We recall that dessins d'enfants may be considered as one of the following equivalent classes of objects. Let G be an appropriate "cartographical" group. (We may take $G = \Gamma(2)$ for general dessins, $G = \Gamma_0(2)$ for pre-clean dessins, and $G = \mathrm{PSL}_2(\mathbb{Z})$ for pre-clean dessins whose valencies at the "other" type of vertex divide 3.)

1. Coverings of the Riemann sphere, branched (in a manner consistent with G) only above $\{0, 1, \infty\}$, along with a choice of (unramified) basepoint.

2. Drawings (structured in a manner consistent with G) on connected oriented surfaces.

3. Subgroups of G.

4. Transitive permutation representations of G, along with a choice of basepoint.

The correspondences among these classes of objects have been explained many times, so we refer the reader to Birch [2], Jones and Singerman [5], and Schneps [7] for details, mentioning only one piece of terminology: In describing the correspondence between (3) and (4), we call the basepointed transitive permutation representation of G corresponding to a subgroup $G_1 \leq G$ the *coset representation of G_1 as a subgroup of G.*

This article describes some computational methods for working with dessins and similar structures. Briefly, one might say that the questions examined here arise from subgroups (viewpoint (3)), and the answers come from looking at their coset representations (viewpoint (4)). We hope that the methods presented here can be used to address other questions about dessins by looking at them in terms of their coset representations.

Throughout, we will describe coset representations of subgroups of $G =$

1991 Mathematics Subject Classification. Primary 11F06, 20H05; Secondary 20B05

Key words and phrases. Classical modular group, dessins d'enfants, finite permutation groups

$\mathrm{PSL}_2(\mathbb{Z})$ using the following notation. Let

$$\gamma_0 = \begin{pmatrix} 1 & 1 \\ -1 & 0 \end{pmatrix}, \quad \gamma_1 = \begin{pmatrix} 0 & -1 \\ 1 & 0 \end{pmatrix}, \quad \gamma_\infty = \begin{pmatrix} 1 & 1 \\ 0 & 1 \end{pmatrix}.$$

γ_0, γ_1, and γ_∞ satisfy the relations

$$1 = \gamma_0^3 = \gamma_1^2 = \gamma_0 \gamma_1 \gamma_\infty$$

in G, and in fact, given the generators γ_0, γ_1, and γ_∞, (1.2) is a set of defining relations for G. Therefore, to specify a basepointed transitive permutation representation of G, it is enough to specify the images σ_0, σ_1, and σ_∞ of γ_0, γ_1, and γ_∞. In other words, to obtain a basepointed transitive permutation representation of G, we need only choose transitive permutations σ_0, σ_1, and σ_∞ which respect (1.2). (By convention, we take the point numbered 1 as our basepoint.)

For instance,

$$\sigma_0 = (1\ 2\ 3), \quad \sigma_1 = (1\ 2), \quad \sigma_\infty = (2\ 3)$$

is the coset representation of $\Gamma_0(2)$ as a subgroup of G.

By convention, we draw the dessin corresponding to the coset representation of a subgroup of G by saying that σ_0 describes how edges are arranged counterclockwise around vertices marked with a $\circ$, σ_1 describes how edges are arranged counterclockwise around vertices marked with a $|$, and σ_∞ how edges are arranged counterclockwise around a 2-cell. For example, the dessin corresponding to (1.3) is shown in Figure 1.

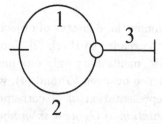

Figure 1. Dessin for $\Gamma_0(2)$

§2. Identifying congruence subgroups

Let $G = \mathrm{PSL}_2(\mathbb{Z})$. In this section, we describe an algorithmic answer to the following problem.

- Given the coset representation of a subgroup $G_1 \leq G$ of finite index, determine if G_1 is a congruence subgroup of G. That is, determine if G_1 contains $\Gamma(N)$ for some value of N.

We begin with the result often known as Wohlfahrt's Theorem [13], for which we need the following definition.

Definition 2.1. Let G_1 be a modular subgroup of finite index, and consider the coset representation of G_1 in G, using the notation of §1. The (generalized) level of G_1 is defined in one of the following equivalent ways, roughly corresponding to the viewpoints listed in §1.

1. The least common multiple of the cusp widths of the quotient of the upper half-plane by G_1.

2. The least common multiple of the numbers of sides of the 2-cells of the dessin corresponding to G_1. (Note that number of sides of a 2-cell is half the number of edges which touch it, counting multiplicities.)

3. The smallest value of N such that all conjugates of γ_∞^N are contained in G_1.

4. The order of the permutation σ_∞.

Theorem 2.2. (Wohlfahrt's Theorem) *Let G_1 be a modular subgroup of generalized level N. Then G_1 is a congruence subgroup if and only if G_1 contains $\Gamma(N)$.*

The following can then be shown. (See [4] for a proof, in slightly different notation.)

Theorem 2.3. *Let G_1 be a modular subgroup of level N, and let*

$$(2.1) \qquad \langle \sigma_0, \sigma_1, \sigma_\infty \mid r_1(\sigma_0, \sigma_1, \sigma_\infty), \ r_2(\sigma_0, \sigma_1, \sigma_\infty), \ \ldots \rangle$$

be a presentation for $\mathrm{SL}(\mathbb{Z}/N)/\{\pm I\}$ which is fulfilled by taking $\sigma_0 = \gamma_0$ (mod N), $\sigma_1 = \gamma_1$ (mod N), and $\sigma_\infty = \gamma_\infty$ (mod N). Then G_1 is a congruence subgroup if and only if the coset representation of G_1 respects the relations $\{r_i\}$.

Implementation. To apply theorem 2.3 effectively, we need a uniform method of presenting $\mathrm{SL}(\mathbb{Z}/N)/\{\pm I\}$. See [4] for one such method. Note that with the uniform method given there, one can write a program to determine if a modular subgroup is congruence by checking no more than 12 relations.

Example 2.4. We illustrate theorem 2.3 by examining two subgroups of index 18 and generalized level 6, one of which is congruence, and the other of which is not. Let G_1 be the modular subgroup with coset representation

$$\sigma_0 = (1\ 13\ 8)(2\ 7\ 16)(3\ 15\ 10)(4\ 9\ 18)(5\ 17\ 12)(6\ 11\ 14),$$

$$(2.3) \qquad \sigma_1 = (1\ 7)(2\ 15)(3\ 9)(4\ 17)(5\ 11)(6\ 13)(8\ 18)(10\ 14)(12\ 16),$$

$$\sigma_\infty = (1\ 2\ 3\ 4\ 5\ 6)(7\ 8\ 9\ 10\ 11\ 12)(13\ 14\ 15\ 16\ 17\ 18),$$

and let G_2 be the modular subgroup with coset representation

$$\sigma_0 = (1\ 15\ 17)(2\ 16\ 6)(3\ 5\ 7)(4\ 12\ 8)(9\ 11\ 13)(10\ 18\ 14),$$

(2.3) $$\sigma_1 = (1\ 16)(2\ 5)(3\ 12)(4\ 7)(6\ 15)(8\ 11)(9\ 18)(10\ 13)(14\ 17),$$

$$\sigma_\infty = (1\ 2\ 3\ 4\ 5\ 6)(7\ 8\ 9\ 10\ 11\ 12)(13\ 14\ 15\ 16\ 17\ 18).$$

The dessins for (2.2) and (2.3) are shown in Figures 2 and 3. Since both dessins have genus 1, we have drawn them as polygons with identified boundaries; identifications are indicated by edge numbers which appear twice.

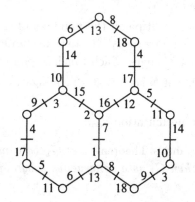

Figure 2. Dessin for G_1

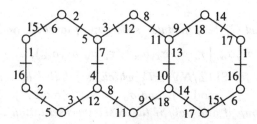

Figure 3. Dessins for G_2

It can be shown by coset enumeration that $\mathrm{SL}(\mathbb{Z}/6)/\{\pm I\}$ has a presentation with generators σ_0, σ_1, and σ_∞, and defining relations

(2.4) $$1 = \sigma_0^3 = \sigma_1^2 = \sigma_0\sigma_1\sigma_\infty = \sigma_\infty^6,$$

(2.5) $$1 = [\sigma_\infty^2, \sigma_1^{-1}\sigma_\infty^{-3}\sigma_1],$$

where $[x,y] = x^{-1}y^{-1}xy$. However, the coset representation of a subgroup of generalized level 6 satisfies (2.4) automatically. Therefore, for groups of level 6, theorem 2.3 boils down to testing (2.5). For G_1, we have

(2.6) $$[\sigma_\infty^2, \sigma_1^{-1}\sigma_\infty^{-3}\sigma_1] = (),$$

and for G_2, we have

$$(2.7) \quad [\sigma_\infty^2, \sigma_1^{-1}\sigma_\infty^{-3}\sigma_1] = (1\ 13\ 7)(2\ 8\ 14)(3\ 15\ 9)(4\ 10\ 16)(5\ 17\ 11)(6\ 12\ 18),$$

so G_1 is congruence and G_2 is noncongruence.

§3. Enlarging subgroups

As a natural extension to the results in the previous section, in this section, we solve the following family of problems. Let $G = \mathrm{PSL}_2(\mathbb{Z})$.

1. Given an arbitrary subgroup G_1 of finite index in G, find the coset representation of $\langle G_1, \Gamma(N) \rangle$ (the closure of the two groups in G).

2. Given an arbitrary subgroup G_1 of finite index in G, find the coset representation of the smallest congruence subgroup $H \leq G$ containing G_1. H is known as the *congruence closure* of G_1. Note that Wohlfahrt's Theorem reduces this problem to problem (1).

3. Given an arbitrary subgroup G_1 of finite index in G, find the coset representation of the smallest subgroup of G which contains G_1 and has generalized level dividing N. In other words, if $U(N)$ is the normal closure of γ_∞^N in G, find the coset representation of $\langle G_1, U(N) \rangle$.

We can generalize these three problems as:

4. Given an arbitrary subgroup G_1 of finite index in G and a "known" subgroup Γ of G, find the coset representation of $\langle G_1, \Gamma \rangle$.

Note that this new formulation of the problem is no longer restricted to the case where $G = \mathrm{PSL}_2(\mathbb{Z})$, and in fact becomes quite a general problem on dessins and related structures.

The goal of this section is to show how we may employ a slight variation of an algorithm which is well-known in another context to solve the above problems efficiently. In doing so, we will also show how to solve the following problem.

5. Given an arbitrary dessin D, find the largest dessin which is covered by D in a manner which identifies two chosen flags of D.

For the reader who may be less familiar with permutation groups, we present a brief summary (after Wielandt [11,§7]) of the theory of imprimitivity, using the language of *lattices*. (For a standard reference, see Crawley and Dilworth [3].) Meet and join in a lattice are denoted by $\wedge$ and $\vee$, respectively. The partial order in a lattice is denoted by $\leq$. Let Ω be a set. A *partition* of Ω is a set P of nonempty disjoint subsets of Ω (the *members* of P) whose union is Ω. We use $\mathrm{Part}(\Omega)$ to denote the set of partitions of

Ω. Part(Ω) has a natural lattice structure (Crawley and Dilworth [3, Ch. 12]) which we describe with the following examples.

Example 3.1. Set

$$\Omega = \{1, 2, 3, 4, 5, 6\},$$
$$E = \{\{1\}, \{2\}, \{3\}, \{4\}, \{5\}, \{6\}\},$$
$$P = \{\{1, 2\}, \{3\}, \{4\}, \{5, 6\}\},$$
$$Q = \{\{1\}, \{2\}, \{3, 4\}, \{5\}, \{6\}\},$$
$$B = \{\{1, 2\}, \{3, 4\}, \{5, 6\}\}.$$

Then $E, P, Q, B \in \mathrm{Part}(\Omega)$, $E \le P \le B \le \{\Omega\}$, $E \le Q \le B$, $P \vee Q = B$, and $P \wedge Q = E$.

Convention 3.2. When Ω is understood, we write partitions of Ω using the conventions for writing cycle shapes of permutations. That is, we omit one-element subsets, commas between subsets, and the outermost pair of brackets, denoting the partition of Ω into one-element subsets by empty brackets. For instance, in the above example, we write $E = \{\}$ and $P = \{1, 2\}\{5, 6\}$.

We also recall that Part(Ω) is a *complete* lattice (Crawley and Dilworth [3, Ch. 12]). In particular, in every sublattice L of Part(Ω), the meet of all elements of L is the unique minimal element of L.

Now suppose a group G acts transitively on Ω. We say that a partition B of Ω is a *system of imprimitivity for G on Ω* if the action of G permutes the members of B. We use $\mathrm{Imp}_G(\Omega)$ to denote the set of systems of imprimitivity for G on Ω. Note that $\mathrm{Imp}_G(\Omega)$ is a sublattice of Part(Ω).

Example 3.3. Let Ω and B be as in Example 3.1, and let G be the group generated by the permutations $(1\ 3\ 5)(2\ 4\ 6)$ and $(1\ 4)(2\ 3)(5\ 6)$. Then $B \in \mathrm{Imp}_G(\Omega)$.

Finally, let 1 be an element of Ω, and let $G_1 = \mathrm{Stab}_G(1)$ (the stabilizer of 1 in G).

Definition 3.4. We define a map Φ_1 from $\mathrm{Imp}_G(\Omega)$ to subgroups of G by saying that, for $B \in \mathrm{Imp}_G(\Omega)$, $\Phi_1(B)$ is the stabilizer of the member of B containing 1.

Theorem 3.5. Φ_1 *is an isomorphism (of lattices) between* $\mathrm{Imp}_G(\Omega)$ *and the lattice of subgroups of G containing G_1.*

Proof. See Wielandt [11, Thm. 7.5] or Aschbacher [1,(5.18)]. $\Diamond$

For the rest of this section, let G be a group acting transitively on a set Ω, let 1 be an element of Ω, let $G_1 = \mathrm{Stab}_G(1)$, let Γ be an arbitrary subgroup of G, and let Γ^G denote the normal closure of Γ in G. For $H \leq G$ and $x \in \Omega$, we use $O_H(x)$ to denote the partition which has $\mathrm{Orb}_H(x)$ (the orbit of x under the action of H) as one of its members, and the single-element subsets of Ω not contained in $\mathrm{Orb}_H(x)$ as its other members. We use O_H to denote the partition of Ω into its H-orbits.

We are now ready to solve problems (1)-(4) in terms of the theory of imprimitivity. The key is the following lemma.

Lemma 3.6. *The isomorphism* Φ_1 *(Definition 3.4) sends the lattice of all* $B \in \mathrm{Imp}_G(\Omega)$ *such that* $O_\Gamma(1) \leq B$ *onto the lattice of all subgroups of* G *containing both* G_1 *and* Γ.

Proof. Suppose $B \in \mathrm{Imp}_G(\Omega)$, and b is the member of B containing 1. If $O_\Gamma(1) \leq B$, then every element of Γ fixes b, that is, $\Gamma \leq \mathrm{Stab}_G(b)$. Conversely, if $\Gamma \leq \mathrm{Stab}_G(b)$, then $\mathrm{Orb}_\Gamma(1)$ must be contained in b, that is, $O_\Gamma(1) \leq B$. $\diamond$

Theorem 3.7. *Let* B_O *be the smallest* $B \in \mathrm{Imp}_G(\Omega)$ *such that* $O_\Gamma(1) \leq B$, *and let* b_O *be the member of* B_O *containing 1. Then* $\mathrm{Stab}_G(b_O) = \langle G_1, \Gamma \rangle$.

Proof. The theorem follows from lemma 3.6 because the minimal elements of isomorphic lattices must correspond. $\diamond$

To prove the analogous theorem for Γ^G, we need a lemma about base-points.

Lemma 3.8. *Suppose* $B \in \mathrm{Imp}_G(\Omega)$. *Then for any* $g \in G$, $O_{g\Gamma g^{-1}}(1) \leq B$ *if and only if* $O_\Gamma(1g) \leq B$.

Proof. For $x \in \Omega$, let b_x be the member of B containing x. We first note that since $B \in \mathrm{Imp}_G(\Omega)$, $b_1 g = b_{1g}$ for all $g \in G$. Now, $\mathrm{Orb}_{g\Gamma g^{-1}}(1)$ is contained in b_1 if and only if $1g\gamma g^{-1} \in b_1$ for all $\gamma \in \Gamma$. However, the latter is true if and only if $(1g)\gamma \in (b_1 g)$ for all $\gamma \in \Gamma$, that is, if and only if $\mathrm{Orb}_\Gamma(1g)$ is contained in b_{1g}. $\diamond$

Theorem 3.9. *Let* B_O *be the smallest* $B \in \mathrm{Imp}_G(\Omega)$ *such that* $O_\Gamma \leq B$, *and let* b_O *be the member of* B_O *containing 1. Then* $\mathrm{Stab}_G(b_O) = \langle G_1, \Gamma^G \rangle$.

Proof. Lemma 3.6 implies that Φ_1 maps the lattice of all subgroups of G containing $\langle G_1, \Gamma^G \rangle$ to the lattice of all $B \in \mathrm{Imp}_G(\Omega)$ such that $O_{g\Gamma g^{-1}}(1) \leq$

B for all $g \in G$. However, using lemma 3.8 and the fact that G is transitive,

$$\{B \in \mathrm{Imp}_G(\Omega) \mid O_{g\Gamma g^{-1}}(1) \le B, \ \forall g \in G\} =$$
$$\{B \in \mathrm{Imp}_G(\Omega) \mid O_\Gamma(1g) \le B, \ \forall g \in G\} =$$
$$\{B \in \mathrm{Imp}_G(\Omega) \mid O_\Gamma \le B\}$$

The theorem then follows from the correspondence of minimal elements. $\Diamond$

When $G = \mathrm{PSL}_2(\mathbb{Z})$, it turns out that theorem 3.9 is particularly useful for solving the motivating problems of §1. For problem (3), we may apply the theorem with $\Gamma = \langle \gamma_\infty^N \rangle$. For problem (1), we may let S be the set of relators of a presentation for $\mathrm{SL}(\mathbb{Z}/N)/\{\pm I\}$, and apply the theorem with Γ equal to the group generated by the elements of S.

For comparison, it is worth noting that the congruence test of §2 is a trivial case of theorem 3.9. To be precise, let S be a set of relators of a presentation for $\mathrm{SL}(\mathbb{Z}/N)/\{\pm I\}$ (i.e., a set of normal generators of $\Gamma(N)$ in $\mathrm{PSL}_2(\mathbb{Z})$), and let $\Gamma = \langle S \rangle$. Then if G_1 has generalized level N, G_1 is congruence if and only if O_Γ is the partition of Ω into its one-element subsets. More generally, the congruence subgroups of $\mathrm{PSL}_2(\mathbb{Z})$ which contain G_1 correspond (under the map Φ_1) precisely with those $B \in \mathrm{Imp}_G(\Omega)$ such that $O_\Gamma \le B$.

Implementation. Examination of Theorems 3.7 and 3.9 shows that we have reduced our original problems to the following problem.

- Given a finitely generated group G which acts transitively on a finite set Ω, and given a partition O of Ω, determine the smallest system of imprimitivity $\ge O$. (In theorem 3.7, $O = O_\Gamma(1)$, and in theorem 3.9, $O = O_\Gamma$.)

Consideration of Theorem 3.7 also shows that solving this problem is exactly what we need to solve problem (5) above; take O to be the partition $\{a, b\}$, where a and b are the flags which are to be identified.

It turns out that the above problem has been studied extensively in computational group theory. For instance, the `BlocksSeed` function for permutation groups in GAP [9] is a readily available algorithm for solving the above problem, and Sims' COINCIDENCE algorithm [10,Ch. 4] is an effectively linear-time algorithm for doing the same thing. Note that both algorithms must be modified slightly to take an arbitrary partition as their input. (An appropriately modified version of `BlocksSeed` is available from the author.)

We demonstrate the use of our results with the following examples.

Example 3.10. Let $G = \mathrm{PSL}_2(\mathbb{Z})$, let $\Omega = \{1, \ldots, 18\}$, and let $G_3 \le G$ be the subgroup of index 18 defined by

$$\sigma_0 = (1\ 11\ 6)(2\ 5\ 12)(3\ 15\ 8)(4\ 10\ 16)(7\ 18\ 13)(9\ 14\ 17),$$
$$\sigma_1 = (1\ 5)(2\ 11)(3\ 10)(4\ 15)(6\ 18)(7\ 12)(8\ 14)(9\ 16)(13\ 17),$$
$$\sigma_\infty = (1\ 2)(3\ 4)(5\ 6\ 7)(8\ 9\ 10)(11\ 12\ 13\ 14\ 15\ 16\ 17\ 18).$$

The dessin for G_3 is shown in Figure 4.

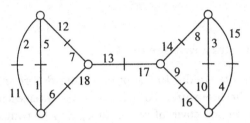

Figure 4. Dessin for G_3

To determine the coset representation of the group $H_1 = \langle G_3, \gamma_\infty \rangle$, we apply theorem 3.7 with $\Gamma = \langle \gamma_\infty \rangle$, which means that $O = O_\Gamma(1) = \{1,2\}$ (using Convention 3.2). A computer calculation shows that the smallest system of imprimitivity $\geq O$ is $\{\Omega\}$, which means that $H_1 = G$.

On the other hand, to determine the coset representation of the group $H_2 = \langle G_3, \gamma_1^{-1}\gamma_\infty\gamma_1 \rangle$, we apply theorem 3.7 with $\Gamma = \langle \gamma_1^{-1}\gamma_\infty\gamma_1 \rangle$, which means that $O = O_\Gamma(1) = \{1,12,18\}$. A computer calculation shows that the smallest system of imprimitivity $\geq O$ is

$$\{\{1,3,12,14,16,18\},\{2,4,11,13,15,17\},\{5,6,7,8,9,10\}\}.$$

Let $a = \{1,3,12,14,16,18\}$, $b = \{2,4,11,13,15,17\}$, and $c = \{5,6,7,8,9,10\}$. We see that σ_0 induces the permutation $(a\ b\ c)$, and σ_1 induces the permutation $(a\ c)$. It is not too hard to see that this implies $H_2 = \gamma_1\Gamma_0(2)\gamma_1^{-1}$. (Compare (1.3).)

Example 3.11. Let $G = \mathrm{PSL}_2(\mathbb{Z})$, let $\Omega = \{1,\ldots,18\}$, and let $G_2 \leq G$ be the subgroup defined by (2.3). Let H be the congruence closure of G_2. Since (2.4) and (2.5) form a set of defining relations for $\mathrm{SL}(\mathbb{Z}/6)/\{\pm I\}$, and (2.4) is satisfied by the coset representation of any group of level 6, to find H, we apply theorem 3.9 with $\Gamma = \langle [\gamma_\infty^2, \gamma_1^{-1}\gamma_\infty^{-3}\gamma_1] \rangle$. However, since

$$[\sigma_\infty^2, \sigma_1^{-1}\sigma_\infty^{-3}\sigma_1] = (1\ 13\ 7)(2\ 8\ 14)(3\ 15\ 9)(4\ 10\ 16)(5\ 17\ 11)(6\ 12\ 18),$$

we just have to find the smallest system of imprimitivity $\geq O$, where

$$O = O_\Gamma = \{\{1,7,13\}\{2,8,14\}\{3,9,15\}\{4,10,16\}\{5,11,17\}\{6,12,18\}\}.$$

However, O is actually invariant under the action of G, so to find the coset representation of H, we just have to look at how G acts on the members of O. Let $a = \{1, 7, 13\}$, $b = \{2, 8, 14\}$, $c = \{3, 9, 15\}$, $d = \{4, 10, 16\}$, $e = \{5, 11, 17\}$, and $f = \{6, 12, 18\}$. Then σ_0 induces $(a\ c\ e)(b\ d\ f)$, σ_1 induces $(a\ d)(b\ e)(c\ f)$, and σ_∞ induces $(a\ b\ c\ d\ e\ f)$. Since the group generated by these three permutations is the abelianization of G, it follows that H is the commutator subgroup of G, a subgroup of index 6, level 6, and genus 1.

§4. Remarks and acknowledgements

Several other results similar to the ones discussed here may be found in various places in the literature. For instance, the intersection of two subgroups (smallest common cover of two dessins) may be computed using the *parallel product*; see Wilson [12]. Also, an efficient algorithm for computing the automorphism group of a dessin (centralizer of a transitive subgroup of the symmetric group) was devised by Kuhn [6, Sect. II], and can be found in GAP [9] as the undocumented function CentralizerTransSymmCSPG.

The author would like to thank A. O. L. Atkin for providing the initial motivation for this article. The author would also like to thank J. H. Conway, K. Harada, R. Lyons, C. C. Sims, R. Solomon, and the anonymous referee, for their help in writing this article. Special thanks goes to L. Schneps, for suggesting that this article be included in this collection.

References

[1]   M. Aschbacher, *Finite group theory*, Cambridge Univ. Press, 1986.

[2]   B. Birch, Noncongruence subgroups, covers, and drawings, in [8], pp. 25–46.

[3]   P. Crawley and R. P. Dilworth, Algebraic theory of lattices, Prentice-Hall, Englewood Cliffs, 1973.

[4]   T. Hsu, Identifying congruence subgroups of the modular group, *Proc. AMS* **124** (1996), no. 5, 1351–1359.

[5]   G. A. Jones and D. Singerman, Maps, hypermaps and triangle groups, in [8], pp. 115–145.

[6]   H. W. Kuhn, On imprimitive substitution groups, *Amer. J. Math.* **26** (1904), 45–102.

[7]   L. Schneps, Dessins d'enfants on the Riemann sphere, in [8], 47-77.

[8]   L. Schneps (ed.), *The Grothendieck theory of dessins d'enfants*, Cambridge Univ. Press, 1994.

[9]   M. Schönert et al., GAP: Groups, algorithms and programming, Lehrstuhl D für Mathematik, RWTH Aachen, April 1992, Version 3.1.

[10] C. C. Sims, *Computation with finitely presented groups*, Cambridge Univ. Press, 1994.

[11] H. Wielandt, *Finite permutation groups*, Academic Press, 1964.

[12] S. E. Wilson, Parallel products in groups and maps, *J. Alg.* **167** (1994), 539–546.

[13] K. Wohlfahrt, An extension of F. Klein's level concept, Ill. J. Math. **8** (1964), 529–535.

Department of Mathematics
Univ. of Michigan
Ann Arbor, MI 48109

e-mail: timhsu@math.lsa.umich.edu

Dessins d'enfants en genre 1

Leonardo Zapponi

§0. Introduction.

Cette introduction est un bref résumé de l'article, visant à dégager la philosophie générale des méthodes et résultat exposés. En gros, le but est celui de mettre en marche les "engrenages" (certains seulement) de ce "mécanisme" qui transforme des données combinatoires (provenant de l'aspect purement topologico-combinatoire des dessins d'enfants) en données arithmétiques (telles que les corps de définition des courbes algébriques associées). La démarche classique, qui permet de relier ces deux aspects, utilise les groupes cartographiques des dessins pour réaliser les revêtements via la théorie de Galois. Une telle interprétation amène à l'étude du groupe fondamental arithmétique d'une courbe définie sur un corps de nombres et peut donc s'appuyer sur des résultats et des techniques profondes de la géométrie algébrique. Malheureusement, la nature générale des objets étudiés dans ce contexte porte souvent à délaisser le côté intuitif du concept de dessin d'enfant (qui en justifie d'ailleurs le nom), en perdant souvent en cours de route des informations combinatoires simples mais utiles.

C'est cette concrétisation si élémentaire qui peut inspirer des méthodes d'analyse aussi diverses que riches. Dans cette optique, avec l'idée en tête que les données combinatoires déterminent complètement celles de nature arithmétiques, nous allons étudier des dessins d'enfants en genre 1. Une attention particulière sera attachée à l'action du groupe de Galois absolu $\Gamma = \mathrm{Gal}(\overline{\mathbb{Q}}/\mathbb{Q})$ sur de telles structures.

Au §1, la famille de dessins est formée de classes de valence particulièrement simples pour lesquelles les paires de Belyi peuvent être déterminées en introduisant les polynômes de Tchebychev (et en jouant sur un automorphisme canonique qui ramène l'etude à celle en genre 0). L'action de Γ est alors facilement explicitée et se réduit à l'action cyclotomique. Il est possible en particulier d'introduire un invariant galoisien "absolu" (i.e. qui détermine complètement les classes de conjuguaison sous Γ). Cet exemple ne porte en lui aucun intérêt majeur, si ce n'est celui de guider et orienter les généralisations et discussions ultérieures.

Au §2, les dessins sont en effet proches, comme nature, de ces derniers et cette analogie est renforcée en reliant la théorie des dessins d'enfants à celle de la décomposition cellulaire de l'espace des modules des courbes algébriques en genre 1 (par le biais des différentiels de Strebel, aspect que nous n'aborderons pas ici). Comme précédemment, le problème peut être reconduit à la détermination de paires de Belyi sur la sphère de Riemann, mais la démarche plus fructueuse et intéressante est une autre, liée à la nature "orientable" (ou "bicoloriable") des dessins considérés. Le premier pas est alors celui de démontrer que les deux pôles de l'application de Belyi (tous les dessins considérés ont deux faces) sont des points de torsion de la courbe, opposés l'un de l'autre – et c'est là le premier argument réellement propre aux courbes elliptiques [1]. L'ordre n de ces points est évidemment un invariant galoisien. On retrouve ici un lien ultérieur avec la première partie où il était question de points de torsion non pas sur une courbe elliptique mais sur $\mathbb{C}^*$. La détermination de cet invariant emprunte des méthodes analytiques complexes (transcendantes !) telles que l'uniformisation (en voyant la courbe comme quotient de $\mathbb{C}$ par rapport à un réseau $\Lambda_\tau = \mathbb{Z} \oplus \tau\mathbb{Z}$). Ceci permet non seulement d'utiliser les fonctions elliptiques mais aussi de représenter de façon simple un point de n-torsion par l'intermédiaire de ses "coordonnées" a, b (modulo n) dans le parallélogramme fondamental. Le fait essentiel est que la simple connaissance du point de torsion P (l'orientation du dessin amène à une distinction entre les deux pôles) détermine complètement le revêtement. Pour obtenir une paire de Belyi, P devra vérifier des relations supplémentaires bien "rigides", qui se traduisent dans l'interprétation analytique (et c'est là qu'elle se montre réellement adéquate) par des relations entre τ et les coordonnées a, b de P.

Plus explicitement, une paire $(\mathbb{C}/\Lambda_\tau, P)$ détermine une paire de Belyi si et seulement si

$$\zeta\left(\frac{a+b\tau}{n}\right) - \eta\left(\frac{a+b\tau}{n}\right) = 0$$

où ζ est la fonction zéta de Weierstrass et η est la fonction de quasi-période. Cette relation est le point central de l'article car elle peut être interprétée comme l'annulation de fonctions de la variable τ indexées (et déterminées) par a, b. De telles fonctions sont en fait des formes modulaires de poids 1 pour des conjugués (dans le groupe modulaire $SL_2(\mathbb{Z})$) du sous-groupe de congruence $\Gamma_1(n)$. Une classe d'entre elles (les fonctions $F_{m,n}$) sont des formes modulaires pour $\Gamma_1(n)$ lui-même et définissent des fonctions méromorphes sur la courbe modulaire $X_1(n) = \Gamma_1(n)\backslash\mathcal{H}$ où $\mathcal{H}$ est le demi-plan supérieur. Les points de $X_1(n)$ paramétrisent de façon naturelle les paires

---

[1] Des résultats analogues, dans un contexte plus général (mettant en jeu les points de n-division sur la jacobienne associée à une courbe hyperelliptique) sont présentés dans le preprint de F. Pakovitch (cf. [P])

(C, P) où C est une courbe elliptique et $P \in C$ est un point de torsion d'ordre exact n. L'annulation des fonctions $F_{m,n}$ caractérise alors les paires amenant à des dessins d'enfants. C'est dans ce sens que nous parlons d' "interprétation modulaire".

Dans la dernière partie du §2, l'attention est centrée sur les invariants modulaires des courbes elliptiques associées aux dessins considérés. Nous allons voir en effet que ces derniers sont les racines de polynômes à coefficients rationnels provenant (et déterminés) par des formes modulaires "globales" Θ_n pour $\mathrm{SL}_2(\mathbb{Z})$; exprimées simplement en termes des fonctions introduites précédemment.

C'est là un point de départ pour une investigation plus générale et systématique, exprimant assez bien d'ailleurs la philosophie de cette deuxième partie. Elle se base sur cette simple considération: Pour une courbe elliptique E, il est possible de construire une paire de Belyi (E, β) si et seulement si l'invariant modulaire $J_0 = J(E)$ est un nombre algébrique. Dans ce cas, si $P(X) \in \mathbb{Q}[X]$ est le polynôme minimal de J_0, alors $f(\tau) = P(J(\tau))$ est une fonction modulaire qui multipliée par une puissance adéquate de $\Delta(\tau)$ devient une forme modulaire.

L'idée est alors celle d'associer à des familles $\mathcal{F}$ de dessins[2] des courbes modulaires [3] $X_{\mathcal{F}}$ déterminées complètement par les caractéristiques combinatoires; les dessins de $\mathcal{F}$ correspondant alors aux zéros de fonctions méromorphes sur $X_{\mathcal{F}}$ définies de façon "canonique".

Dans la théorie générale, les points des courbes modulaires paramétrisent des "formes", i.e. des paires (C, S) où C est comme d'habitude une courbe en genre 1 et S est une "structure" supplémentaire sur laquelle Γ opère de façon naturelle. C'est cette structure qui devrait être déterminée par les caractéristiques de $\mathcal{F}$. Une telle interprétation est particulièrement utile quand on s'intéresse à l'action de Γ, qui se réduit essentiellement à l'action sur la courbe (sur son invariant modulaire) d'une part, et à l'action sur S de l'autre.

Le but ultime est alors celui de traduire l'action de Γ sur la famille $\mathcal{F}$ par l'action de ce dernier sur les points $\overline{\mathbb{Q}}$-rationnels de la courbe modulaire [4]. En particulier, le revêtement canonique $X_{\mathcal{F}} \xrightarrow{J} \mathbb{P}^1$ permet de distinguer les deux différents aspects de cette action.

---

[2] Familles qui pourraient mériter l'appellation de "familles modulaires".

[3] Quotients du demi-plan supérieur par rapport à l'action de sous-groupes d'indice fini du groupe modulaire.

[4] La présentation canonique de $\mathrm{PSL}_2(\mathbb{Z})$ induit sur toute courbe modulaire X une structure de dessin d'enfant réalisée par la paire $(X, J(\tau))$. En particulier, X est définie sur un corps de nombres...

Le §3 aborde le problème de l'action de Γ sur les dessins d'enfants d'une façon différente. Il est question ici de réaliser des groupes de Galois arithmétiques (infinis) comme groupes d'automorphismes des groupes cartographiques (profinis), en accord avec [G] ou [I]. La technique utilisée est simple et se réduit à composer des applications de Belyi avec des isogénies. Malheureusement, de telles méthodes ne s'appliquent qu'à des dessins réguliers, et en genre 1 il en existe essentiellement deux types, correspondant aux courbes elliptiques d'invariant $J = 0$ et $J = 1728$. Il est possible néanmoins de dégager quelques résultats généraux. Le point essentiel est celui de réduire l'action de Γ à l'action cyclotomique d'une part, et à l'action sur le module de Tate de la courbe de l'autre.

Dans cet article nous avons adopté la terminologie de [S] et [SV]; textes auxquels nous renvoyons la lectrice ou le lecteur ayant peu de familiarité avec la théorie des dessins d'enfants de Grothendieck.
Dans la suite, nous ne considérerons que des dessins pré-propres (et le plus souvent propres). De plus, un dessin D sera dit de type $[a, b, c]$ si le p.p.c.m. des valences des sommets (resp. arêtes, resp. faces) est égal à a (resp. b, resp. c). Dire que les dessins sont (pré-)propres revient à affirmer que $b = 2$.
Il existe une caractérisation plus fine pour les dessins propres que nous utiliserons dans la suite: Si N est le nombre de drapeaux, on peut alors construire deux partitions $N = n_1 + 2v_2 + ... + kn_k + ... = m_1 + 2m_2 + ... + km_k$ où n_i (resp. m_i) est le nombre de sommets (resp. de faces) de valence i. On utilisera la notation $P_V = [1]^{n_1}[2]^{n_2}....[k]^{n_k}...$ et $P_F = [1]^{m_1}[2]^{m_2}....[k]^{m_k}...$ pour indiquer de telles partitions. Deux dessins (propres) induisent les mêmes partitions si et seulement s'ils appartiennent à la même classe de valence.

§1. Un exemple paramétrique.

Nous allons commencer avec un exemple particulièrement simple qui ne fait intervenir que les polynômes de Tchebychev. Notre but est celui de déterminer complètement l'action de $\mathrm{Gal}(\overline{\mathbb{Q}}/\mathbb{Q})$.

1.1. Description combinatoire.

Pour tout $N > 1$ on définit $\mathcal{F}_N$ comme étant la famille de dessins propres ayant $2N$ drapeaux et induisant les partitions

$$P_V = [2]^{N-2}[4]^1 \quad \text{et} \quad P_F = [2N]^1$$

Tous ces dessins sont de type $[4, 2, 2N]$. En fixant une valeur de N, on obtient une classe de valence. De plus (cf. caractéristique d'Euler-Poincaré) tous ces dessins ont genre 1. Dans la figure suivante sont représentés les éléments de $\mathcal{F}_5$.

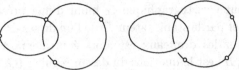

Il existe une bijection entre $\mathcal{F}_N$ et $\mathcal{D}_N = \{1, ..., n\}$ où $n = \frac{N}{2}$ si N est pair et $n = \frac{N-1}{2}$ sinon (il suffit d'associer à k le dessin ayant k arêtes sur l'une des deux "boucles", le nombre d'arêtes sur l'autre boucle étant alors automatiquement déterminé). Nous indiquerons donc par D_k le dessin de $\mathcal{F}_N$ correspondant à k.

Pour décrire le groupe cartographique de $D_k \in \mathcal{F}_N$, nous allons nous limiter à donner les permutations de S_{2N} induites par une numérotation des drapeaux. Si l'on indique par l et f les permutations relatives aux arêtes et aux faces, alors

$$l = (1 \ N + k)(2 \ N + k - 1) \cdots (k \ N + 1)(k + 1 \ 2N) \cdots (N \ N + k + 1)$$

$$f = (1 \ 2 \cdots 2N)$$

(la permutation v relative aux sommets est déterminée par la relation $vlf = 1$). On peut remarquer que tous ces dessins ont un groupe d'automorphismes non trivial (mais pratiquement toujours réduit à deux éléments). Il existe en particulier une involution ϕ donnée par

$$\phi = (1 \ N + 1)(2 \ N + 2) \cdots (k \ N + k) \cdots (N \ 2N)$$

Cet automorphisme possède quatre points fixes et est la traduction combinatoire de l'involution canonique de la courbe elliptique correspondante. Notre but est alors de déterminer le dessin quotient obtenu en faisant agir ϕ. Il ne dépend que de N et pas de k (figure suivante). Pour cette étude nous allons reprendre le discours de façon différente.

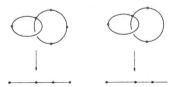

1.2. Groupes diédraux et polynômes de Tchebychev.

Soit $\Delta_N = <a, b \mid ab = b^{-1}a, \ a^2 = b^N = 1\}$ le groupe diédral d'ordre $2N$ ($\Delta_N \cong \mathbb{Z}/2\mathbb{Z} \ltimes \mathbb{Z}/N\mathbb{Z}$), et considérons la présentation $\Delta_N = <v, l, f>$

84   Leonardo Zapponi

où $v = a$, $l = ba$ et $f = b$. On obtient alors un dessin régulier. On rappelle qu'un dessin ("épinglé" en suivant la terminologie de [G]) est déterminé par son groupe cartographique G muni d'une présentation $G =< v, l, f \mid vlf = 1 >$ et par le stabilisateur H de l'un de ses drapeaux. L'action (à droite) de G sur les drapeaux est alors équivalente à celle de G sur G/H. De plus (G, H) est le quotient du dessin régulier $(G, 1)$ par le groupe d'automorphismes H...cf. [Z]. Dans notre cas, posons $H =< a >\cong \mathbb{Z}/2\mathbb{Z}$. Nous rappelons que le N-ième polynôme de Tchebychev T_N est défini par la relation $T_N(\cos(\theta)) = \cos(N\theta)$.

Lemme. *Une paire de Belyi associée au dessin $(\Delta_N, \mathbb{Z}/2\mathbb{Z})$ est donnée par $(\mathbb{P}^1, T_N)$ où T_N est le N-ième polynôme de Tchebychev.*

Démonstration. On a le "diagramme" commutatif

$$(\Delta_N, \mathbb{Z}/2\mathbb{Z}) \longleftarrow \Delta_N$$

$$(1) \longleftarrow \mathbb{Z}/2\mathbb{Z}$$

où $\Delta_N \to \mathbb{Z}/2\mathbb{Z}$ est la projection naturelle. D'un point de vue "graphique" on obtient (dans la figure on a indiqué le cas $N = 5$)

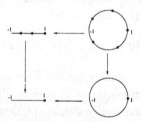

Finalement, vu que tous ces dessins ont genre 0, on arrive au diagramme commutatif

$$\begin{array}{ccc} \mathbb{P}^1 & \xleftarrow{p_2} & \mathbb{P}^1 \\ {\scriptstyle T_N}\downarrow & & \downarrow{\scriptstyle F_N} \\ \mathbb{P}^1 & \xleftarrow{p_1} & \mathbb{P}^1 \end{array}$$

Tout d'abord, F_N admet comme modèle $F_N(Z) = Z^N$ (en ayant imposé $F_N(\pm 1) = \pm 1$, ce qui n'est pas une limitation...cf [Sh]). p_1 est un revêtement double de la sphère de Riemann ayant deux points critiques ($Z = \pm 1$). On

[0]

déduit facilement que $p_1(Z) = p_2(Z) = \frac{1}{2}(Z + \frac{1}{Z})$. Par la commutativité du diagramme on obtient

$$T_N\left(\frac{1}{2}(Z + \frac{1}{Z})\right) = \frac{1}{2}(Z^N + \frac{1}{Z^N})$$

En particulier, si $Z = e^{i\theta}$, $p(Z) = Re(Z) = \cos(\theta)$ et $T_N(\cos(\theta)) = \cos(N\theta)$. T_N est donc le N-ième polynôme de Tchebychev. $\diamond$

1.3. Modèles pour les éléments de $\mathcal{F}_N$.

Revenons finalement aux dessins présentés au §1. Soit $D_k \in \mathcal{F}_N$ et posons $\lambda_k = Re(e^{i\pi\frac{k}{N}}) = \cos(\pi\frac{k}{N})$. Du fait que $0 < k \leq \frac{n}{2}$, on en déduit que $0 \leq \lambda_k < 1$. On peut alors considérer la courbe elliptique réelle définie par

$$C_k : Y^2 = (X^2 - 1)(X - \lambda_k)$$

Si $\pi(X, Y) = X$ est le revêtement canonique de C_k, alors (par construction) (C_k, f_k) est une paire de Belyi, où $f_k = (-1)^k T_N \pi$. De plus le dessin correspondant est un élément de $\mathcal{F}_N$ (en effet $\mathcal{F}_N$ est une classe de valence et l'on voit immédiatement que les partitions induites sont bien $[2]^N[4]^1$ et $[2n]^1$). La dynamique du revêtement π montre facilement que ce dessin est bien D_k. On a ainsi le

Lemme. *Considérons la courbe elliptique $C_k : Y^2 = (X^2 - 1)(X - \lambda_k)$ où $k \in \mathcal{D}_N$ et $\lambda_k = \cos(\pi\frac{k}{N})$. Soit de plus $f_k = (-1)^k T_N \pi$ où π est le revêtement canonique de la courbe. Alors la paire (C_k, f_k) est de Belyi et le dessin correspondant est $D_k \in \mathcal{F}_N$.*

Remarque. Si le dessin est donné par (G, H), alors $[G : H] = 2N$ et Δ_N est quotient de G.

1.4. Action du groupe de Galois.

Le groupe $\Gamma = \mathrm{Gal}(\overline{\mathbb{Q}}/\mathbb{Q})$ opère d'une façon naturelle sur les paires de Belyi: $(C, f) \longmapsto (C^\sigma, f^\sigma)$. Pour les dessins de $\mathcal{F}_N$, nous venons de présenter un modèle qui permet de décrire complètement cette action. Nous pouvons finalement démontrer la

Proposition. *La famille (classe de valence) $\mathcal{F}_N$ se décompose en $d(N) - 1$ orbites sous l'action de Γ. Soit $D \in \mathcal{F}_N$ et posons $c(D) = \frac{N}{(n,N)}$ où n est le nombre d'arêtes sur une des boucles. Alors $c(D)$ est un invariant galoisien*

absolu (i.e. D et D' sont conjugués si et seulement si $c(D) = c(D')$). Si J est l'invariant modulaire de la courbe associée à D, alors $\mathbb{Q}(J)/\mathbb{Q}$ est une extension abélienne de degré $\frac{1}{2}\phi(c(D))$.

Démonstration. Dans le modèle (C_k, f_k), le revêtement f_k est défini sur $\mathbb{Q}$. L'action de Γ se réduit alors à celle sur la courbe, et donc à l'action sur λ_k. Soit $\mathbb{Q}_N$ le corps (réel) obtenu en rajoutant λ_k à $\mathbb{Q}$ (pour $k \in \mathcal{D}_N$). Si $\mathbb{Q}(\zeta_N)/\mathbb{Q}$ est l'extension cyclotomique de degré N, alors, $\mathbb{Q}_N$ est une sous-extension de $\mathbb{Q}(\zeta_{2N})$ (en effet λ_k est la partie réelle d'une racine $2N$-ième de l'unité) . En particulier λ_k et λ_m sont conjugués si et seulement s'il existe un automorphisme de $\mathbb{Q}(\zeta_{2N})/\mathbb{Q}$ transformant $e^{i\pi\frac{k}{N}}$ en $e^{i\pi\frac{m}{N}}$. Ceci revient à dire que $(m, N) = (k, N)$ (où (a, b) indique le p.g.c.d de a et de b).

Pour $D = D_k \in \mathcal{F}_N$ définissons alors l'entier $c(D) = \frac{N}{(k,N)}$. On peut remarquer que $(k, N) = (N - k, N)$ et $c(D) = \frac{N}{(d,N)}$ où d est le nombre d'arêtes sur une quelconque des deux boucles.

Deux dessins D et D' sont donc conjugués si et seulement si $c(D) = c(D')$. En d'autres termes, c est un invariant galoisien absolu. On en déduit immédiatement que $\mathcal{F}_N$ se décompose en $d(N) - 1$ orbites sous l'action de Γ (il faut éliminer le cas $c = 1$).

Intéressons-nous finalement à l'invariant modulaire: Pour D_k on obtient

$$J = 32\frac{\left(7 + \cos(2\pi\frac{k}{N})\right)^3}{\left(1 - \cos(2\pi\frac{k}{N})\right)^2}$$

Si $D \in \mathcal{F}_N$, le corps $\mathbb{Q}(J)$ est contenu dans $\mathbb{Q}(\zeta_{c(D)})$ (où $\zeta_{c(D)}$ est une racine primitive $c(D)$-ième de l'unité). De plus, l'automorphisme σ qui transforme $\zeta_{c(D)}$ en $\zeta_{c(D)}^{-1}$ est le seul qui fixe J. On en déduit que $\mathbb{Q}(J)/\mathbb{Q}$ est galoisien, de groupe de Galois $(\mathbb{Z}/c(D)\mathbb{Z})^* / \pm 1$. En particulier

$$[\mathbb{Q}(J) : \mathbb{Q}] = \frac{1}{2}\phi(c(D))$$

si $c(D) \neq 2$ (ici ϕ est la fonction indicatrice d'Euler. Pour le cas $c(D) = 2$ l'invariant est rationnel). $\diamond$

L'invariant $c(D)$ mesure donc la "complexité" arithmétique de $J = J(D)$. On peut en particulier se demander quand $J(D)$ est rationnel. Le premier cas correspond à $c(D) = 2$ et donc $J = 1728$ (on aurait pu remarquer que $c(D) = 2$ implique que le nombre d'arêtes est le même sur les deux boucles. Dans ce cas le dessin admet un automorphisme d'ordre quatre qui fixe un point. Or il existe une seule courbe elliptique admettant un tel isomorphisme: $J = 1728$). En général, ceci revient à résoudre $\phi(c(D)) = 2$ et donc $c(D) \in \{3, 4, 6\}$. Pour $c(D) = 3$ on obtient $J = \frac{35152}{9}$, pour $c(D) = 4$, $J = 10976$ et enfin pour $c(D) = 6$, $J = 54000$.

Remarque. Tous ces dessins ont tous une seule face. En fait, les dessins propres ayant une seule face pour lesquels tous les sommets ont valence > 1 sont de deux types: ceux décrits plus haut et ceux induisant les partitions $P_V = [3]^2[2]^{N-3}$ et $P_F = [2N]^1$. Cette distinction est liée à la décomposition cellulaire de l'espace des modules des surfaces de Riemann en genre 1 avec un point marqué (cf [Z]). Les dessins que nous avons étudiés sont *tous* les dessins en genre 1 ayant une seule face tels la courbe associée admette une équation du type $Y^2 = (X - e_1)(X - e_2)(X - e_3)$ où $e_1, e_2, e_3 \in \mathbb{R}$. Le cas général est bien plus délicat. Dans la deuxième partie de l'article, nous allons étudier des dessins qui, tout en ayant deux faces, sont intimement liés à ce problème (cf. [Z]).

1.5. Exemples.

Nous allons terminer cette section en donnant quelques exemples numériques:

-$N = 5$. $\mathcal{F}_5$ contient deux dessins et possède une seule orbite sous l'action de $\mathbb{\Gamma}$ ($d(5) - 1 = 1$). Les deux dessins seront donc conjugués et le résultat est présenté dans la figure suivante.

$$J = \frac{71224 + 26664\sqrt{5}}{5} \qquad\qquad J = \frac{71224 - 26664\sqrt{5}}{5}.$$

-$N = 10$. Dans ce cas $\mathcal{F}_{10}$ contient 5 dessins. $d(10) = 4$ et donc on retrouve 3 orbites.

Tout d'abord on a le dessin D_5 ($C(D) = 2$), qui, pour des raisons de symétrie correspond à $J = 1728$.

Pour $c(D) = 5$ on obtient $\frac{1}{2}\phi(5) = 2$ dessins conjugués. Ce sont les mêmes courbes que dans le cas $N = 5$.

Finalement, les deux derniers dessins vérifient $c(D) = 10$. Leurs invariants appartiennent à une extension quadratique de $\mathbb{Q}$ (figure).

$$J = 211688 + 92168\sqrt{5} \qquad\qquad J = 211688 - 92168\sqrt{5};$$

§2. Dessins en genre 1, points de torsion et formes modulaires.

Nous allons à présent étudier une deuxième classe de valence pour laquelle les dessins sont en quelque sorte une généralisation de ceux décrits précédemment – on rajoute une "boucle". Ici il est bien plus difficile de donner une solution paramétrique; on pourra néanmoins introduire un invariant galoisien calculable de façon combinatoire. En utilisant des méthodes analytiques (en passant par le revêtement universel de la courbe associée au dessin) on donnera ensuite une caractérisation "modulaire" de ces dessins. Dans son article, F. Pakovitch [P] présente des résultats analogues, en utilisant des méthodes différentes.

2.1. Description combinatoire.

Soit $N > 2$ un entier. La famille $\mathcal{E}_N$ est formée par les dessins propres ayant $2N$ drapeaux (N arêtes, vu que les dessins sont propres) et induisant les partitions

$$P_V = [2]^{N-3}[6]^1 \quad \text{et} \quad P_F = [N]^2$$

En d'autres termes, $D \in \mathcal{E}_N$ a un sommet de valence 6, tous les autres sommets de valence 2 et deux faces de valence N. Ici aussi $\mathcal{E}_N$ est une classe de valence. Ces dessins sont obtenus en prenant le dessin régulier de groupe $G = \mathbb{Z}/6\mathbb{Z}$ muni de la présentation $v = 1$, $l = 3$, $f = 2$ et en lui "rajoutant" des sommets de valence 2 (figure).

Comme précédemment, tous ces dessins possèdent une involution avec quatre points fixes qui permute les deux faces. L'ordre cyclique des drapeaux sortant du sommet de valence 6 induit un ordre des "boucles" du dessin. $D \in \mathcal{E}_N$ est alors déterminé par un triplet d'entiers positifs (N_1, N_2, N_3)

(avec $N_1 + N_2 + N_3 = N$) modulo une permutation cyclique des indices (N_i est le nombre d'arêtes sur la i-ième boucle).

On pourrait étudier $\mathcal{E}_N$ en suivant les méthodes utilisées pour $\mathcal{F}_N$: Soit en effet $D_{N_1,N_2,N_3} \in \mathcal{E}_N$. L'involution σ fixe le sommet de valence 6 du dessin. Si N_i est pair alors σ fixe le sommet central de la boucle correspondante; s'il est impair, c'est l'arête centrale qui est fixée. Le dessin quotient que l'on obtient est un *arbre* ayant un seul sommet de valence 3. Il est toujours pré-propre (propre si N_1, N_2 et N_3 sont pairs).

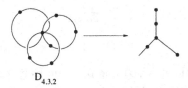

$$D_{4,3,2}$$

Les paires de Belyi associées à ces arbres sont du type $(\mathbb{P}^1_{\mathbb{C}}, p)$ òu p est un polynôme de degré N ayant deux valeurs critiques sur $\mathbb{C}$. p est appelé *polynôme de Tchebychev généralisé* [Sh]. Supposons qu'on connaît p, et considérons les points e_1, e_2, e_3 relatifs aux "extrémités" de l'arbre, et e, le sommet de valence 3. Soit finalement $\mathcal{C}$ la courbe elliptique définie par

$$\mathcal{C} : Y^2 = (X - e)(X - e_1)(X - e_2)(X - e_3)$$

et $\pi(X, Y) = X$ le revêtement canonique. $(\mathcal{C}, p \circ \pi)$ est alors une paire de Belyi associée à D_{N_1,N_2,N_3} (avec des modifications élémentaires il est possible de donner un modèle non singulier de $\mathcal{C}$).

Obtenir des polynômes de Tchebychev généralisés n'est pas simple et en général ils ne sont pas définis sur $\mathbb{Q}$ (pour un exemple numérique on renvoie la lectrice ou le lecteur à [S], p. 75). C'est en jouant sur une autre caractéristique de ces dessins (qui sera introduite dans le paragraphe suivant) que nous pourrons obtenir des résultats plus tangibles.

2.2. Dessins orientables.

Avant de continuer nous allons définir le concept de dessin orientable, qui permet de simplifier les calculs. En effet, pour un tel dessin, l'application de Belyi est la composition de deux revêtements.

Un dessin orienté est simplement un dessin pour lequel chaque arête a été orientée. Evidemment, nous pouvons choisir 2^n orientations différentes (où n est le nombre d'arêtes). Une orientation est dite *admissible* si toutes les arêtes appartenant à une même face sont orientées de façon cohérente. On démontre facilement que si elle existe, une telle orientation est unique (au signe près).

Un dessin est *orientable* s'il existe une orientation admissible. Ce concept est équivalent à celui de dessin "bicoloriable".

Soit D un dessin de groupe cartographique $G = < v, l, f >$ et $B < G$ le stabilisateur d'un drapeau (notation: v (resp. l, resp. f) est l'élément de G correspondant aux sommets (resp. aux arêtes, resp. aux faces)). On démontre alors le critère suivant [Z]:

Critère. *Un dessin est orientable si et seulement s'il existe un homomorphisme*

$$G \xrightarrow{Or} \mathbb{Z}/2\mathbb{Z}$$

tel que $Or(v) = Or(l) = -1$, $Or(f) = 1$ *et* $B < Ker(Or)$.

Soit donc D orientable et $(\mathcal{C}, \beta)$ une paire de Belyi correspondante (nous supposons que β ramifie au dessus de ± 1 et ∞). Si $2N = [G : B]$ est le nombre de drapeaux, l'homomorphisme Or induit alors un revêtement $\mathcal{C} \xrightarrow{\beta_0} \mathbb{P}^1_\mathbb{C}$ de degré N tel que $\beta = \frac{1}{2}\left(\beta_0 + \frac{1}{\beta_0}\right)$.

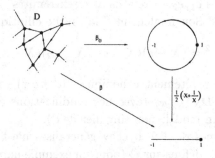

Tous les dessins ne sont pas nécessairement orientables; mais ils possèdent toujours un revêtement double, ramifié au dessus des sommets de valence impaire pour lequel le dessin correpondant est orientable cf. [Z]. Un tel concept a une traduction analytique: Le différentiel de Strebel associé à un dessin est le carré d'un différentiel ordinaire si et seulement si le dessin est orientable.

Pour revenir à $\mathcal{E}_N$, on voit facilement que tous les dessins de $\mathcal{E}_N$ sont orientables (figure suivante).

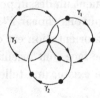

2.3. Un invariant galoisien pour les dessins de $\mathcal{E}_N$.

Il est évident que la valence N des deux faces d'un dessin de $\mathcal{E}_N$ est un invariant galoisien. Notre but est celui d'introduire un invariant plus fin. Dans la suite, quand on associera à un dessin de $\mathcal{E}_N$ une paire de Belyi $(\mathcal{C}, \beta)$, nous supposerons toujours que *le sommet de valence 6 correspond à l'élément neutre de la courbe $\mathcal{C}$*, vue comme groupe abélien. Ceci n'est pas une limitation et revient à composer à droite β avec une translation. Pour l'instant nous allons démontrer la

Proposition. *Soit $D \in \mathcal{E}_N$ et $(\mathcal{C}, \beta)$ une paire de Belyi correspondante. Alors les deux pôles de β sont des points de torsion de $\mathcal{C}$ opposés l'un de l'autre (en voyant $\mathcal{C}$ comme groupe abélien) et leur ordre n_D divise $2N$.*

Remarque. n_D est évidemment un invariant galoisien.

Démonstration. $D \in \mathcal{E}_N$ est orientable; l'application de Belyi peut donc être exprimée par $\beta = p\beta_0$ où $p(z) = \frac{1}{2}(z + \frac{1}{z})$. Le diviseur de β_0 est particulièrement simple:

$$\text{div}(\beta_0) = N[P^+] - N[P^-]$$

où P^+ et P^- sont les deux pôles de β. Un théorème classique en théorie des courbes elliptiques affirme que si f est une fonction méromorphe et $\text{div}(f) = n_1[P_1] + \cdots + n_k[P_k]$ alors $n_1 P_1 + \cdots + n_k P_k = 0$. Pour β_0 on obtient $N(P^+ - P^-) = 0$.

L'involution du dessin se traduit en une involution analytique σ de la courbe $\mathcal{C}$ fixant 4 points. En particulier, elle fixe l'élément neutre (qui par hypothèse correspond au sommet de valence 6). On en déduit immédiatement que $\sigma(P) = -P$. Le fait que σ permute les deux faces se traduit par $P^+ = -P^-$ et donc $2NP^+ = 0 \Rightarrow P^+, P^- \in \mathcal{C}[2N]$. $\diamond$

2.4. Expression combinatoire de l'invariant n_D.

Pour le calcul de l'ordre n_D de P^+ nous utiliserons des méthodes analytiques complexes. C'est celui-ci le premier pas pour pouvoir ensuite donner une "caractérisation modulaire" de la famille $\mathcal{E}_N$.

Soit $D \in \mathcal{E}_N$ et $(\mathcal{C}, \beta)$ une paire de Belyi associée. Considérons la décomposition $\beta = p\beta_0$ induite par l'orientation du dessin. Nous allons exprimer le revêtement β_0 en utilisant des fonctions elliptiques.

Choisissons une valeur $\tau \in \mathcal{H} = \{z \in \mathbb{C} \mid Im(z) > 0\}$ telle que $\mathcal{C}$ soit isomorphe à $\mathbb{C}/\Lambda$ où $\Lambda = \Lambda_\tau = \mathbb{Z} \oplus \tau\mathbb{Z}$ est le réseau engendré par 1 et τ.

Les fonctions sigma et zéta de Weierstrass sont définies par

$$\sigma(z) = \sigma(z, \tau) = z \exp\left(-\sum_{n>1} \frac{G_{2n}(\tau)}{2n} z^{2n}\right)$$

$$\zeta(z) = \zeta(z, \tau) = \frac{1}{z} - \sum_{n>0} G_{2n+2}(\tau) z^{2n+1}$$

où $G_{2n}(\tau)$ est la $2n$-ième série de Eisenstein. On a alors les relations suivantes (qui seront utiles dans la suite):

$$\wp(z) = -\zeta'(z) \qquad \zeta(z) = \frac{\sigma'(z)}{\sigma(z)}$$

($\wp(z) = \wp(z, \tau)$ est la fonction $\wp$ de Weierstrass).

$$\begin{cases} \zeta(z+1) &= \zeta(z) + 2\eta_1 \\ \zeta(z+\tau) &= \zeta(z) + 2\eta_2 \end{cases} \qquad \begin{cases} \sigma(z+1) &= -\sigma(z)e^{2\eta_1\left(z+\frac{1}{2}\right)} \\ \sigma(z+\tau) &= -\sigma(z)e^{2\eta_2\left(z+\frac{\tau}{2}\right)}. \end{cases}$$

avec $\eta_1 = \eta_1(\tau) = \zeta(\frac{1}{2})$ et $\eta_2 = \eta_2(\tau) = \zeta(\frac{\tau}{2})$ constantes vérifiant

$$\eta_1 \tau - \eta_2 = \pi i$$

La fonction sigma nous intéresse particulièrement . Elle n'est pas elliptique mais toute fonction elliptique peut être exprimée grâce à elle de façon particulièrement simple. On a en effet le

Theorème. *Soit $f(z)$ une fonction elliptique ayant 1 et τ comme périodes réduites ($Im(\tau) > 0$). Soit $\operatorname{div} f = n_1[z_1] + \cdots + n_k[z_k] - m_1[p_1] - \cdots - m_l[p_l]$ le diviseur de f avec $n_i, m_j > 0$. Alors*

$$\omega = n_1 z_1 + \cdots + n_k z_k - m_1 p_1 - \cdots - m_l p_l \in \Lambda$$

et il existe une constante $c \in \mathbb{C}$ telle que

$$f(z) = c \frac{\sigma(z - z_1)^{n_1} \cdots \sigma(z - z_k)^{n_k}}{\sigma(z - p_1)^{m_1} \cdots \sigma(z - p_l)^{m_l - 1} \sigma(z - p_l - \omega)}$$

Démonstration. [Si] proposition 5.5, p.45   ◇

Nous allons appliquer ce théorème à β_0 en démontrant le

Lemme. *Soit $D \in \mathcal{E}_N$ et $(\mathcal{C}, \beta)$ une paire de Belyi correspondante telle que l'élément neutre de $\mathcal{C}$ corresponde au sommet de valence 6. Considérons*

une valeur $\tau \in \mathcal{H}$ associée à C et $\overline{\beta}_0 : \mathbb{C} \to C \xrightarrow{\beta_0} \mathbb{P}^1_{\mathbb{C}}$ la fonction elliptique correspondante (où $\beta = p\beta_0$). Si $z^+ = \frac{a}{2N} + \frac{b}{2N}\tau$ est le zéro de $\overline{\beta}_0$ dans le parallélogramme fondamental, alors

$$\overline{\beta}_0(z) = (-1)^N \frac{\sigma(z-z^+)^N}{\sigma(z+z^+)^N} e^{2z(a\eta_1+b\eta_2)}$$

Démonstration. Dans la suite, le parallélogramme fondamental sera celui déterminé par les points $0,1,\tau$ et $1+\tau$. En ayant supposé que le sommet de valence 6 corresponde à l'élément neutre de C, on en déduit que $\overline{\beta}_0(0) = 1$. D'après la proposition précédente, le diviseur de $\overline{\beta}_0$ est donné par $\mathrm{div}(\overline{\beta}_0) = N[z^+] - N[-z^+]$. De plus, $2Nz^+ = a + b\tau$. En appliquant le théorème on obtient

$$\overline{\beta}_0(z) = c\frac{\sigma(z-z^+)^N}{\sigma(z+z^+)^{N-1}\sigma(z+z^+-a-b\tau)}$$

où $c \in \mathbb{C}$ est une constante. Or $\sigma(z+z^+-a-b\tau) = \sigma(z+z^+)e^{-2z(a\eta_1+b\eta_2)+d}$ où d est une constante dépendant de η_1,η_2,τ,a,b et N. On arrive donc à l'expression

$$\overline{\beta}_0(z) = c\frac{\sigma(z-z^+)^N}{\sigma(z+z^+)^N} e^{2z(a\eta_1+b\eta_2)+d}$$

Finalement, $\overline{\beta}_0(0) = 1 \Rightarrow c\frac{\sigma(-z^+)^N}{\sigma(z^+)} e^d = (-1)^N c e^d = 1$ car σ est impaire. On en déduit immédiatement que $c = (-1)^N e^{-d}$ et le lemme est démontré.$\diamond$

Avant de continuer nous allons montrer le lien entre $\overline{\beta}_0$ et la fonction de hauteur locale de Néron (cf. [Si] Ch.6):

Corollaire. $\log|\overline{\beta}_0(z)| = N[\lambda(z+z^+) - \lambda(z-z^+)]$ *où λ est la fonction de hauteur locale de Néron.*

Démonstration. L'expression de $\overline{\beta}_0$ présentée dans le lemme précédent peut être réécrite comme

$$\overline{\beta}_0(z) = \left[-\frac{\sigma(z-z^+)}{\sigma(z+z^+)} e^{2z\eta(z^+)}\right]^N$$

où η est la fonction de quasi-période ([S] p. 465). En effet $2z(a\eta_1+b\eta_2) = 2Nz\left(\frac{a}{N}\zeta\left(\frac{1}{2}\right) + \frac{b}{N}\zeta\left(\frac{\tau}{2}\right)\right) = 2Nz\left(\frac{a}{2N}\eta(1) + \frac{b}{2N}\eta(\tau)\right) = 2Nz\,\eta\left(\frac{a+b\tau}{2N}\right)$.

Or, $\frac{1}{2}(z+z^+)\eta(z+z^+) - \frac{1}{2}(z-z^+)\eta(z-z^+) = z\eta(z^+) + z^+\eta(z)$ (par $\mathbb{R}$-linéarité) et $z^+\eta(z) - z\eta(z^+) = il$ avec $l \in \mathbb{R}$ ([Si] prop. 3.1, p. 465)$\Rightarrow$

$$\frac{1}{2}(z+z^+)\eta(z+z^+) - \frac{1}{2}(z-z^+)\eta(z-z^+) = 2z\eta(z^+) + il \Rightarrow$$

$$\overline{\beta}_0(z) = \left[-\frac{\sigma(z - z^+)e^{-\frac{1}{2}(z-z^+)\eta(z-z^+)}}{\sigma(z + z^+)e^{-\frac{1}{2}(z+z^+)\eta(z+z^+)}} e^{-il} \right]^N.$$

Finalement, une expression pour la fonction de hauteur locale de Néron est donnée par ([Si] Th. 3.2, p.466)

$$\lambda(z) = -\log \left| \sigma(z)\Delta(\tau)^{\frac{1}{12}} e^{-\frac{1}{2}z\eta(z)} \right| \Rightarrow$$

$$\log |\overline{\beta}_0(z)| = N\left[\lambda(z + z^+) - \lambda(z - z^+) \right] \qquad\qquad \Diamond$$

Corollaire. *Soit $(\mathbb{C}/\Lambda, \overline{\beta})$ une paire de Belyi associée à un dessin de $\mathcal{E}_N$ telle que le sommet de valence 6 corresponde au point $z = 0$. Alors*

$$\overline{\beta}^{-1}([-1,1]) = \left\{ z \in \mathbb{C} \mid \lambda(z - z^+) = \lambda(z + z^+) \right\}$$

Démonstration. Immédiate en utilisant le corollaire précédent. $\qquad \Diamond$

Ce dernier corollaire permet de donner une description en termes de la fonction de hauteur locale de Néron du complexe K immergé dans la courbe elliptique, déterminant la structure de dessin. En voyant λ comme une "distance", K est alors l'ensemble des points "équidistants" de z^+ et z^-.

Nous allons à présent exprimer l'invariant n_D de façon combinatoire:

Proposition. *Soit $D = D_{N_1,N_2,N_3} \in \mathcal{E}_N$, alors*

$$n_D = \frac{2N}{(N_1 + N_2, N_2 + N_3, N_3 + N_1)}$$

Démonstration. Choisissons une orientation admissible de D_{N_1,N_2,N_3} (parmi les deux possibles). Les "boucles" deviennent alors des chemins fermés orientés γ_1, γ_2 et γ_3 (numérotés comme dans la figure qui suit) tels que $\gamma_1 + \gamma_2 + \gamma_3 = 0$ (en genre 1, $\pi_1(C)$ est commutatif).

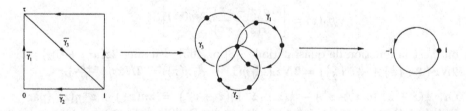

De plus $\pi_1(C) = < \gamma_1, \gamma_2 >$ et l'on peut choisir une valeur de $\tau \in \mathbb{C}$ telle que dans le parallélogramme fondamental γ_1 et γ_2 soient respectivement homotopes à $\overline{\gamma}_1(t) = t\tau$ et $\overline{\gamma}_2(t) = 1 - t$, avec $\tau \in [0,1]$ (les deux côtés). Dans ce cas pour γ_3 on obtient $\overline{\gamma}_3(t) = \tau + t(1 - \tau)t$ (la diagonale ne passant pas par l'origine). Supposons de plus que pour ce choix de τ on ait $z^+ = \frac{a}{2N} + \frac{b}{2N}\tau$.

Considérons sur $\mathbb{P}^1_{\mathbb{C}}$ le différentiel méromorphe $\omega = \frac{1}{N}\frac{dX}{X}$ et le chemin $\gamma(t) = e^{\pi i t}$, $t \in [0,1]$. Posons en général $\gamma_n(t) = e^{\pi i t}$, $t \in [0,n]$. γ peut être relevé via $\overline{\beta}_0$ de trois façons différentes sur $\mathbb{C}$ en choisissant l'origine comme point base. Supposons de relever γ le long de $\overline{\gamma}_1$. On peut alors relever γ_{N_1} et son extrémité est $\frac{\tau}{2}$ (point fixe de la "boucle" par l'automorphisme du dessin). On a alors

$$
e^{\pi i \frac{N_1}{N}} = \exp\left(\int_{\gamma_{N_1}} \omega\right) = \exp\left(\int_0^{\frac{\tau}{2}} \overline{\beta}_0^* \omega\right) = -\frac{\sigma(\frac{\tau}{2} - z^+)}{\sigma(\frac{\tau}{2} + z^+)} e^{\frac{a\eta_1 + b\eta_2}{N}\tau}
$$

$$
= -\exp\left(\frac{a\eta_1 + b\eta_2}{N}\tau - 2\eta_2 z^+\right) = -\exp\left(\frac{a\eta_1 + b\eta_2}{N}\tau - \eta_2 \frac{a + b\tau}{N}\right)
$$

$$
= -e^{\pi i \frac{a}{N}}
$$

car $\sigma(\frac{\tau}{2} + z^+) = e^{2\eta_2 z^+}\sigma(\frac{\tau}{2} - z^+)$ et $\eta_1\tau - \eta_2 = i\pi$. On obtient donc

$$
e^{\pi i \frac{N_1}{N}} = e^{i\pi + \pi i \frac{a}{N}} \Rightarrow
$$

$$
i\pi \frac{N_1}{N} \equiv i\pi + i\pi \frac{a}{N} \quad \mathrm{mod}(2\pi i) \Rightarrow a \equiv N + N_1 \quad \mathrm{mod}(2N).
$$

De façon tout à fait analogue, $b \equiv N + N_2 \quad \mathrm{mod}(2N)$. L'ordre de z^+ est alors égal à $\frac{2N}{d}$ où $d = (N + N_1, N + N_2, 2N)$ et on démontre facilement que $(N + N_1, N + N_2, 2N) = (N_1 + N_2, N_2 + N_3, N_3 + N_1)$. $\qquad\qquad \Diamond$

Remarque. Dans la démonstration on donne explicitement les valeurs de a et b en fonction des données combinatoires. Le point de torsion est donc uniquement déterminé, en ayant choisi une orientation admissible.

2.5. Formes modulaires.

2.5.1. Inversion du problème.

Soit $\mathcal{C} = \mathbb{C}/\Lambda_\tau$ une courbe elliptique avec $\tau \in \mathcal{H}$. Considérons $0 \leq a, b < 2N$, a, b entiers et posons $z^+ = \frac{a}{2N} + \frac{b}{2N}\tau$. La fonction $\overline{\beta}_0$ définie comme dans le lemme précédent est alors elliptique et a le "bon" diviseur. Il ne reste qu'à vérifier les ramifications.

Proposition. *La fonction $\overline{\beta}_0$ est de Belyi si et seulement si*

$$a\zeta\left(\frac{1}{2}\right) + b\zeta\left(\frac{\tau}{2}\right) = N\zeta\left(\frac{a}{2N} + \frac{b}{2N}\tau\right)$$

Démonstration. Nous devrons avoir

$$\operatorname{div}\overline{\beta}_0' = (N-1)[z^+] + 2[0] - (N+1)[-z^+].$$

Par un calcul direct nous obtenons

$$\frac{\overline{\beta}_0'(z)}{\overline{\beta}_0(z)} = N\left[\zeta(z - z^+) - \zeta(z + z^+)\right] + 2a\eta_1 + 2b\eta_2$$

$$= N\left[\zeta(z - z^+) - \zeta(z + z^+) + 2\zeta(z^+)\right] + 2a\eta_1 + 2b\eta_2 - 2N\zeta(z^+).$$

Pour transformer la dernière expression utilisons le lemme suivant (cf. [C], p. 55):

Lemme. $\dfrac{\wp'(v)}{\wp(u) - \wp(v)} = \zeta(u - v) - \zeta(u + v) + 2\zeta(v).$

Nous arrivons à l'égalité

$$\frac{\overline{\beta}_0'(z)}{\overline{\beta}_0(z)} = N\frac{\wp'(z^+)}{\wp(z) - \wp(z^+)} + 2a\eta_1 + 2b\eta_2 - 2N\zeta(z^+) \Rightarrow$$

$$\overline{\beta}_0'(z) = N\overline{\beta}_0(z)\frac{\wp'(z^+)}{\wp(z) - \wp(z^+)} + \overline{\beta}_0(z)\left(2a\eta_1 + 2b\eta_2 - 2N\zeta(z^+)\right)$$

Or $\operatorname{div}\dfrac{1}{\wp - \wp(z^+)} = 2[0] - [z^+] - [-z^+] \Rightarrow$

$$\operatorname{div}\frac{\overline{\beta}_0}{\wp - \wp(z^+)} = (N-1)[z^+] + 2[0] - (N+1)[-z^+]$$

qui est le diviseur cherché. Nous en déduisons immédiatement que $a\eta_1 + b\eta_2 = N\zeta(z^+)$. $\diamond$

Remarque: En utilisant la fonction de quasi-période on peut reformuler la proposition en disant que $\overline{\beta}_0$ *est de Belyi si et seulement si*

$$\zeta(z^+) = \eta(z^+)$$

2.5.2. Le sous-groupes de congruence $\Gamma_1(n)$ et le caractère Ψ_n.

Ouvrons à présent une parenthèse en décrivant une famille de sous-groupes de congruence de $\Gamma(1) = \mathrm{SL}_2(\mathbb{Z})$ qui seront très utiles dans la suite.

Soit $R_0(n) = \{(a,b) \in \mathbb{Z}/n\mathbb{Z} \oplus \mathbb{Z}/n\mathbb{Z} \mid\ <a,b>= \mathbb{Z}/n\mathbb{Z}\}$. Un résultat classique affirme que la cardinalité de $R_0(n)$ est égale à $n^2 \prod_{p|n}(1 - \frac{1}{p^2})$. On définit une action à droite de $\Gamma(1)$ sur $R_0(n)$ en posant

$$(a,b)^M = (a',b') \quad \text{où} \quad \begin{pmatrix} b' \\ a' \end{pmatrix} = {}^t M \begin{pmatrix} b \\ a \end{pmatrix}$$

avec $M \in \Gamma(1)$ (ici ${}^t M$ indique la matrice transposée de M). Posons

$$H(a,b) = Stab_{\Gamma(1)}(a,b)$$

$H(a,b)$ est un sous-groupe de congruence. En effet on vérifie immédiatement l'inclusion $\Gamma(n) \subset H(a,b)$ (où $\Gamma(n) = \ker(\Gamma(1) \to \mathrm{SL}_2(\mathbb{Z}/n\mathbb{Z}))$).
Pour $b = 0$ on a $H(a,0) = H(1,0) = \Gamma_1(n)$ où

$$\Gamma_1(n) = \left\{ M \in \Gamma(1) \mid M \equiv \begin{pmatrix} 1 & * \\ 0 & 1 \end{pmatrix} \bmod n \right\}$$

$H(a,b)$ est alors le conjugué de $\Gamma_1(n)$ par un élément de $\Gamma(1)$ (en effet l'action est transitive).

Considérons sur $R_0(n)$ la relation d'équivalence $(a,b) \sim (-a,-b)$.
$\Gamma(1)$ opère sur $R(n) = R_0(n)/\sim$ (l'action est bien définie) et nous indiquons par $[a,b] = \{(a,b);(-a,-b)\}$ la classe d'équivalence de (a,b).
Trivialement, $H[a,b] = Stab_{\Gamma(1)}([a,b]) =< Stab(a,b), -I >$. Tout ce qui a été dit pour $H(a,b)$ peut se transposer à $H[a,b]$. En particulier c'est un sous-groupe de congruence et $\Gamma[n] =< \Gamma(n), -I >\subset H[a,b]$. Comme avant, pour $b = 0$ on a $H[a,0] = H[1,0] = \Gamma_1[n] =< \Gamma_1(n), -I >$.

Soit $\Gamma(1) = \bigcup_i \Gamma_1[n]M_i$ une décomposition en classes à droite du groupe modulaire. De cette décomposition on en déduit une pour $\Gamma_1(n)$, donnée par

$$\Gamma(1) = \bigcup_i \left(\Gamma_1(n)M_i \cup -\Gamma_1(n)M_i \right).$$

Posons $m = [\Gamma(1), \Gamma_1[n]] = \frac{1}{2}n^2 \prod_{p|n}(1 - \frac{1}{p^2})$.
On a alors $\Gamma_1(n)M_iM = \chi_i(M)\Gamma(1)M_{\sigma_M(i)}$ où $\sigma \in S_m$ et $\chi_i(M) \in \{-1,1\}$ vérifient

$$\chi_i(M'M) = \chi_i(M)\chi_{\sigma_M(i)}(M')$$

Soit $\Psi_n : \Gamma(1) \to \mathbb{Z}/2\mathbb{Z}$ défini par $\Psi_n(M) = \prod_i \chi_i(M)$. Alors Ψ_n est un caractère. En effet

$$
\begin{aligned}
\Psi_n(M'M) &= \prod_i \chi_i(M)\chi_{\sigma_M(i)}(M') \\
&= \prod_i \chi_i(M) \prod_i \chi_{\sigma_M(i)}(M') \\
&= \prod_i \chi_i(M) \prod_j \chi_j(M') \\
&= \Psi_n(M)\Psi_n(M')
\end{aligned}
$$

Remarque. Si Ψ_n est non trivial alors son noyau est un sous-groupe d'indice 2 (donc distingué) de $\Gamma(1)$. Or un tel sous-groupe est unique et égal à $H = \Gamma(1)^2$ (cf. [L] p.359). Une décomposition en classes à droite est donnée par $\Gamma(1) = H \cup HU$ où $U = \begin{pmatrix} 1 & 1 \\ 0 & 1 \end{pmatrix}$. Il suffira donc d'étudier l'action de U pour vérifier la trivialité de Ψ_n. Par exemple, Ψ_4 est non trivial.

2.5.3. Les fonctions $F_{a,b}(\tau)$ et $F_m(\tau)$.

En reprenant les notations du 2.5.1 nous allons étudier à fond la condition

$$
N\zeta\left(\frac{a}{2N} + \frac{b}{2N}\tau\right) = a\zeta\left(\frac{1}{2}\right) + b\zeta\left(\frac{\tau}{2}\right)
$$

Soit $n > 1$ et $(a,b) \in R_0(n)$ (cf. 2.5.2). Posons

$$
F_{a,b}(\tau) = \frac{1}{i\pi}\left(n\zeta\left(\frac{a}{n} + \frac{b}{n}\tau\right) - 2a\zeta\left(\frac{1}{2}\right) - 2b\zeta\left(\frac{\tau}{2}\right)\right)
$$

Pour tous $a, b \in \mathbb{Z}$, $F_{a+n,b}(\tau) = F_{a,b+n}(\tau) = F_{a,b}(\tau)$. $F_{a,b}$ ne dépend que de l'image de a et b dans $\mathbb{Z}/n\mathbb{Z}$, elle est donc bien définie.
Dans la suite, nous écrirons plus simplement $\eta_1(\tau) = \zeta\left(\frac{1}{2}\right)$ et $\eta_2(\tau) = \zeta\left(\frac{\tau}{2}\right)$.

Proposition. $F_{a,b}(\tau)$ *est une forme modulaire de poids 1 pour* $H(a,b)$.

Démonstration. $F_{a,b}(\tau)$ est holomorphe (ce n'est pas difficile à démontrer en utilisant des critères de convergence). L'expression de la fonction ζ présentée au 2.4 permet d'obtenir l'égalité

$$
\zeta\left(z, \frac{A\tau + B}{C\tau + D}\right) = (C\tau + D)\zeta\left((C\tau + D)z, \tau\right) \qquad \forall\, M = \begin{pmatrix} A & B \\ C & D \end{pmatrix} \in \Gamma(1)
$$

En particulier,

$$\zeta\left(\frac{a + b\frac{A\tau+B}{C\tau+D}}{n}, \frac{A\tau+B}{C\tau+D}\right) = (C\tau+D)\zeta\left(\frac{aD+bB}{n} + \frac{bA+aC}{n}\tau, \tau\right)$$

De même,

$$\begin{cases} \zeta\left(\frac{1}{2}, \frac{A\tau+B}{C\tau+D}\right) & = (C\tau+D)\zeta\left(\frac{C\tau+D}{2}, \tau\right) = \frac{1}{2}(C\tau+D)\eta(C\tau+D) \\ \zeta\left(\frac{1}{2}\frac{A\tau+B}{C\tau+D}, \frac{A\tau+B}{C\tau+D}\right) & = (C\tau+D)\zeta\left(\frac{A\tau+B}{2}, \tau\right) = \frac{1}{2}(C\tau+D)\eta(A\tau+B) \end{cases}$$

où η est la fonction de quasi-période.

Finalement, en combinant ces expressions, on arrive à

$$F_{a,b}(M\tau) = (C\tau+D)F_{a',b'}(\tau) \quad \text{avec} \quad \begin{pmatrix} b' \\ a' \end{pmatrix} = {}^tM\begin{pmatrix} b \\ a \end{pmatrix}$$

En accord avec l'action de $\Gamma(1)$ définie au 2.5.2 on peut donc écrire $(a', b') = (a, b)^M$. En prenant $M \in H(a,b)$ on montre que $F_{a,b}$ est une *fonction faiblement modulaire* de poids 1. Pour terminer la démonstration il faut étudier le comportement à l'infini, question qui est reportée au prochain paragraphe.$\Diamond$

Corollaire. *Soit* $m \in (\mathbb{Z}/n\mathbb{Z})^*$ *et posons* $F_m(\tau) = F_{m,0}(\tau)$. *Alors* F_m *est une forme modulaire de poids 1 pour* $\Gamma_1(n)$.

2.5.4. Comportement à l'infini. La fonction théta.

La fonction θ est proche (comme nature) de la fonction σ. Elle est souvent utilisée pour des raisons numériques. En effet elle admet un développement en série à convergence rapide. Nous allons l'introduire par l'expression suivante (en produit infini):

$$\theta(z) = \theta(z, \tau) = 2c\,q^{\frac{1}{8}}\sin(\pi z)\prod_{n>0}(1 - q^n e^{2\pi iz})\prod_{n>0}(1 - q^n e^{-2\pi iz})$$

avec $c = \prod_{n>0}(1 - q^n)$ et $q = e^{2i\pi\tau}$. Grâce à ce produit on pourra étudier le comportement de F_n à l'infini.

En utilisant les propriétés élémentaires de θ (cf. [C], p.60) on obtient:

$$(1)\ \sigma(z,\tau) = \frac{\theta(z,\tau)}{\theta'(0,\tau)}e^{\eta_1 z^2}$$

(2) $\overline{\beta_0}(z) = \dfrac{\theta(z - z^+)^n}{\theta(z + z^+)^n} e^{-2\pi i z}$

(3) $\zeta(z) = \dfrac{\theta'(z)}{\theta(z)} + 2\eta_1 z$

La troisième relation permet d'obtenir

$$n\zeta(\frac{a + b\tau}{n}) = n\frac{\theta'(\frac{a+b\tau}{n})}{\theta(\frac{a+b\tau}{n})} + 2a\eta_1 + 2b\eta_1\tau \Rightarrow$$

$$n\zeta(\frac{a + b\tau}{n}) - 2a\eta_1 - 2b\eta_2 = n\frac{\theta'(\frac{a+b\tau}{n})}{\theta(\frac{a+b\tau}{n})} + 2b\eta_1\tau - 2b\eta_2 = n\frac{\theta'(\frac{a+b\tau}{n})}{\theta(\frac{a+b\tau}{n})} + 2b\pi i$$

car $\eta_1\tau - \eta_2 = \pi i \Rightarrow$

$$F_{a,b}(\tau) = \frac{n}{\pi i}\frac{\theta'(\frac{a+b\tau}{n})}{\theta(\frac{a+b\tau}{n})} + 2b.$$

L'expression de θ en produit infini amène à l'égalité

$$\frac{1}{\pi i}\frac{\theta'(z)}{\theta(z)} = \frac{e^{2\pi i z} + 1}{e^{2\pi i z} - 1} - 2e^{2\pi i z}\sum_{m>0}\frac{q^m}{1 - q^m e^{2\pi i z}} + 2e^{-2\pi i z}\sum_{m>0}\frac{q^m}{1 - q^m e^{-2\pi i z}}$$

En supposant que a et b sont des entiers avec $a, b < n$,

$$\frac{1}{\pi i}\frac{\theta'(\frac{a+b\tau}{n})}{\theta(\frac{a+b\tau}{n})} = \frac{\alpha q^{\frac{b}{n}} + 1}{\alpha q^{\frac{b}{n}} - 1} - 2\sum_{m>0}\frac{\alpha q^{m+\frac{b}{n}}}{1 - \alpha q^{m+\frac{b}{n}}} + 2\sum_{m>0}\frac{\alpha^{-1}q^{m-\frac{b}{n}}}{1 - \alpha^{-1}q^{m-\frac{b}{n}}}$$

avec $\alpha = e^{2\pi i \frac{a}{n}}$. Posons $p = e^{2\pi i \frac{\tau}{n}} = q^{\frac{1}{n}}$; la dernière expression devient alors

$$n\frac{1}{\pi i}\frac{\theta'(\frac{a+b\tau}{n})}{\theta(\frac{a+b\tau}{n})} + 2b = n\frac{\alpha p^b + 1}{\alpha p^b - 1} - 2n\sum_{m>0}\frac{\alpha p^{mn+b}}{1 - \alpha p^{mn+b}}$$

$$+ 2n\sum_{m>0}\frac{\alpha^{-1}p^{mn-b}}{1 - \alpha^{-1}p^{mn-b}} + 2b$$

$$= c_0 + c_1 p + c_2 p^2 + \cdots.$$

où $c_i \in \mathbb{C}$. De plus, $\alpha \in \overline{\mathbb{Q}}_{ab}$ et donc $c_i \in \overline{\mathbb{Q}}_{ab}$. F_n peut donc être exprimé en série de Puiseux:

$$F_{a,b}(\tau) \in \overline{\mathbb{Q}}_{ab}[[q^{\frac{1}{n}}]]$$

ce qui montre que c'est effectivement une *forme* modulaire et la démonstration de la proposition du 2.5.3 est terminée.

Plus explicitement, pour $b = 0$

$$F_m(\tau) = n\frac{\alpha + 1}{\alpha - 1} - 2n\sum_{k>0}\frac{\alpha q^k}{1 - \alpha q^k} + 2n\sum_{k>0}\frac{\alpha^{-1}q^k}{1 - \alpha^{-1}q^k}$$

$$= -in(c_0 + c_1 q + c_2 q^2 + \cdots)$$

où $c_0 = i \dfrac{e^{2\pi i \frac{m}{n}}+1}{e^{2\pi i \frac{m}{n}}-1} = \cotg\left(\pi\dfrac{m}{n}\right)$ et $c_k = 4\sum_{d|k} sin\left(2\pi\frac{md}{n}\right)$.

2.5.5. Quelques propriétés des formes F_m.

Dans ce paragraphe nous allons nous limiter au cas $b = 0$ (et $H(a,b)$ sera donc le sous-groupe de congruence $\Gamma_1(n)$) pour étudier certaines propriétés des formes F_m, liées en particulier aux formes de $\Gamma_0(n)$ avec caractère et aux courbes modulaires.

En général, si $M_k(\Gamma_1(n))$ désigne le $\mathbb{C}$-espace vectoriel des formes modulaires de poids 1 pour $\Gamma_1(n)$, on a l'isomorphisme (cf. [K] Prop. 28, p. 137)

$$M_k(\Gamma_1(n)) \cong \bigoplus_\chi M_k(n,\chi)$$

la somme directe étant étendue à tous les caractères $\chi : (\mathbb{Z}/n\mathbb{Z})^* \to \mathbb{C}^*$ et $M_k(n,\chi)$ désignant l'espace vectoriel des formes modulaires de poids 1 et caractère χ pour $\Gamma_0(n)$. On rappelle que $\Gamma_0(n)$ est le sous-groupe de $\Gamma(1)$ formé par les matrices du type $\begin{pmatrix} a & b \\ c & d \end{pmatrix}$ avec $c \equiv 0 \bmod n$. On a alors $\Gamma_1(n) \lhd \Gamma_0(n)$ et $\Gamma_0(n)/\Gamma_1(n) \cong (\mathbb{Z}/n\mathbb{Z})^*$. Un tel isomorphisme peut être obtenu en considérant l'homomorphisme $\mu : \Gamma_0(n) \to (\mathbb{Z}/n\mathbb{Z})^*$ défini par $\mu(M) = d$. Pour tout caractère χ de $(\mathbb{Z}/n\mathbb{Z})^*$, on en obtient un de $\Gamma_0(n)$ en posant $\overline{\chi}(M) = \chi(d) = \chi\mu(M)$. $M_k(n,\chi)$ est alors l'ensemble des éléments $f \in M_k(\Gamma_1(n))$ tels que $f(M\tau) = \overline{\chi}(M)(c\tau+d)^k f(M) \ \forall M \in \Gamma_0(n)$.

Les fonctions F_m avec $(m,n) = 1$ et $2m < n$ définies dans le corollaire du 2.5.3 forment un sous-espace vectoriel V_n de $M_1(\Gamma_1(n))$ (on remarquera que $F_{-m} = -F_m$). De plus, si $M = \begin{pmatrix} a & b \\ c & d \end{pmatrix} \in \Gamma_0(n)$ alors

$$F_m(\tau)|_{[M]_1} := (c\tau+d)^{-1} F_m(M\tau) = F_{m\mu(M)}(\tau) = F_{md}(\tau)$$

En particulier elles n'appartiennent pas à $M_1(n,\chi)$ pour tout χ.

Notre but est celui de construire des éléments de $M_1(n,\chi)$ en partant des fonctions $f_{m,n}$. Ceci est lié au fait que les éléments de $M_1(n,\chi)$ sont des fonctions propres pour l'opérateur $T_{N,N}$ (i.e. $T_{N,N}f = \frac{\chi(N)}{N}f$ avec $(N,n) = 1$ cf. [K] Prop. 35 p. 160 et p. 156 pour la définition de $T_{N,N}$). Soit donc $\chi : (\mathbb{Z}/n\mathbb{Z})^* \to \mathbb{C}^*$ un caractère, $\overline{\chi} = \chi\mu$ le caractère induit pour $\Gamma_0(n)$ et posons en général

$$g(\tau) = \sum_{(m,n)=1} c_m F_m(\tau)$$

Alors on voit que $g(\tau)|_{[M]_1} = \overline{\chi}(M)g(\tau) \Leftrightarrow$

$$\overline{\chi}(M)g(\tau) = \sum_{(n,m)=1} c_m F_{dm}(\tau) = \sum_{(n,m)=1} c_{d^{-1}m} F_m(\tau) \Leftrightarrow$$

$$\chi(d)c_m = c_{d^{-1}m} \;\forall\, d \in (\mathbb{Z}/n\mathbb{Z})^*$$

En posant $c_1 = 1$ on obtient finalement $c_m = \chi^{-1}(m)$.

Proposition. *Soit $\chi : (\mathbb{Z}/n\mathbb{Z})^* \to \mathbb{C}^*$ un caractère. Alors $V_n \cap M_1(n, \chi) = \mathbb{C}g_\chi$ avec*

$$g_\chi(\tau) = \sum_{(n,m)=1,\; 2m<n} \chi(m)^{-1} F_m(\tau).$$

Soient maintenant $\Gamma_i[n] = <\Gamma_i(n), -I>$, $i = 0,1$ et considérons les courbes modulaires

$$X_i(n) = \Gamma_i[n]\backslash\overline{\mathcal{H}}$$

(ici $\overline{\mathcal{H}} = \mathcal{H} \cup \mathbb{Q} \cup \{i\infty\}$ est la compactification ordinaire du demi-plan supérieur). $X_0(n)$ paramétrise de façon naturelle les couple (E, S) où E est une courbe elliptique et S est un sous-groupe de E d'ordre n. De même $X_1(n)$ paramétrise les couples (E, P) où E est une courbe elliptique et $P \in E$ est un point d'ordre (exactement) n (voir par exemple [K] §5 p.153). La théorie générale des courbes modulaires affirme alors que $X_i(n)$ est une courbe projective sur $\mathbb{C}$ (et même sur $\overline{\mathbb{Q}}_{ab}$). On indiquera par $K_i(n)$, $i = 0, 1$ les corps des fonctions respéctifs. $K_i(n)$ s'identifie avec le corps des formes modulaires *méromorphes* de poids 0 pour $\Gamma_i(n)$.
$\Gamma_1(n)$ est un sous-groupe distingué de $\Gamma_0(n)$, ce qui induit un revêtement galoisien

$$X_1(n) \xrightarrow{\pi} X_0(n)$$

de degré $\frac{1}{2}\phi(n)$ et groupe de Galois isomorphe à $(\mathbb{Z}/n\mathbb{Z})^*/\{\pm 1\}$[6].
Le corps $K_1(n)$ est donc une extension galoisienne de $K_0(n)$. Dans ce contexte, l'action du groupe de Galois sur une fonction f est simplement donnée par $f(\tau) \longmapsto f(M\tau)$ pour tout $M \in \Gamma_0(n)$. Les formes modulaires F_m sont particulièrement intéressantes car elles réalisent de façon pratique cette extension:

Proposition. *Soit $F \in M_1(\Gamma_0(n))$ une forme modulaire quelconque (non-identiquement nulle) de poids 1 pour $\Gamma_0(n)$. Considérons $0 < m < \frac{n}{2}$ avec $(m, n) = 1$ et posons $f_m(\tau) = \frac{F_m(\tau)}{F(\tau)}$. Alors*

$$K_1(n) = K_0(n)(f_m).$$

---

[6] $\Gamma_0[n] = \Gamma_0(n)$ tandis que $\Gamma_1[n] \cong \Gamma_1(n) \times \mathbb{Z}/2\mathbb{Z}$ pour $n > 2$.

En d'autres termes, le polynôme

$$p(X) = \prod_{(m,n)=1,\ 2m<n} (X - f_m) \in K_0(n)[X]$$

est irréductible et $K_1(n) \cong K_0(n)[X]/(p)$.

Démonstration. Tout d'abord, $f_{-m} = -f_m$ et f_m est évidemment un élément de $K_1(n)$. La démonstration est immédiate en utilisant le fait que $f_m(M\tau) = f_{\mu(M)m}(\tau)$. L'action étant transitive on en déduit que $P(X)$ est un élément irréductible de $K_0(n)[X]$ (cf. [GI] Th. III.29 p. 181). De plus, son degré est égal à $\frac{1}{2}\phi(n) \Rightarrow K_0(n)(f_m) = K_1(n)$ $\diamondsuit$

Remarque. On peut reformuler la proposition précédente de façon plus explicite en remplaçant F_m par F_m^2 et en posant

$$F = \sum_{(m,n)=1,\ 2m<n} F_m^2 \quad \text{et} \quad f_m = \frac{F_m^2}{F}$$

F n'est pas identiquement nulle; en effet si $F(\tau) = a_0 + a_1 q + \cdots$. alors $a_0 = -n^2 \sum_{(m,n)=1,\ 2m<n} \cotg(\pi\frac{m}{n})^2 < 0$ (cf. 2.5.4).

Pour revenir à la famille de dessins $\mathcal{E}_N$, $\frac{F_m^{12}(\tau)}{\Delta(\tau)}$ définit une fonction méromorphe g_m sur $X_1(n)$ et holomorphe sur $X_1(n)^0 = \Gamma_1[n]\backslash\mathcal{H}$ (ouvert de $X_1(n)$ obtenu en enlevant les "cusps"). Par construction, les zéros de $g_m(\tau)$ correspondent aux valeurs τ pour lesquelles l'application $\overline{\beta}_0$ définie au 2.4 est de Belyi. On en déduit une caractérisation "modulaire" de $\mathcal{E}_N$:

Proposition. *Il existe une bijection entre les zéros de g_m dans $X_1(n)^0$ et les dessins $D \in \mathcal{E}_N$ d'invariant $n_D = n$.*

Remarque. Il est possible de remplacer g_m par la fonction méromorphe $h_m(\tau) = \frac{f_m(\tau)^4}{G_4(\tau)}$ ou même par f_m (définie dans la remarque précédente.

Nous allons terminer par l'étude d'une forme modulaire pour $\mathrm{SL}_2(\mathbb{Z})$ qui permet de donner une caractérisation des invariants modulaires des courbes associées aux dessins de $\mathcal{E}_N$.

2.5.6. La forme modulaire Θ_n.

Soit $\Gamma(1) = \bigcup_{i=1}^{r} \Gamma_1[n]M_i$ une décomposition latérale et posons

$$\Theta_n(\tau) = \prod_i F_1(\tau)|_{[M_i]_1}$$

NB. Evidemment, Θ_n dépend du choix des représentants dans la décomposition en classes à droite de $\Gamma(1)$ (et même de m). Changer de représentants n'affecte que le signe. Ce qui nous intéresse réellement, ce sont les zéros de Θ_n; donc dans la suite nous n'indiquerons pas le choix des M_i.
En accord avec les notations du 2.5.2, Θ_n vérifie

$$\Theta_n(M\tau) = \Psi_n(M)(C\tau + D)^r \Theta_n(\tau)$$

Si Ψ_n est trivial (hypothèse que nous supposerons vérifiée dans la suite), alors les considérations précédentes permettent d'affirmer que Θ_n est une forme modulaire de poids $r = \frac{1}{2}n^2 \prod_{p|n}(1 - \frac{1}{p^2})$. On démontre facilement que Θ_n est non-parabolique pour n impair et parabolique avec un zéro d'ordre $\frac{1}{2}\phi(\frac{n}{2})$ si n est pair (en utilisant le développement en série du 2.5.4)
.

Considérons la courbe modulaire $X = \Gamma(1)\backslash\overline{\mathcal{H}}$ et $X^0 = \Gamma(1)\backslash\mathcal{H}$. Dans ce cas $X \cong \mathbb{P}^1_{\mathbb{C}}$ et $X^0 \cong \mathbb{C}$, l'isomorphisme étant réalisé par la fonction $J(\tau)$. Si $n > 4$ alors $r \equiv 0 \mod 12$ et $\frac{\Theta_n}{\Delta^{\frac{r}{12}}}$ est une *fonction modulaire*. De plus elle est régulière sur $\mathcal{H}$ et possède un pôle d'ordre $d = \frac{r}{12}$ en $i\infty$ si n est impair et $d = \frac{r}{12} - \frac{1}{2}\phi\left(\frac{n}{2}\right)$ si n est pair. On en déduit que

$$\frac{\Theta_n(\tau)}{\Delta^{\frac{r}{12}}(\tau)} = P_n(J(\tau)) \in \mathbb{C}[J]$$

Proposition. *Si Ψ_n est trivial, alors il existe une bijection entre les zéros de $P_n(X) \in \mathbb{C}[X]$ et les invariants modulaires des courbes associées aux dessins $D \in \mathcal{E}_N$ tels que $n_D = n$. En particulier, le degré de P_n est égal à $\frac{r}{12}$ si n est impair et à $\frac{r}{12} - \frac{1}{2}\phi\left(\frac{n}{2}\right)$ si n est pair.*

Remarque. Le cas $n = 3$ est étudié dans la section 2.5.8 et pour $n = 4$, Ψ_n est non-trivial.

La seule chose qui reste à faire est de démontrer que, à multiplication par une constante près, $P_n(X)$ est un élément de $\mathbb{Q}[X]$.

2.5.7. Les coefficients de Θ_n.

Soit $\Theta_n(\tau) = \sum_{k\geq 0} a_k q^k$ où $q = e^{2\pi i\tau}$.
D'après l'expression de $F_{a,b}(\tau)$ en série de Puiseux (cf. 2.5.4), les coefficients appartiennent tous à une extension cyclotomique de $\mathbb{Q}$ (de degré n). Nous pouvons donc considérer l'action de $\mathrm{Gal}(\mathbb{Q}_n/\mathbb{Q})$ (où $\mathbb{Q}_n = \mathbb{Q}(e^{\frac{2\pi i}{n}})$) sur F_n induite par l'action sur les coefficients.

Identifions $\text{Gal}(\mathbb{Q}_n/\mathbb{Q})$ avec $(\mathbb{Z}/n\mathbb{Z})^*$ ($\sigma \longmapsto \chi(\sigma)$ où χ est le caractère cyclotomique). On a alors

$$^{\sigma}F_{a,b}(\tau) = F_{\chi(\sigma)a,b}(\tau) \quad \forall \, \sigma \in \text{Gal}(\mathbb{Q}_n/\mathbb{Q})$$

En particulier l'action préserve les formes du type F_m.
Nous obtenons une action sur Θ_n:

$$^{\sigma}\Theta_n(\tau) = \mathcal{G}_n(\sigma)\Theta_n(\tau)$$

$\mathcal{G}_n$ est un caractère $\text{Gal}(\mathbb{Q}_n/\mathbb{Q}) \to \mathbb{Z}/2\mathbb{Z}$. Si $\mathcal{G}_n$ est trivial, alors les coefficients a_k sont rationnels, sinon il existe une constante C_n appartenant à une extension quadratique de $\mathbb{Q}$ telle que $C_n a_k \in \mathbb{Q}$ pour tout k. Ceci découle de l'action "globale" de $\text{Gal}(\mathbb{Q}_n/\mathbb{Q})$ sur Θ_n. C_n est effectivement calculable en déterminant a_0 dans le développement en série. En particulier, $\mathcal{G}_n$ est trivial si et seulement si n n'est pas premier et $\phi(n) \equiv 0 \bmod 4$. Si n n'est pas premier mais $\phi(n) \equiv 2 \bmod 4$ alors $C_n = i$. Finalement, pour $n = p$ premier, $C_p = \sqrt{(-1)^{\frac{p-1}{2}}p}$. Le caractère $\mathcal{G}_n$ est obtenu en considérant l'homomorphisme $\text{Gal}(\overline{\mathbb{Q}}/\mathbb{Q}) \longrightarrow \text{Gal}(\mathbb{Q}(C_n)/\mathbb{Q})$. En résumant, nous obtenons la

Proposition. *Supposons que Ψ_n soit trivial. Alors il existe une constante c_n (ne dépendant que de n) telle que les coefficients de $c_n\Theta_n$ sont rationnels.*

Remarque. On peut s'arranger pour que les coefficients soient *entiers*.

Corollaire. $P_n(X) \in \mathbb{Q}[X]$ *(à multiplication par une constante près).*

Démonstration. Nous supposerons, en appliquant la proposition précédente, que les coefficients de Θ_n sont rationnels. En suivant la remarque du paragraphe précédent, nous avons l'égalité (on suppose que Ψ_n est trivial, donc en particulier $n > 4$)

$$\frac{\Theta_n(\tau)}{\Delta^{\frac{r}{12}}(\tau)} = P_n(J(\tau))$$

En développant en série les deux termes, on voit que les coefficients de P_n peuvent être obtenus en resolvant des équations linéaires à coefficients rationnels (car les coefficients de Δ, J et Θ_n sont rationnels). $\Diamond$

2.5.8. Exemples.

Etudions le cas $N = 3$; le plus simple. Il existe un seul dessin et pour des raisons de symétrie il correspond à la courbe d'invariant modulaire $J = 0$. D'un point de vue "modulaire" on retrouve la démarche suivante:

$\mathcal{E}_3 = \{D_{1,1,1}\}$; $N_1 = N_2 = N_3 = 1$. On aura donc $(2,2,2) = 2 \Rightarrow n_D = 3$. $R(3)$ est composé de 4 éléments. De plus Ψ_3 est trivial et la forme modulaire est non parabolique (car n est impair). Par un théorème classique d'unicité, on en déduit immédiatement que

$$\Theta_3(\tau) = cg_2(\tau)$$

avec $c \in \mathbb{C}$ et l'on retrouve $J = 0$.

$N = 4$. Dans ce cas aussi nous avons un seul dessin: $D_{1,1,2}$. On obtient $n = 8$ et il faut donc étudier $\Theta_8(\tau)$. C'est une forme modulaire parabolique de poids 24 ayant un zéro simple en $i\infty$. On a alors

$$\Theta_8(\tau) = a\Delta^2(\tau)(J(\tau) - J_0)$$

où a est une constante et J_0 est l'invariant modulaire de la courbe associée au dessin.

En développant en série (Maple V) on obtient

$$\Theta_8(\tau) = 549755813888q(6561 + 4358810q + \cdots)$$

$$a\Delta^2(\tau)(J(\tau) - J_0) = aq(1 + (696 - J_0)q + \cdots)$$

En identifiant les coefficients on détermine donc J_0:

$$J_0 = \frac{210646}{6561}$$

Remarque. Le dessin n'ayant pas de conjugué sous l'action de Γ et le point de torsion étant uniquement déterminé (en ayant choisi une orientation admissible), on en déduit que la courbe elliptique d'invariant $J_0 = \frac{210646}{6561}$ possède un point de torsion d'ordre 8 défini au pire sur une extension quadratique de $\mathbb{Q}$ (ceci car il faut considérer l'éventuelle action $P \longmapsto -P$ de Γ).

$N = 5$. Ici on retrouve deux dessins $D_{1,2,2}$ et $D_{3,1,1}$, avec respectivement $n = 10$ et $n = 5$. Ces deux dessins ne sont donc pas conjugués. Etudions de plus près $D_{3,1,1}$: Θ_5 est une forme modulaire non parabolique de poids 12 et l'on obtient

$$\Theta_5(\tau) = a\Delta(\tau)(J(\tau) - J_0)$$

Comme dans l'exemple précédent, en développant en série on arrive aux égalités

$$\Theta_5(\tau) = 5\sqrt{5}(243 + 154480q + \cdots)$$
$$a\Delta(\tau)(J(\tau) - J_0) = a(1 + (720 - J_0)q + \cdots)$$

et finalement

$$J_0 = \frac{20480}{243}$$

§3. Dessins d'enfants en genre 1 et isogénies. Un exemple.

Cette dernière partie est consacrée à la réalisation de groupes de Galois (infinis) comme groupes d'automorphismes de groupes profinis. Notre but est celui de fournir des exemples explicites décrivant le théorème général (action de $\mathbb{\Gamma} = \mathrm{Gal}(\overline{\mathbb{Q}}/\mathbb{Q})$ sur $\widehat{F}_2$) présenté dans [I].

3.1. Réalisation de $\mathbb{\Gamma}_{ab} = \mathrm{Gal}(\overline{\mathbb{Q}}_{ab}/\mathbb{Q})$.

Ce premier paragraphe est équivalent au §1: Il n'est pas d'un intérêt fondamental mais permet de mettre en évidence certaines caractéristiques qui se retrouvent dans des généralisations ultérieures. $\mathbb{\Gamma}_{ab}$ est bien connu et isomorphe à $\widehat{\mathbb{Z}}^*$. Dans ce contexte, sa réalisation comme groupe d'automorphismes perd un peu d'intérêt (car pour le faire il n'est réellement pas nécessaire de passer par les dessins d'enfants). Nous allons néanmoins décrire une telle construction qui met en jeu les groupes diédraux et les points d'ordre fini sur $\mathbb{C}^*$ (dans la suite il sera question de "groupes diédraux généralisés" et de points de torsion sur des courbes elliptiques).

Nous avons vu dans le 1.2, que la paire de Belyi $(\mathbb{P}^1, f_n)$ (où $f_n = \frac{1}{2}(z^n + \frac{1}{z^n})$) est un modèle pour le dessin régulier associé au groupe diédral

$$\Delta_n =< a, b \mid ab = b^{-1}a, \ a^2 = b^n = 1 >$$

muni de la présentation $v = a$, $l = ba$ et $f = b$. Nous indiquerons par $\mathrm{Aut}(f_n)$ le groupe d'automorphismes du revêtement ("deck transformations"). Le choix d'un drapeau de référence amène alors à un isomorphisme[7]

$$\Delta_n \xrightarrow{\phi} \text{Aut}(f_n)$$

On peut poser ici $\phi(f)(z) = \zeta z$ et $\phi(v)(z) = \frac{1}{z}$ avec $\zeta = e^{\frac{2\pi i}{n}}$. L'action de Γ se traduit alors par l'action sur un tel isomorphisme, et elle est donc cyclotomique.

Indiquons par $\chi : \Gamma \to \widehat{\mathbb{Z}}^*$ le caractère cyclotomique et par $\phi \longmapsto {}^\sigma\phi$ l'action de Γ sur ϕ. On a alors

$$\begin{cases} {}^\sigma\phi(f)(z) &= \zeta^{\chi(\sigma)}z = \phi(f^{\chi(\sigma)})(z) \\ {}^\sigma\phi(v)(z) &= \frac{1}{z} = \phi(v^{\chi(\sigma)})(z) \end{cases}$$

Finalement, on peut considérer le groupe (profini)

$$\Delta_\infty = \varprojlim \Delta_n \cong \mathbb{Z}/2\mathbb{Z} \ltimes \widehat{\mathbb{Z}}$$

muni de la présentation canonique V, L, F provenant de celles des Δ_n. L'action de Γ est alors explicitée par les relations

$$\begin{cases} V &\longmapsto V^{\chi(\sigma)} \\ L &\longmapsto L^{\chi(\sigma)} \\ F &\longmapsto F^{\chi(\sigma)} \end{cases}$$

et permet de caractériser les automorphismes de Δ_∞ provenant de l'action de Γ [8]. Ces relations sont en accord avec le théorème énoncé dans [I], et se réduisent au cas le plus simple.

3.2. Réalisation de $\text{Gal}(\overline{\mathbb{Q}(i)}_{ab}/\mathbb{Q})$.

Nous allons à présent nous intéresser à un groupe de Galois non-commutatif dans le but de rendre moins "triviale" l'action de Γ. La construction présentée passe par l'étude des dessins obtenus en composant une application de Belyi avec une isogénie. Dans le cas présent nous nous limiterons aux dessins réguliers en genre 1.

Soit $\mathcal{C}$ la courbe elliptique d'équation

$$C : Y^2 = 4X(X^2 - 1)$$

---

[7] qui induit une structure de Δ_n-revêtement.

[8] On remarquera que $\Gamma_{ab} \to \text{Aut}(\Delta_\infty)$ est *injectif*.

et considérons le revêtement $f : \mathcal{C} \to \mathbb{P}^1_{\mathbb{C}}$ défini par $f(X, Y) = X^{-2}$.

Les points critiques de f sont $0, 1$ et ∞ donc $(\mathcal{C}, f)$ est une paire de Belyi (f a un zéro d'ordre 4 dans le point à l'infini de $\mathcal{C}$, un pôle d'ordre 4 en $X = 0$, et $f - 1$ a deux zéros doubles en $X = \pm 1$). Le revêtement est galoisien et le groupe d'automorphismes est cyclique, isomorphe à $\mathbb{Z}/4\mathbb{Z}$ et engendré par $\theta : \mathcal{C} \to \mathcal{C}$ avec $\theta(X, Y) = (-X, iY)$.[9] La paire $(\mathcal{C}, f)$ est définie sur $\mathbb{Q}$.

Le choix $\tau = i$ comme valeur du module associé à la courbe détermine le revêtement universel $\mathbb{C} \xrightarrow{\pi} \mathcal{C}$. On obtient alors l'isomorphisme analytique $\mathcal{C} \cong \mathbb{C}/\Lambda$ où $\Lambda = \mathbb{Z} \oplus i\mathbb{Z}$. Dans le parallélogramme fondamental (déterminé par les points $0, i, 1, 1 + i$), le zéro de f correspond à $z = 0$; $z = \frac{1+i}{2}$ est le pôle, et les deux autres points de ramification sont $z = \frac{1}{2}$ et $z = \frac{i}{2}$. Nous retrouvons donc les points de $\mathcal{C}[2]$. Sur $\mathbb{C}$ l'automorphisme θ s'exprime simplement par $\theta(z) = iz$ et est donc un morphisme (en voyant $\mathcal{C}$ comme groupe abélien).

Considérons à présent le revêtement non-ramifié (endomorphisme)

$$\mathcal{C} \xrightarrow{\phi_n} \mathcal{C}$$

correspondant à la multiplication par $n \in \mathbb{N}$. Si l'on pose $f_n = f \, \phi_n$, alors la paire $(\mathcal{C}, f_n)$ est de Belyi et nous obtenons un nouveau dessin D_n. Cette méthode est effectivement utilisable pour tout dessin en genre 1 mais dans ce cas les choses se simplifient particulièrement. Notre but est celui d'étudier à fond D_n.

L'application f_n a n^2 zéros d'ordre 4 correspondant aux points de $\mathcal{C}[n]$, n^2 zéros d'ordre 4; de plus $f_n - 1$ possède exactement $2n^2$ zéros doubles et ceux-ci sont les seuls points de ramification de f_n.

Le revêtement est galoisien de degré $4n^2$. Son groupe de Galois $\mathrm{Aut}(f_n)$ est engendré par θ et $T_n = \{\sigma \in \mathrm{Aut}(\mathcal{C}) \mid \sigma(P) = P + P_n \text{ où } P_n \in \mathcal{C}[n]\}$. On a évidemment $T_n \cong \mathcal{C}[n]$.

Pour décrire $\mathrm{Aut}(f_n)$ il est comode de passer par le revêtement universel. Nous allons identifier $\mathcal{C}[n]$ avec $(\mathbb{Z}/n\mathbb{Z})^2$ (en tant que groupes) en considérant l'application

$$(\mathbb{Z}/n\mathbb{Z})^2 \longrightarrow \mathbb{C}/\Lambda$$
$$(a, b) \longmapsto \frac{a}{n} + i\frac{b}{n}$$

Dans cette optique, $\theta(a, b) = (-b, a)$.

Pour tout $P \in \mathcal{C}[n]$, indiquons par σ_P l'automorphisme de $\mathcal{C}$ défini par $\sigma_P(Q) = Q + P$. On a alors $\theta \, \sigma_P \, \theta^{-1} = \sigma_{\theta(P)} \Rightarrow T_n \lhd G_n$. De plus

---

[9] On retrouve ici un des dessins d'enfants (le plus simple) présentés au §1.

$T_n \cap\ <\theta> = 1$, et G_n est donc le produit semi-direct de $<\theta>$ et T_n. On aura alors

$$\mathrm{Aut}(f_n) \cong G_n = (\mathbb{Z}/n\mathbb{Z})^2 \rtimes_\theta \mathbb{Z}/4\mathbb{Z} \cong \mathcal{C}[n] \rtimes_\theta \mathrm{Aut}(\mathcal{C})$$

Remarque. La courbe $\mathcal{C}$ possède des isogénies non-triviales (l'anneau des endomorphismes est isomorphe à $\mathbb{Z}[i]$). On peut donc appliquer la construction précédente pour un morphisme quelconque correspondant à $\alpha = a + ib$. Si nous ne l'avons pas fait c'est que ces derniers peuvent être obtenus comme quotients de ceux décrits plus haut. En effet, si l'on pose $n = a^2 + b^2$, alors le revêtement $\mathcal{C} \xrightarrow{\phi_n} \mathcal{C}$ se factorise en $\mathcal{C} \xrightarrow{\phi_\alpha} \mathcal{C} \xrightarrow{\phi_{\overline{\alpha}}} \mathcal{C}$.

Choisissons pour G_n la présentation $v = (0,0,1)$, $l = (1,0,2)$, $f = (1,0,1)$ et un drapeau de référence b_0 donné (sur le revêtement universel) par le "segment" $]0, \frac{1}{n}[$ orienté de gauche à droite (figure suivante).

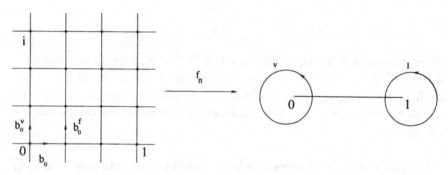

Dans ce cas,

$$\begin{cases} b_0^v &= \theta(b_0) \\ b_0^l &= \sigma_0\,\theta^2(b_0) \\ b_0^f &= \sigma_0\,\theta(b_0) \end{cases}$$

où σ_0 est la translation déterminée par le point de $\mathcal{C}[n]$ correspondant à $z = \frac{1}{n}$ sur le revêtement universel.

On définit ainsi, en accord avec le paragraphe précédent, un isomorphisme

$$G_n \xrightarrow{\phi} \mathrm{Aut}(f_n)$$

qui induit une structure de G_n-revêtement, exprimé par $g_v = \phi(v) = \theta$, $g_l = \phi(l) = \sigma_0\theta^2$ et $g_f = \phi(f) = \sigma_0\theta$ avec $g_v g_l g_f = 1$.

On peut finalement passer à l'action de Γ. $(\mathcal{C}, f_n)$ étant définie sur $\mathbb{Q}$, une telle action n'intervient que sur l'isomorphisme $\phi : G_n \rightarrow \mathrm{Aut}(f_n)$.

Pour tout $\tau \in \mathbb{\Gamma}$ on aura ainsi un isomorphisme

$$^\tau\phi : G_n \to \mathrm{Aut}(f_n)$$
$$g \longmapsto \tau(\phi(g))$$

Tout d'abord, $\theta(X, Y) = (-X, iY) \Rightarrow \tau(\theta) = \theta^{\chi(\tau)}$ où χ est le caractère cyclotomique. On peut aussi écrire $\tau(\theta) = \theta^{\chi_4(\tau)}$ où $\chi_4 : \mathbb{\Gamma} \to (\mathbb{Z}/4\mathbb{Z})^*$ est l'homomorphisme correspondant à l'extension $\mathbb{Q}(i)/\mathbb{Q}$.

Pour les translations, en ayant choisi deux générateurs de $\mathcal{C}[n]$ (ce que nous avons fait en identifiant $\mathcal{C}[n]$ avec $(\mathbb{Z}/n\mathbb{Z})^2$) on obtient l'homomorphisme de groupes

$$\mathbb{\Gamma} \to \mathrm{GL}_2(\mathbb{Z}/n\mathbb{Z})$$

$$\tau \longmapsto M_\tau = \begin{pmatrix} a_\tau & b_\tau \\ c_\tau & d_\tau \end{pmatrix}$$

et sur $\mathbb{C}$, l'action de $\mathbb{\Gamma}$ sur $\mathcal{C}[n]$ s'exprime par

$$\tau\left(\frac{N}{n} + i\frac{M}{n}\right) = \frac{a_\tau N + b_\tau M}{n} + i\frac{c_\tau N + d_\tau M}{n}$$

Pour σ_P on obtient $\tau(\sigma_P) = \sigma_{\tau(P)}$ et $^\tau\phi$ est déterminé par

$$\begin{cases} ^\tau g_v &= {^\tau\phi(v)} = \theta^{\chi(\tau)} \\ ^\tau g_f &= {^\tau\phi(f)} = {^\tau\sigma_0}\theta^{\chi(\tau)} \end{cases}$$

En général on aura $\sigma_P^{-1}\theta\sigma_P = \sigma_{\theta(P)-P}\theta$.

Soit P_0 (resp. P_1) le point de $\mathcal{C}[n]$ correspondant au point $z = \frac{1}{n}$ (resp. $z = \frac{i}{n}$) sur le revêtement universel ($P_1 = \theta(P_0)$). Notre but est celui de déterminer un point $P = \frac{N}{n} + i\frac{M}{n} \in \mathcal{C}[n]$ (dans le parallélogramme fondamental) tel que

$$(*) \quad \begin{cases} \theta(P) - P &= {^\tau P_0} - P_0 \quad \text{si} \quad \chi_4(\tau) \equiv 1 \mod(4) \\ \theta(P) - P &= {^\tau P_0} - P_1 \quad \text{si} \quad \chi_4(\tau) \equiv 3 \mod(4) \end{cases}$$

Intéressons-nous avant toute chose au cas $\chi_4(\tau) \equiv 1 \mod(4)$: On arrive au système d'équations

$$\begin{cases} 2N &\equiv 1 - a_\tau + c_\tau \mod(n) \\ 2M &\equiv 1 - a_\tau - c_\tau \mod(n) \end{cases}$$

Si n est impair on obtient une solution unique. Pour n pair il suffit de vérifier que a_τ et c_τ n'ont pas la même parité. Pour cela remarquons que $\mathcal{C}[2] \subset \mathcal{C}(\mathbb{Q})$ et $\mathbb{\Gamma}$ opère donc trivialement sur ces points. En particulier $M_\tau \begin{pmatrix} \frac{n}{2} \\ 0 \end{pmatrix} = \begin{pmatrix} \frac{n}{2} \\ 0 \end{pmatrix}$, qui se traduit par $a_\tau \equiv 1 \mod(2)$ et $c_\tau \equiv 0 \mod(2)$.

De tels arguments peuvent être appliqués au cas $\chi_4(\tau) \equiv 3 \mod(4)$ et $(*)$ admet donc toujours une solution.

Par construction, $^\tau\phi(f) = \sigma_P^{-1}\phi(f^{\chi(\tau)})\sigma_P$. Posons $g_1 = fv^{-1}$, $g_2 = v^{-1}f$ et

$$(**) \qquad g_\tau = g_1^N g_2^M \in G_n$$

On vérifie immédiatement que $\phi(g_\tau) = \sigma_P$.

Proposition. *Soit $G_n = (\mathbb{Z}/n\mathbb{Z})^2 \rtimes_\theta \mathbb{Z}/4\mathbb{Z}$ muni de la présentation $v = (0,0,1)$, et $f = (1,0,1)$. Soit P un point de $\mathcal{C}[n]$ vérifiant $(*)$. Alors l'action de Γ sur G_n est donnée par*

$$\begin{cases} v & \longmapsto v^{\chi(\tau)} \\ f & \longmapsto g_\tau^{-1} f^{\chi(\tau)} g_\tau \end{cases}$$

où g_τ est défini comme en $(**)$.

Remarque. Si n est pair, $N = N_\tau$ et $M = M_\tau$ ne sont pas uniquement déterminés. Les éléments g_1 et g_2 ne dépendent pas de $\tau \in \Gamma$. De plus $g_1 g_2 = g_2 g_1$ et on vérifie immédiatement que $\{g_\tau\}$ définit un cocycle; on obtient donc une classe de cohomologie dans le groupe $H^1(\Gamma, \mathcal{C}[n])$.

Pour obtenir un énoncé mettant en jeu des groupes profinis remarquons tout d'abord que l'action de Γ se réduit à l'action de $\mathrm{Gal}(K/\mathbb{Q})$ où K est la limite directe des corps K_n obtenus en rajoutant à $\mathbb{Q}(i)$ les coordonnées des points de n-torsion. On démontre de plus ([Si] chap II) que $K = \overline{\mathbb{Q}(i)}_{ab}$.

En posant

$$\mathcal{G} = \varprojlim G_n \cong \widehat{\mathbb{Z}}^2 \rtimes_\theta \mathbb{Z}/4\mathbb{Z} \cong \widehat{T}(\mathcal{C}) \rtimes \mathrm{Aut}(\mathcal{C})$$

et en considérant la présentation induite, on peut reformuler la proposition en remplaçant G_n par $\mathcal{G}$ et en considérant l'homomorphisme

$$\Gamma \to \mathrm{GL}_2(\widehat{\mathbb{Z}})$$

provenant de l'action sur le module de Tate de la courbe. L'application

$$\mathrm{Gal}(\overline{\mathbb{Q}(i)}_{ab}/\mathbb{Q}) \to \mathrm{Aut}(\mathcal{G})$$

ainsi définie est *injective*.

3.3. Applications.

Dans ce qui précède nous avons étudié une famille de dessins d'enfants galoisiens. Dans la suite nous allons nous intéresser aux quotients de ceux-ci et en particulier aux dessins de genre 1 obtenus par ce procédé.

D_n est le seul dessin galoisien (propre) en genre 1 ayant n^2 faces et sommets de valence 4. Ceci revient à dire que le groupe $\text{Aut}(G_n)$ opère transitivement sur l'ensemble des présentations $G_n =< v, l, f \mid vlf = 1, \; o(f) = 2o(l) = o(v) = 4 >$.

Soit $E(4, 2, 4)$ l'ensemble des classes d'équivalence de paires de Belyi (prépropres) $(\mathcal{D}, \beta)$ où l'ordre des zéros et de pôles de β divise 4 (ce sont tous les dessins de type $[4, 2, 4]$... cf. l'introduction).

Posons de même $E^1(4, 2, 4) = \{(\mathcal{D}, \beta) \in E(4, 2, 4) \mid \mathcal{D} \text{ a genre } 1\}$.

Pour toute paire $(\mathcal{D}, \beta) \in E(4, 2, 4)$, il existe un entier positif n et une suite de revêtements

$$\mathcal{C} \xrightarrow{\phi_D} \mathcal{D} \xrightarrow{\beta} \mathbb{P}^1$$

telle que $\beta \, \phi_D = f_n$. $\mathcal{D}$ est alors le quotient de $\mathcal{C}$ par un sous-groupe de $\text{Aut}(f_n)$[10]. En particulier, vu que les éléments de ce dernier son tous définis sur $K_n \subset \overline{\mathbb{Q}(i)}_{ab}$, on obtient la

Proposition. *Tout dessin de $E(4, 2, 4)$ peut être défini sur une extension abélienne de $\mathbb{Q}(i)$.*

Passons maintenant aux éléments de $E^1(4, 2, 4)$. Ceux-ci correspondent à des courbes elliptiques isogènes à $\mathcal{C}$. On peut alors décomposer $E^1(4, 2, 4)$ en "strates" $E^1(4, 2, 4)_n = \{(\mathcal{D}, \beta) \in E^1(4, 2, 4) \mid deg\beta = 4n\}$[11]. Décrire les éléments de $E^1(4, 2, 4)$ revient à déterminer tous les sous-réseaux d'indice n de $\Lambda = \mathbb{Z} \oplus i\mathbb{Z}$ à homothéties près. C'est le problème des homothéties qui rend les calculs compliqués et fastidieux (déterminer le nombre de sous-réseaux d'indice n est relativement facile). En travaillant avec les groupes G_n on peut simplifier cette démarche. Voici la méthode générale:

$\mathcal{D}$ est le quotient de $\mathcal{C}$ par un sous-groupe d'automorphismes $H < T_n$ (translations) de degré n. Le dessin d'enfant est alors déterminé par la paire (G_n, H) (cf. 1.2.).

Il faut donc déterminer toutes les classes de conjugaison dans G_n de sous-groupes d'indice n de T_n (on remarquera que H possède au plus un conjugué donné par $\theta^{-1}H\theta$).Pour des raisons de simplicité nous allons nous borner au cas $n = p$ premier. Ceci n'est pas vraiment une limitation.

---

[10] Le groupe profini $\mathcal{G}$ défini plus haut est l'équivalent (pour les dessins de $E(4, 2, 4)$) de $\widehat{F_2}$ dans le cas général.

[11] Ceci revient à dire qu'il existe une isogénie $\mathcal{C} \to \mathcal{D}$ de degré n.

Si $n = p$ est premier, alors tout sous-groupe (non-trivial) de T_p (que l'on identifiera dans la suite avec $\mathcal{C}[p] \cong \mathbb{F}_p^2$) est cyclique engendré par σ_Q avec $Q \in \mathcal{C}[p]$. Il existe donc une bijection entre l'ensemble des sous-groupes d'indice p de T_p et les points de $\mathbb{P}^1_{\mathbb{F}_p}$. On aura donc $p + 1$ choix possibles. Si $H = <\sigma_Q>$, alors $H^\theta = \theta^{-1}H\theta = <\sigma_{\theta(Q)}>$.

En posant $Q = (x, y) \in \mathbb{F}_p$,

$$H = H^\sigma \iff x^2 + y^2 \equiv 0 \mod(p)$$

En particulier, si $p \equiv 3 \mod(4)$, alors H ne peut pas être distingué dans G_n et on obtient donc $\frac{p+1}{2}$ dessins non isomorphes.

Si $p \equiv 1 \mod(4)$, alors on obtient deux sous-groupes distingués distincts et $E^1(2, 4, 4)_p$ est constitué de $\frac{p+3}{2}$ dessins dont $\frac{p-1}{2}$ non-galoisiens et 2 galoisiens. De plus, pour les deux dessins galoisiens on obtient un automorphisme d'ordre 4 qui fixe un point. Les courbes correspondantes sont donc isomorphes à $\mathcal{C}$, et les revêtements obtenus correspondent à des isogénies non triviales (autres que la multiplication par n...cf. remarque du 3.2) . En résumé, on obtient la

Proposition. *Soit p premier; alors*

$$\left| E^1(2, 4, 4)_p \right| = \begin{cases} \frac{p+1}{2} & si\ p \equiv 3 \mod(4) \\ \frac{p+3}{2} & si\ p \equiv 1 \mod(4) \end{cases}$$

En particulier, si $p \equiv 3 \mod(4)$, alors les dessins admettent un modèle défini sur une extension abélienne de $\mathbb{Q}(i)$, de degré inférieur ou égal à $\frac{p+1}{2}$. Si $p \equiv 1 \mod(4)$, alors on obtient 2 dessins réguliers et tous les autres peuvent être définis sur une extension abélienne de $\mathbb{Q}(i)$ de degré au plus $\frac{p-1}{2}$.

Exemple. Nous allons reprendre les exemples présentés dans $[SS]$: On s'intéresse ici au cas $p = 3, 5$. Avant toute chose, nous allons utiliser le théorème 1 p.3 qui affirme que la courbe est définie sur $\mathbb{Q}$ seulement dans deux cas: $J = 1728$ et $J = 287496$ qui correspondent respectivement à $\tau = i$ et $\tau = 2i$. En particulier, pour $p \neq 2$, les dessins non galoisiens de $E^1(2, 4, 4)_p$ ne sont pas définis sur $\mathbb{Q}$.

Pour $p = 3$ on obtient deux dessins qui sont donc conjugués.

Pour $p = 5$ les dessins sont 4; deux sont galoisiens et les deux autres sont conjugués et définis sur une extension quadratique de $\mathbb{Q}$ (pour un exemple "graphique" voir $[SS]$ p.6).

Ces résultats sont en accord avec ceux de $[SS]$, tout en utilisamt des méthodes différentes.

3.4. Conclusion. Remarques.

Les dessins réguliers propres en genre 1 (et on le déduit par le calcul de la caractéristique d'Euler-Poincaré) sont du type $[4, 2, 4]$ ou $[3, 2, 6]$. Le premier cas a été étudié à fond dans les paragraphes précédents. Pour le deuxième cas, on peut suivre exactement la même démarche en remplaçant la courbe $\mathcal{C}$ par celle d'invariant modulaire $J = 0$. Les résultats obtenus sont alors une simple transcription de ceux présentés ici, avec des modifications moindres (dans la définition des groupes G_n ou des corps mis en jeu). On obtient en particulier une réalisation du groupe $\mathrm{Gal}(\overline{\mathbb{Q}(\rho)}_{ab}/\mathbb{Q})$ où $\rho^3 = 1$.

§4 Remerciements.

Une grande partie des résultats, et en particulier pour ce qui regarde les formes modulaires, provient de ma thèse de "Laurea". Un premier remerciement va donc à E. Arbarello qui en premier m'a adressé vers la théorie des dessins d'enfants. Le développement ultérieur n'aurait jamais été possible sans les conseils et le soutien de Leila Schneps et de Pierre Lochak que je remercie entre autre pour m'avoir fourni les preprints de F. Pakovicth et de D.Singerman et R.I.Syddall. Un remerciement particulier est adressé à l'Istituto Nazionale Di Alta Matematica "F. Severi" (INDAM) qui a rendu possible mon séjour à Paris.

References

[C]   K. Chandrasekharan, Elliptic functions, Springer-Verlag, 1980.

[G]   A. Grothendieck, Esquisse d'un programme, *Geometric Galois Actions*, volume I.

[GI]   M. Girardi, G. Israel, *Teoria dei campi*, Feltrinelli, 1976.

[I]   Y.Ihara, On the embedding of $\mathrm{Gal}(\overline{\mathbb{Q}}/\mathbb{Q})$, in *The Grothendieck theory of dessins d'enfants*, L. Schneps, ed., London Mathematical Society Lecture Note Series **200**, Cambridge University Press, 1994.

[K]   N. Koblitz, *Introduction to elliptic curves and modular forms*, GTM **97**, 1993.

[L]   J. Lehner, Discontinuous groups and automorphic functions, American Mathematical Society, 1964.

[P]   F. Pakovitch, Arbres planaires et points d'ordre fini sur les jacobiennes des courbes hyperelliptiques, Prépublication de l'institut Fourier, Grenoble, 1996.

[S]   L. Schneps, Dessins d'enfants on the Riemann sphere, in *The Grothendieck Theory of dessins d'enfants*, L. Schneps, ed. London Mathematical Society Lecture Note Series, 200, Cambridge Univ. Press, 1994.

[Si]   J. Silverman, *Advanced topics in the theory of elliptic curves*, GTM **151**, Springer-Verlag, 1994.

[Sh]   G. Shabat, Plane trees, in *The Grothendieck Theory of dessins d'enfants*, L. Schneps, ed., London Mathematical Society Lecture Note Series, **200**, Cambridge Univ. Press, 1994.

[SS]   D.Singerman, R.I.Syddall, Belyi uniformisation of elliptic curves, preprint, 1996.

[SV]   G. Shabat and V. Voevodsky, Drawing curves over number fields, in *The Grothendieck Festschrift* **3**, Birkäuser, 1990.

[Z]   L. Zapponi, Grafi, differenziali di Strebel e curve, Tesi di Laurea, università degli studi di Roma "La Sapienza", 1995.

Laboratoire de Mathématiques
Faculté des Sciences
25030 Besançon CEDEX
zapponi@dmi.ens.fr   zapponi@math.univ-fcomte.fr

Part II. The Inverse Galois Problem

The Regular Inverse Galois Problem over Large Fields

Pierre Dèbes and Bruno Deschamps

§1. Introduction.

There has been recent progress on the Inverse Galois Problem. Most of it consists of results on the absolute Galois group $G(K(T))$ when K is a field with various good arithmetic properties. This paper is a survey of that recent progress. Our goal is also to try to unify the results and the questions that have arisen in the last few years.

Historically the Inverse Galois Problem (IGP) is: is each finite group the Galois group $G(E/\mathbb{Q})$ of an extension of $\mathbb{Q}$? The modern approach consists in studying rather the Regular Inverse Galois Problem (RIGP): is each finite group the Galois group $G(E/\mathbb{Q}(T))$ of *a Galois extension $E/\mathbb{Q}(T)$ with $E/\mathbb{Q}$ a regular extension*? (As usual, we just say in the sequel *regular Galois extension $E/\mathbb{Q}(T)$*). *Regular* means that $E \cap \overline{\mathbb{Q}} = \mathbb{Q}$, or, equivalently, $G(E/\mathbb{Q}(T)) = G(E\overline{\mathbb{Q}}/\overline{\mathbb{Q}}(T))$. Here are three reasons why considering this problem is more natural:

(1) A positive answer to the RIGP implies a positive answer to the IGP. This classically follows from Hilbert's irreducibility theorem.

(2) Regular Galois extensions $E/\mathbb{Q}(T)$ correspond to Galois covers $f : X \to \mathbb{P}^1$ that are defined over $\mathbb{Q}$ along with their automorphisms. Thus the RIGP essentially consists in studying the action of $G(\mathbb{Q})$ on covers of the projective line. This fits the general feeling that the action of $G(\mathbb{Q})$ on geometric objects is expected to reflect much of its structure.

(3) The RIGP can be formulated more generally for an arbitrary field K (possibly of positive characteristic) in place of $\mathbb{Q}$. Given any field K, each group might well be the Galois group of a regular Galois extension of $K(T)$: at least, no counter-example is known. In contrast with the IGP, the RIGP might not depend on the base field K but rest on some universal property of the field $K(T)$ of rational functions.

Because of the regularity condition, solving the RIGP (in an affirmative way) over a given field automatically solves it over each overfield. Therefore it is sufficient to solve the RIGP over each prime field Q. The RIGP has been solved over each algebraically closed field: the real problem is the descent from $\overline{\mathbb{Q}}$ to Q. Roughly speaking there are then two directions of work. Group theorists fix a group and try to realize it over $Q(T)$ regularly (*i.e.*, as the Galois group of a regular Galois extension of $Q(T)$). Thanks to the so-called rigidity criterion and its developments, there have been since the late 70' many

results in that direction for simple groups in particular. We refer the reader
to the works of Matzat ([Mat], [MatMa]) and his students for that aspect of
the question (see also [De1]). On the other hand, arithmetic geometers try to
realize regularly *all* finite groups over $K(T)$ with K a given algebraic extension
(as small as possible) of Q. There has been more recently some progress in
that direction. We will focus on this second aspect of the problem, which was
developed in particular by Fried, Harbater and Pop. Of course this distinction
is somewhat artificial for in practice there is a strong interaction between the
group theory and the arithmetic.

Basically most of the recent progress can be summarized by saying that
the RIGP is solved if the base field is "large". By large we actually mean
the following precise property, which was introduced by F. Pop: each smooth
geometrically irreducible curve defined over K has infinitely many K-rational
points provided there is at least one. We will study this property more precisely in §3. Haran and Jarden call it "ample" in [HaJa], which expresses
a certain tendency of these fields to develop abundantly through the points
of a variety. We note that AMPLE can also be understood as a property of
"Automatique Multiplication des Points Lisses Existants". We will use this
terminology.

The statement above however does not account for a second series of
results of the same spirit. Similarly, the general RIGP does not account
for a second series of classical conjectures of the area. Those results and
conjectures give, or predict, under certain conditions, the exact structure of
some absolute Galois groups. Historically, the starting problem of this second
circle is the conjecture of Shafarevich: the absolute Galois group $G(\mathbb{Q}^{ab})$ is a
free profinite group (on countably many generators).

These two circles of the area are closely related. They are both concerned
with the structure of absolute Galois groups. The methods use the same kind
of arguments, namely some patching and gluing techniques for analytic covers
and some specialization arguments for ample fields. Still it was our feeling
that the exact connection had never been made completely clear. The goal
of this paper is

(1) to state a conjecture that unifies all classical questions. This conjecture
is the Main Conjecture. We state it in §1 and show the connection with all
other conjectures.

(2) to state a theorem that summarizes most recent results of the area. This
theorem is the Main Theorem. It merely asserts that the Main Conjecture
is true if the base field K is ample. This contains the above statement that
the RIGP is solved if K is ample. The Main Theorem is precisely stated in
§2 where we also show how to deduce most recent results as special cases. A
general proof of the Main Theorem was given by F. Pop [Po4].

(3) to explain the main arguments of the proof of Main Theorem. For simplicity, we will restrict to the solution of the RIGP over large fields.

The necessary background on the arithmetic of fields and the theory of covers can be found in the Fried-Jarden book [FrJa] and in the recent books [MatMa] by Matzat-Malle and [Vo] by Völklein.

We wish to thank M. Fried, D. Haran, D. Harbater, M. Jarden, F. Pop, H. Völklein for very helpful comments and many valuable suggestions.

§2. Conjectures.

2.1. Classical conjectures.

2.1.1 First circle. Recall from the introduction these various conjectures around the Inverse Galois Problem. The Galois group of a Galois extension E/k is denoted by $\mathrm{G}(E/k)$. Given a field K, we denote by K_s (resp. by $\overline{K}$) a separable (resp. algebraic) closure of K and by $\mathrm{G}(K)$ the absolute Galois group $\mathrm{G}(K) = \mathrm{G}(K_s/K)$ of K.

Conjecture. ($\mathbf{RIGP}/_{K(T)}$) — *For every field K and for every finite group G, there exists a regular Galois extension $E_G/K(T)$ such that $\mathrm{G}(E_G/K(T)) = G$.*

Conjecture. ($\mathbf{IGP}/_{K\ hilb.}$) — *For every hilbertian field K and for each finite group G, there exists a Galois extension E_G/K such that $\mathrm{G}(E_G/K) = G$. Or, equivalently, each finite group G is a quotient of the absolute Galois group $\mathrm{G}(K)$ of K.*

Conjecture. ($\mathbf{IGP}$) — *For each finite group G, there exists a Galois extension $E_G/\mathbb{Q}$ such that $\mathrm{G}(E_G/\mathbb{Q}) = G$.*

We have: $$\mathbf{RIGP}/_{K(T)} \Longrightarrow \mathbf{IGP}/_{K\ hilb.} \Longrightarrow \mathbf{IGP}$$

Proof. Indeed, it follows from the hilbertian property that, if $E_T/K(T)$ is a regular Galois extension of Galois group G, then there exists some specialization $t \in K$ of T such that the specialized extension E_t/K is a Galois extension of Galois group G. Whence $\mathbf{RIGP}/_{K(T)} \Rightarrow \mathbf{IGP}/_{K\ hilb.}$. From Hilbert's irreducibility theorem, $\mathbb{Q}$ is a hilbertian field. So $\mathbf{IGP}/_{K\ hilb.} \Rightarrow \mathbf{IGP}$. □

2.1.2 Second circle. Conjectures of this second circle are conjectures about embedding problems. Recall that an *embedding problem* for a group Γ is a diagram of group homomorphisms

$$\Gamma$$
$$\downarrow f$$
$$1 \longrightarrow N \longrightarrow G \xrightarrow{\ \alpha\ } H \longrightarrow 1$$

where the horizontal sequence is exact and the map $f : \Gamma \to H$ is surjective. A *proper solution* is a surjective group homomorphism $g : \Gamma \to G$ such that $\alpha g = f$. Without the condition "g surjective", such a map g is said to be a *weak solution*. The embedding problem is said to be *finite* if G is finite. It is said to be *split* if $\alpha : G \twoheadrightarrow H$ has a group-theoretic section.

A profinite group Γ is said to be *projective* if each finite embedding problem for Γ is weakly solvable. Free profinite groups are some examples of projective groups. Finally recall that, given a weakly solvable embedding problem for a group Γ, there exists a standard procedure that generally allows to reduce to a split situation. More precisely, this procedure (e.g. [Po4;§1 B) 2)]) uses a weak solution to construct a *split* embedding problem for Γ such that existence of a proper solution for it implies existence of a proper solution for the original one. We will use this reduction in several occasions. For simplicity, we call it the *weak→split* reduction.

Conjecture. (Fried-Völklein [FrVo2]) — *Let K be a hilbertian countable field such that* $G(K)$ *is projective (e.g. cohdim$(K) \leq 1$), then* $G(K)$ *is pro-free.*

Conjecture. (Shafarevich) — $G(\mathbb{Q}^{ab})$ *is pro-free.*

These conjectures are more or less classical. We will denote them respectively by **FrVo** and **SHA**. The latter is actually a special case of the former. Indeed, from a result of Kuyk, $\mathbb{Q}^{ab}$ is hilbertian, see [FrJa,Th.15.6]; and $\mathbb{Q}^{ab}$ is of cohomological dimension ≤ 1: this follows from the fact that any division algebra over every number field has a cyclotomic splitting field [CaFr]. In turn, **FrVo** is a consequence of the two following equivalent conjectures about embedding problems.

Conjecture. (Split EP$/_{K(T)}$) — *Let K be an arbitrary field. Then each split embedding problem for* $G(K(T))$ *has a proper solution.*

Conjecture. (Split EP$/_{K \ hilb.}$) — *Let K be a hilbertian field. Then each split embedding problem for* $G(K)$ *has a proper solution.*

We call these conjectures the *Split Embedding Problem* conjecture over

$K(T)$ (resp. over hilbertian fields). We have:

$$\textbf{Split EP}/_{K(T)} \Longleftrightarrow \textbf{Split EP}/_{K \ hilb.} \Longrightarrow \textbf{FrVo} \Longrightarrow \textbf{SHA}$$

Proof of **Split EP**/$_{K(T)}$ ⇔ **Split EP**/$_{K\ hilb.}$. (⇐) follows from the fact that for every field K, the field $K(T)$ is hilbertian [FrJa;Th.12.10].

(⇒): Given a split embedding problem for $G(K)$ with K hilbertian, consider the embedding problem for $G(K(T))$ obtained by composition with the map $G(K(T)) \to G(K)$. The **Split EP**/$_{K(T)}$ conjecture provides a proper solution. Then, as in the proof of **RIGP**/$_{K(T)}$ ⇒ **IGP**/$_{K\ hilb.}$ above, use the hilbertian property to specialize this proper solution to a proper solution of the original embedding problem for $G(K)$. □

Proof of **Split EP**/$_{K\ hilb.}$ ⇒ **FrVo**. We will show that any given embedding problem for $G(K)$ has a proper solution. Conclusion "$G(K)$ pro-free" will follow then from Iwasawa's theorem recalled below. Since $G(K)$ is assumed to be projective, the given embedding problem has a weak solution. From the weak→split reduction, one may assume that the embedding problem is split. Therefore, from the **Split EP**/$_{K\ hilb.}$ conjecture, this split embedding problem has a proper solution. □

Theorem 2.1. (Iwasawa [Iw],[FrJa;Cor.24.2]) *Let K be a countable field, $G(K)$ is pro-free if and only if each finite embedding problem for $G(K)$ has a proper solution.*

2.1.3 Conclusion. Finally we have **Split EP**/$_{K\ hilb.}$ ⇒ **IGP**/$_{K\ hilb.}$. Indeed realizing a group G over K amounts to solving the split embedding problem for $G(K)$ in which the exact sequence is $1 \to G \to G \to 1 \to 1$. The following diagram summarizes §2.1.

$$\textbf{RIGP}/_{K(T)} \quad \Longrightarrow \quad \textbf{IGP}/_{K\ hilb.} \quad \Longrightarrow \quad \textbf{IGP}$$
$$\Uparrow$$
$$\textbf{Split EP}/_{K(T)} \quad \Longleftrightarrow \quad \textbf{Split EP}/_{K\ hilb.} \quad \Longrightarrow \quad \textbf{FrVo} \quad \Longrightarrow \quad \textbf{SHA}$$

Remark 2.2. Conjectures of the second circle seem to be stronger since they give results on the structure of absolute Galois groups whereas those of the first circle only deal with the finite quotients of absolute Galois groups. But there is in fact no other obvious implication than **Split EP**/$_{K(T)}$ ⇒ **IGP**/$_{K(T)}$ between both these sets of conjectures. For example there is no obvious implication between **FrVo** and **IGP** [1], nor there is between **Split EP**/$_{K(T)}$ and **RIGP**/$_{K(T)}$. The motivation of the following section is to state a more general conjecture that unifies both these circles of the area. The difference between the two conjectures **Split EP**/$_{K(T)}$ and **RIGP**/$_{K(T)}$ will then become clear (see Remark 2.4).

2.2 Main Conjecture.

2.2.1 Statement of the Main Conjecture.
Suppose given a commutative diagram of group homomorphisms

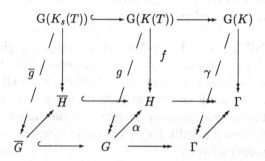

- where the sequences in which the arrows are lined up are exact,
- where $\hookrightarrow$ means that the homomorphism in question is injective,
- where $\twoheadrightarrow$ means that the homomorphism in question is surjective.

Such a diagram is called an *embedding problem of exact sequences over* $K(T)$. A *proper solution* is a triple $(\bar{g}, g, \gamma)$ of surjective maps as in the diagram above, such that the enlarged diagram commutes. If the condition "surjective" is removed, the triple $(\bar{g}, g, \gamma)$ is said to be a *weak solution*.

Main Conjecture. — *Let K be an arbitrary field. Then for each embedding problem EP of exact sequences over $K(T)$, if EP has a weak solution, then EP has a proper solution.*

Remark 2.3. (a) Weak solutions $(\bar{g}, g, \gamma)$ actually correspond in a one-one way with the maps $g : G(K(T)) \to G$ such that $\alpha g = f$. Indeed, given g, take for $\bar{g}$ the restriction of g to $G(K_s(T))$. The containment $g(G(K_s(T))) \subset \overline{G}$ holds because the two down-right groups in the diagram are equal (to Γ). There exists then a unique map γ that makes the diagram commute. The triple $(\bar{g}, g, \gamma)$ is a weak solution.

---

[1] The main point is that $G(\mathbb{Q})$ is not projective (since $G(\mathbb{Q})$ has elements of order 2) and so **FrVo** cannot be applied directly to $G(\mathbb{Q})$.

(b) We will say that the embedding problem of exact sequences over $K(T)$ is split if the map α splits. In that case, the embedding problem has a weak solution. Conversely, it is an exercise to show if an embedding problem of exact sequences has a weak solution, then there exists a *split* embedding problem of exact sequences such that existence of a proper solution for it implies existence of a proper solution for the original one: the weak→split reduction mentioned above for classical embedding problems generalizes with no difficulties. Therefore, in the Main Conjecture, the condition that the embedding problem has a weak solution can be replaced by the condition that it is split.

2.2.2 Relation with other conjectures. The Main Conjecture contains all conjectures of §2.1. In summary we have

$$\begin{array}{ccccccccc}
 & \Longrightarrow & \mathbf{RIGP}/_{K(T)} & \Longrightarrow & \mathbf{IGP}/_{K\ hilb.} & \Longrightarrow & \mathbf{IGP} & & \\
\mathbf{Main\ Conj.} & & & & \Uparrow & & & & \\
 & \Longrightarrow & \mathbf{Split\ EP}/_{K(T)} & \Longleftrightarrow & \mathbf{Split\ EP}/_{K\ hilb.} & \Longrightarrow & \mathbf{FrVo} & \Longrightarrow & \mathbf{SHA}
\end{array}$$

Proof of **Main Conj.** $\Rightarrow$ **RIGP**/$_{K(T)}$. From the Main Conjecture, the following embedding problem of exact sequences over $K(T)$,

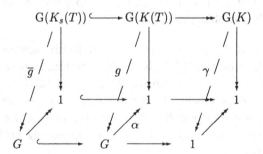

which is obviously split, has a proper solution. This exactly means that there exists an extension $E_G/K(T)$ such that $G(E_G/K(T)) = G(E_G K_s/K_s(T)) = G$.

Proof of **Main Conj.** $\Rightarrow$ **Split EP**/$_{K(T)}$. Consider a split embedding problem

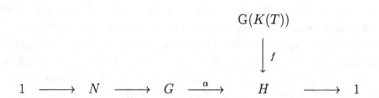

Denote the group $f(G(K_s(T)))$ by $\overline{H}$, the natural map $H \to H/\overline{H}$ by p and the kernel of the surjective map $p\alpha$ by $\overline{G}$. It is readily checked that α maps surjectively $\overline{G}$ onto $\overline{H}$. Thus we can form the following embedding problem of exact sequences over $K(T)$

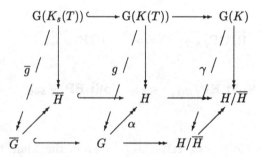

This embedding problem is split. From the Main Conjecture there exists a proper solution $(\bar{g}, g, \gamma)$. In particular the map g is a proper solution of the original embedding problem for $G(K(T))$.   □

Remark 2.4. Roughly speaking the Main Conjecture asserts that each split embedding problem for $G(K(T))$ given with a certain constraint over K_s has a proper solution. Conjectures **Split EP**$/_{K(T)}$ and **RIGP**$/_{K(T)}$ correspond to the following special cases. The **Split EP**$/_{K(T)}$ conjecture is obtained by leaving out the constraint over K_s. While the **RIGP**$/_{K(T)}$ has the constraint over K_s but is concerned only with the split embedding problem for which the quotient is $H = 1$.

2.2.3 Further comments. (a) The Main Conjecture can be generalized in two directions. First the field $K(T)$ can be replaced by the function field $K(C)$ of any smooth projective K-curve. That is, the upper exact sequence of the diagram in §2.2.1 is replaced by the exact sequence

$$1 \to G(K_s(C)) \to G(K(C)) \to G(K) \to 1$$

Harbater [Har4] and Pop [Po2] work in this more general context. An extra difficulty is that this exact sequence need not be split in general, in contrast with the special case $C = \mathbb{P}^1$.

A second generalization is to replace the absolute Galois group $G(K(C))$ by the algebraic fundamental group $\pi_1(C-D)$ where D is a reduced $G(K)$-invariant divisor of C. That is, the upper exact sequence of the diagram in §2.2.1 is replaced by the exact sequence

$$1 \to \pi_1(\overline{C}-\overline{D}) \to \pi_1(C-D) \to G(K) \to 1$$

where $\overline{C}-\overline{D} = (C-D) \otimes_K K_s$. However, with that change, the conjecture is false. Take $K = \mathbb{C}$, $C = \mathbb{P}^1$ and $|D| > 0$. Since $\pi_1(\mathbb{P}^1-D)$ is pro-free of rank $|D| - 1$, an embedding problem

$$\pi_1(\mathbb{P}^1-D)$$

$$\downarrow$$

$$1 \longrightarrow N \longrightarrow G \stackrel{\alpha}{\longrightarrow} H \longrightarrow 1$$

with G of rank $\geq |D|$ cannot have any proper solution. But the Main Conjecture predicts that the embedding problem has a proper solution if some extra branch points are allowed. More precisely, the Main Conjecture can be rephrased as follows:

(*) given a split embedding problem for the exact sequence above of fundamental groups, there exists a finite set $D' \subset C$ such that the embedding problem for the exact sequence

$$1 \to \pi_1(\overline{C}-(\overline{D} \cup \overline{D'})) \to \pi_1(C-(D \cup D')) \to G(K) \to 1$$

obtained by composing with the map $\pi_1(C-(D \cup D')) \to \pi_1(C-D)$ has a proper solution.

The equivalence "Statement (*) $\Leftrightarrow$ **Main Conj.**" follows from:

$$G(K(C)) = \varprojlim_{D} \pi_1(C-D)$$

Further questions of interest then arise: how should the extra branch points be selected? how many of them are needed?, etc.

(b) Conjecturally, if K is an arbitrary field, both the Split Embedding Problem conjecture and the Inverse Galois Problem hold over $K(T)$. This is not true anymore if $K(T)$ is replaced by K. Specifically, there exist fields K such that each finite group is a Galois group over K but for which there exist some split embedding problems for $G(K)$ with no proper solution. An example was given by Fried and Völklein [FrVo2;§2 Example] (see also [Ja1;Ex.3.5]).

Fried and Völklein say a field K is RG-hilbertian if the Hilbert special-ization property is true for all polynomials $P(T, Y) \in K(T)[Y]$ such that the associated extension of $K(T)$ is Galois and regular. They showed that there exists a countable field K of characteristic 0 that is RG-hilbertian but not hilbertian and that is PAC (Cf. §3.1 Ex.1). It follows from Cor.3.5 below and the RG-hilbertianity that each finite group is a Galois group over K. (For PAC fields of characteristic 0, the RG-hilbertianity and the property that each finite group is a Galois group over K are actually equivalent [FrVo2;Th.B]).

Assume now that each split embedding problem for $G(K)$ has a proper solution. Then from the weak→split reduction, each weakly solvable embed-ding problem for $G(K)$ has a proper solution. But from a result of Ax, since K is PAC, $G(K)$ is projective [FrJa,Th.10.17] and each embedding problem for $G(K)$ is weakly solvable. Finally, this proves that any embedding problem for $G(K)$ has a proper solution. From Iwasawa's theorem, $G(K)$ is pro-free. But Roquette showed that a PAC field such that $G(K)$ is pro-free is necessar-ily hilbertian [FrJa;Cor.24.38]. A contradiction. Conclude that there exists some split embedding problem for $G(K)$ with no proper solution.

§3. Results.

3.1. Ample fields.

Definition 3.1 — *A field K is said to be ample if for each geometrically irreducible smooth curve C defined over K we have*

$$C(K) \neq \emptyset \Rightarrow C(K) \text{ infinite}$$

This definition is due to F. Pop who calls it *large*. We prefer to follow Jarden who calls it *ample* and to keep the word *large* for informal contexts. Before giving some examples of ample fields, we state a nice observation of F. Pop, which is used in the proof of the Main Theorem (§4): ample fields are *existentially closed* in the field $K((x))$ of formal Laurent series over K.

Definition 3.2. *Let Ω/K be a regular extension. The field K is said to be* *existentially closed in Ω if the following equivalent properties hold:*

(1) For each smooth geometrically irreducible K-variety V,

$$V(\Omega) \neq \emptyset \Rightarrow V(K) \neq \emptyset$$

(2) For each smooth geometrically irreducible K-variety V,

$$K(V) \subset \Omega \Rightarrow V(K) \neq \emptyset$$

(3) For each smooth geometrically irreducible K-variety V,

$$V(\Omega) \text{ Zariski-dense} \Rightarrow V(K) \text{ Zariski-dense}$$

[(3) $\Rightarrow$ (1): Pick $\xi \in V(\Omega)$. Consider the Zariski closure of ξ in the K-variety V: it is an irreducible closed subscheme W of V. It follows from $K(W) = K(\xi) \subset \Omega$ and the regularity of the extension Ω/K that the extension $K(W)/K$ is regular. Hence W is a geometrically irreducible sub-variety of V. It follows from (3) that $W(K) \neq \emptyset$ and so $V(K) \neq \emptyset$.

(1) $\Rightarrow$ (2): If $K(V) \subset \Omega$, the generic point ξ of V is in $V(\Omega)$ (since $K(\xi) = K(V)$). It follows from (1) that $V(K) \neq \emptyset$.

(2) $\Rightarrow$ (3): Pick $\xi \in V(\Omega)$. Consider the Zariski closure W of ξ in the K-variety V. As explained above, W is a geometrically irreducible sub-variety of V. It follows from (2) that $W(K) \neq \emptyset$ and so $V(K) \neq \emptyset$. The same argument holds with V replaced by an open subset of V. Conclude that $V(K)$ is Zariski-dense.]

Theorem 3.3. (Pop, [Po4;Prop.1.1]) *A field K is ample if and only if it is existentially closed in its formal Laurent series field $K((x))$.*

Examples: (1) PAC fields are ample. This follows from the definition of PAC: a field K is P(seudo) A(lgebraically) C(losed) if $V(K) \neq \emptyset$ for each geometrically irreducible variety defined over K) [FrJa;Ch.10]. Clearly, algebraically closed fields are PAC. So are separably closed fields. There are many other examples of PAC fields, even inside $\overline{\mathbb{Q}}$. For example, from Pop's theorem stated below (Th.3.4), the field $\mathbb{Q}^{tr}(\sqrt{-1})$ is PAC, where $\mathbb{Q}^{tr}$ is the field of totally real algebraic numbers.

(2) Complete valued fields (*e.g.* $K = \mathbb{Q}_p$, $\mathbb{R}$, $k((x))$, etc.) are ample. This follows immediately from the Implicit Function Theorem. The following argument shows directly (*i.e.*, without Th.3.3) that a complete valued field (K, v) is existentially closed in $K((x))$. Let V be a variety defined over K. Assume that condition (2) of the definition holds, *i.e.*, that $K(V) \subset K((x))$. Denote the Henselian closure of $K(x)$ in $K((x))$ by $\widetilde{K(x)}$. The field $\widetilde{K(x)}$ is known to be existentially closed in $K((x))$ [Ja1; Lemmas 2.2 and 2.3]. Therefore $V(\widetilde{K(x)}) \neq \emptyset$. Now $\widetilde{K(x)}$ is the algebraic closure of $K(x)$ in $K((x))$. Consequently elements of $\widetilde{K(x)}$ are formal power series with a positive radius of

convergence (*e.g.* [De3;Prop.p.387]). Pick a point $M_X \in V(\widetilde{K(x)})$. Specializing x to some element $\xi \neq 0$ in the disk of convergence of the series involved in the coordinates of M_x yields a point $M_\xi \in V(K)$.

(3) For each prime number p, denote by $\mathbb{Q}^{tp}$ the field of totally p-adic algebraic numbers (*i.e.*, such that all conjugates over $\mathbb{Q}$ lie in a given copy of $\mathbb{Q}_p$). We use the notation $\mathbb{Q}^{tr}$ for the field of totally real algebraic numbers, which corresponds to the case $p = \infty$. Then for each prime number p (including $p = \infty$), the field $\mathbb{Q}^{tp}$ is ample. This immediately follows from the preceding example and the following result of Pop [Po5] (see also [GrPoRo], [Po4;App.I], [Ja3]).

Theorem 3.4. (Pop) *Let V be a smooth geometrically irreducible variety defined over $\mathbb{Q}^{tp}$, then $V(\mathbb{Q}^{tp}) \neq \emptyset$ provided that $V(\mathbb{Q}_p^\sigma) \neq \emptyset$ for each $\sigma \in G(\mathbb{Q}_p)$. (This last condition can be replaced by $V(\mathbb{Q}_p) \neq \emptyset$ if V is defined over $\mathbb{Q}$).*

This example can be generalized to consider the field K^S of totally S-adic elements of a global field K. Here S is a finite set of places of K and K^S is the subfield of K_s of all elements such that, for all $v \in S$, all conjugates over K lie in a given copy of the completion K_v. Pop's result is shown to hold in this generality. The field K^S is ample.

(4) The field $\mathbb{Q}$ is not ample. More generally, from Faltings's theorem, no number field is ample. Jarden mentioned to us that, using Faltings's theorem and the theorem of Manin and Grauert, one could show that no field which is finitely generated over its prime field is ample.

3.2 Main results

Main Theorem. *The Main Conjecture is true if the base field K is ample.*

The Main Theorem is built from a series of results. The main contributions are due to Fried, Harbater and Pop. The idea of working with families of covers goes back to Fried [Fr]. Harbater [Har2] introduced some very efficient patching and gluing techniques for covers, especially over complete and algebraically closed fields. Pop developed these gluing techniques and the arithmetic aspect of the method. In particular he introduced and studied the notion of ample fields. An equivalent form of the Main Theorem along with a proof can be found in Pop's papers ([Po2],[Po4]). Others took an active part to the Main Theorem, in particular, Dèbes, Jarden, Haran, Liu, Völklein. The Main Theorem has two main consequences (Cor.3.5 and Cor.3.6). We show below that they contain most recent results on this topic.

Corollary 3.5 — *The* **RIGP**$/_{K(T)}$ *conjecture holds if K ample. That is, if K is ample, each finite group is the Galois group of a regular Galois extension of $K(T)$.*

This result contains the following special cases:

- $K = \mathbb{C}$ (Riemann),

- $K = \mathbb{R}$ (Hurwitz (1890)),

- K algebraically closed (Harbater (1984) [Har1]),

- $K = \mathbb{Q}_p$ (Harbater (1985) [Har2]),

- K PAC (Fried-Völklein for *char* $K = 0$ (1991) [FrVo1], Pop in general (1993) [Po4]),

- $K = \mathbb{Q}^{tr}$ (Dèbes-Fried (1991) [DeFr]),

- $K = \mathbb{Q}^{tp}$ (Dèbes (1993) [De2]; Pop (1993) [Po4]),

- K ample (Pop (1995) [Po4]).

Corollary 3.6 — *The* **Split EP**$/_{K(T)}$ *conjecture holds if the base field K is ample. In particular, the Fried-Völklein conjecture* **FrVo** *is true if K is ample.*

Corollary 3.7 — *For every countable field K, $\mathrm{G}(\overline{K}(T))$ is pro-free.*

Proof. From Iwasawa's theorem (Th.2.1), we need to show that each embedding problem for $\mathrm{G}(\overline{K}(T))$ has a proper solution. From Tsen's theorem, $\mathrm{G}(\overline{K}(T))$ is projective. Therefore a given embedding problem for $\mathrm{G}(\overline{K}(T))$ has a weak solution. From the weak→split reduction, one may then assume that the embedding problem is split. But since the algebraically closed field $\overline{K}$ is ample, the Split Embedding conjecture holds over $\overline{K}(T)$. Hence, the given embedding problem has indeed a proper solution.   □

Cor.3.7 is due to:

- Douady in *char* $K = 0$ [Do] (1964),

- Harbater [Har4] and Pop [Po2] in general (1993).

Corollary 3.8 — *For every countable Hilbertian and PAC field K, $\mathrm{G}(K)$ is pro-free.*

Proof. From a result of Ax [FrJa;Th.10.17], if K is PAC then $\mathrm{G}(K)$ is projective. Furthermore PAC fields are ample. The Fried-Völklein conjecture, which is true for ample fields, gives the conclusion.   □

Cor.3.8 is due to:

- Fried-Völklein [FrVo2] in *char* $K = 0$ (1992),
- Pop in general [Po4] (1993).

Remark 3.9. (a) Conjecture (**RIGP**/$_{K(T)}$) predicts in particular that, given a finite group G, then, for every finite field F, G is the Galois group of a regular Galois extension $E/F(T)$. Cor.3.5 allows to show that at least this is known for all but finitely many finite fields F. The proof uses the following model-theoretical argument. Suppose on the contrary that, for some finite group G, the set $\mathfrak{L}$ of all finite fields F for which G is not the Galois group of a regular Galois extension $E/F(T)$ is infinite. Consider a non-principal ultraproduct K of the set of finite fields in $\mathfrak{L}$ [FrJa;Ch.6]. Then [FrJa, Cor.10.6] says that K is a PAC field (essentially this rests on the Lang-Weil estimate for the number of rational points on a curve over a finite field [FrJa;Th.4.9]). So, from Cor.3.5, if G is a finite group, then G can be realized regularly as a Galois group over $K(T)$. But every first order statement true about K is true about all but finitely many F in $\mathfrak{L}$ [FrJa;Cor.6.12]. A contradiction.

This consequence of Cor.3.5 first appeared in the 1991 paper of Fried-Völklein [FrVo1; Cor.2], in a slightly weaker form where only finite prime fields are considered. In that context, the ultraproduct K above is a PAC field of characteristic 0 and the argument works with the special case of Cor.3.5 proved by Fried and Völklein. In their paper they also give an alternate more geometrical exposition of the argument above. The general case of the above argument requires the version of Cor.3.5 for PAC fields of arbitrary characteristic proved by Pop [Po4]. This result of Pop already appeared in some form in a letter of Roquette of 1991.

(b) For $K = \mathbb{Q}^{ab}$, conclusion "G(K) pro-free" of Cor.3.8 would yield Shafarevich's conjecture. But Frey noted that $\mathbb{Q}^{ab}$ is not PAC [FrJa;Cor.10.15]. So Cor.3.8 cannot be applied to $K = \mathbb{Q}^{ab}$. On the other hand, it follows from Cor.3.7 that G($\overline{\mathbb{F}_p}(T)$) is pro-free. This can be regarded as the functional analog of Shafarevich's conjecture: $\overline{\mathbb{F}_p}(T)$) is indeed the maximal cyclotomic extension $\mathbb{F}_p(T)^{cycl}$ of $\mathbb{F}_p(T)$ (just as $\mathbb{Q}^{ab} = \mathbb{Q}^{cycl}$). Similarly the absolute Galois group of the maximal cyclotomic extension of $K = \mathbb{Q}^{tr}$ is pro-free. Indeed $(\mathbb{Q}^{tr})^{cycl} = \mathbb{Q}^{tr}(\sqrt{-1})$; from Cor.3.8, G($\mathbb{Q}^{tr}(\sqrt{-1})$) is pro-free (since $\mathbb{Q}^{tr}(\sqrt{-1})$ is PAC (by Th.3.4) and is hilbertian (by Weissauer's theorem [FrJa;Prop.12.14])).

(c) Cor.3.7 also holds in the uncountable case: for any field K, the group G($\overline{K}(T)$) is pro-free (of rank $card(K)$). This can be proved by the same methods but some adjustments are needed. Namely, one should use the following generalization of Iwasawa's theorem. If m is an infinite cardinal, then an iff condition for a profinite group F to be pro-free of rank m is that each finite embedding problem for F with a non-trivial kernel has exactly m proper solutions (see [Ja2;Lemma 2.1] where this generalization is credited to Z. Chatzidakis). So, for the general case of Cor.3.7, we have to prove that each embedding problem for $G(\overline{K}(T))$ with a non-trivial kernel has $card(K)$ proper solutions, which requires more precise statements than the mere existence result of a proper solution given by the Main Theorem (see [Har4] and [Po2]).

(d) A field K is said to be ω-free if every finite embedding problem for $G(K)$ has a proper solution. From Iwasawa's theorem, "ω-free" is equivalent to "pro-free" if K is countable. M. Jarden [Ja3;Remark 11.3] observed that if the conclusion "$G(K)$ pro-free" is replaced by "$G(K)$ ω-free" then Cor.3.8 also holds for uncountable fields. That is, we have "PAC + hilbertian $\Rightarrow$ ω-free" in general. On the other hand, the implication "PAC + hilbertian $\Rightarrow$ pro-free" is false in general. There are fields which are PAC and Hilbertian but with a non pro-free absolute Galois group [Ja2;Example 3.2].

§4 Main arguments.

In this section we sketch the proof of Cor.3.5. That is, we show that, if K is an ample field, then each finite group G is the Galois group of a regular Galois extension of $K(T)$. The proof of this special case of the Main Theorem contains the main arguments of the method. There are two stages. The first one consists in solving the problem over the field $K((x))$ of formal power series with coefficients in K. The second one uses a specialization argument to descend from $K((x))$ to K.

4.1 The Regular Inverse Galois Problem over $K((x))(T)$. The main tool is this patching and gluing result.

Theorem 4.1. *Let k be a local field. Let G be a finite group generated by two subgroups G_1 and G_2. Assume that for $i = 1, 2$, there exists a regular Galois extension $F_i/k(T)$ of Galois group G_i and with an unramified prime of degree 1. Then there exists a regular Galois extension $F/k(T)$ of Galois group G and with an unramified prime of degree 1.*

This result reduces the problem to the realization of cyclic groups (regularly over $k(T)$ and with an unramified prime of degree 1). In [Har2] Harbater uses some results of Saltman to handle this case (see also [Li] and [Vo;Ch.11]). In characteristic 0, Deschamps recently gave a simpler proof, which, in addition, provides some precise information on the ramification of the constructed extension [Des2;Lemme 2.1.2].

Th.4.1 is due to Harbater [Har2]. The main tools of his proof are a patching and gluing result for formal analytic spaces along with a formal GAGA theorem. Harbater's result was revisited from the point of view of rigid geometry by Liu ([Li], see also [Des1]), following a suggestion of Serre [Se;p.93]. Pop and Harbater developed then independently the patching and gluing procedure: in particular they showed it could be used not only to realize groups but also to solve embedding problems. Pop also managed to keep some arithmetic control on the procedure. That led him to his so-called "1/2-Riemann's existence theorem" [Po1]. The patching and gluing procedure is also an essential ingredient of the proof of Abhyankar's conjecture on Galois groups over curves: the case of the affine line was first proved by Raynaud [Ra]; using formal patching, Harbater could then prove the general case [Har3]; an alternate proof of the general case from Raynaud's result, using rigid patching, was given a little later by Pop [Po3]. There is now an elementary exposition of the proof of Th.4.1 which does not need any geometric background by Haran and Völklein ([HaVo], [Vo;Ch.11]).

4.2 Specialization argument (see also [Ja4;Prop.2.2]). Let Q be a prime field. From the first stage, we know that, for each finite group G, there exists a regular Galois extension $E_X/Q((x))(T)$ of Galois group G.

Let F/Q be an extension of finite type with $F \subset Q((x))$ such that the irreducible polynomial of y over $Q((x))(T)$ lies in $F(T)[Y]$ and such that all conjugates of y over $Q((x))(T)$ lie in $F(T, y)$. Set $E = F(T, y)$. The extension $E/F(T)$ is a regular Galois extension of $F(T)$ such that $G(E/F(T)) = G$.

The containment $F \subset Q((x))$ implies in particular that $F \cap \overline{Q} = Q$. Thus F is the function field of a geometrically irreducible algebraic variety V defined over Q. The equality $G(E/F(T)) = G$ rewrites $G(E/Q(V)(T)) = G$.

The extension $E/Q(V)(T)$ is a regular extension. So the Bertini-Noether theorem (e.g. [FrJa;Prop.8.8]) applies. There exists a Zariski closed subset Z of V such that, for each $v \in V(\overline{Q}) - Z$, the extension $E/Q(V)(T)$ specializes to a regular Galois extension $E_v/Q(v)(T)$ of degree $[E_v : Q(v)(T)] = [E : Q(V)(T)]$. Up to enlarging the Zariski closed subset Z, we may:

- conclude that the specialized extension $E_v/Q(v)(T)$ is also Galois. Then we have $G(E_v/Q(v)(T)) = G$ (for $v \in V(\overline{Q}) - Z$),

- assume that the variety V is smooth.

Finally since $F = Q(V) \subset Q((x))$, the set $V(Q((x)))$ is Zariski-dense. We have proved the following.

Theorem 4.2. *Let Q be a prime field. Then to each finite group G can be attached a smooth irreducible variety V defined over Q such that*

(1) For each field K containing Q, if $V(K)$ is Zariski-dense then the group G is the Galois group of a regular extension of $K(T)$.

(2) $V(Q((x)))$ is Zariski-dense.

Assume now that K is an ample field. From Th.3.3, K is existentially closed in $K((x))$. Therefore, from (2) above, $V(K)$ is Zariski-dense. Conclude from (1) that G is the Galois group of a regular extension of $K(T)$, thus completing the proof of Cor.3.5.

Remark 4.3. From Th.4.1, the extension $E_X/Q((x))(T)$ in the argument above can be taken to have an unramified prime **p** of degree 1. The field F can next be enlarged to contain the field of definition $Q(\mathbf{p}) \subset Q((x))$ of **p** (viewed as a point). The specialization argument then goes through to show this more precise form of Cor.3.5. If K is an ample field, then each finite group is the Galois group of a regular extension of $K(T)$ with an unramified prime of degree 1.

From §3.1, complete valued fields are ample. Thus Th.4.2 has this other consequence.

Corollary 4.4 — *To each finite group G can be attached a smooth irreducible variety V defined over $\mathbb{Q}$ such that*

(1) If $V(\mathbb{Q}) \neq \emptyset$ then G is the Galois group of a regular extension of $\mathbb{Q}(T)$.

(2) $V(\mathbb{Q}_p) \neq \emptyset$ for each prime p (including the prime $p = \infty$).

In condition (2) the field $\mathbb{Q}_p$ can actually be replaced by any complete valued field of characteristic 0. Condition (2) however is not sufficient in general to conclude that $V(\mathbb{Q}) \neq \emptyset$ (see [De2;Ex.4.2]).

Cor.4.4 is due to B. Deschamps [Des2]. His proof uses a different approach based on the Hurwitz space theory of M. Fried [Fr]. This theory had been developed by Fried and Völklein to reduce the Regular Inverse Galois Problem to finding rational points over irreducible varieties ([FrVo1], [Em], [Vo;Ch.10]). More specifically, to a given finite group G and an integer r larger than the rank of G, can be attached a specific smooth algebraic variety $\mathfrak{H} = \mathfrak{H}_{G,r}$ — called Hurwitz space — defined over $\mathbb{Q}$ such that, for every field K of characteristic 0, points in $\mathfrak{H}(K)$ correspond to regular Galois extensions $E/K(T)$ of group G with at most r branch points. Under certain conditions, one has a good control on the field of definition of the irreducible components of the algebraic variety $\mathfrak{H}_{G,r}$. That indeed reduces the problem to finding such components defined over $\mathbb{Q}$ with $\mathbb{Q}$-rational points.

A major application was the original proof of Fried and Völklein of the Regular Inverse Galois Problem over PAC fields [FrVo1] and of the implication "PAC + hilbertian $\Rightarrow$ ω-free" [FrVo2] in characteristic 0. The algebraic variety V constructed in §4 is replaced in the Fried-Völklein method by some irreducible component of some Hurwitz space $\mathfrak{H}_{G,r}$. It is worth noting that the variety V constructed in §4 really appears as a subvariety of some Hurwitz space. So while the techniques are somewhat different, there is some relationship between the two methods.

Deschamps also used the Hurwitz space approach. He managed to show that for some suitably large integer r, there was some irreducible component of $\mathfrak{H}_{G,r}$ defined over $\mathbb{Q}$ and with $\mathbb{Q}_p$-points for all primes p. Furthermore he obtains this extra conclusion in Cor.4.4:

(3) For each prime p, there exists some point in $V(\mathbb{Q}_p)$ such that the corresponding regular extension $E_p/\mathbb{Q}_p(T)$ of group G has the property that its branch points lie in $\mathbb{P}^1(\mathbb{Q}^{ab})$ and are $G(\mathbb{Q})$-invariant as a set.

The Hurwitz space method is more intricate but has the advantage of being much more explicit. The variety V and the Zariski closed subset Z involved in Th.4.2 are precisely described: V is a specific component of a certain Hurwitz space $\mathfrak{H}_{G,r}$ and Z is the branch locus of a certain cover $\mathfrak{H}_{G,r} \rightarrow \mathbb{P}_r$. Some further investigation on these spaces might be the key to a general solution of the Regular Inverse Galois Problem.

References

[CaFR]   J.W.S. Cassels and A. Fröhlich, *Algebraic number theory*, Academic Press, (1967).

[De1]   P. Dèbes, Groupes de Galois sur $K(T)$, *Séminaire de Théorie des Nombres*, Bordeaux **2** (1990), 229–243.

[De2]   P. Dèbes, Covers of $\mathbb{P}^1$ over the p-adics, *Cont. Math.*, **186**, (1995), 217–238.

[De3]   P. Dèbes, G-fonctions et Théorème d'irréductibilité de Hilbert, *Acta Arithmetica*, **47**, n° 4, (1986), 371–402.

[DeFr]   P. Dèbes and M. Fried, Nonrigid constructions in Galois theory, *Pacific J.Math.*, **163** #1, (1994), 81–122.

[Des1]   B. Deschamps, Autour d'un théorème d'Harbater, Mémoire de DEA, Univ. Paris VI, (1993)

[Des2]   B. Deschamps, Existence de points p-adiques pour tout p sur un espace de Hurwitz, *Cont. Math.*, **186**, (1995), 239–247.

[Do]   A. Douady, Détermination d'un groupe de Galois, *C.R. Acad. Sc. Paris*, **258** (1964), 5305–5308.

[Em]   M. Emsalem, Familles de revêtements de la droite projective, *Bull. Soc. Math. France*, **123**, (1995).

[Fr]   M. Fried, Fields of definition of function fields and Hurwitz families–Groups as Galois groups, *Comm. in Alg.* **5(1)** (1977), 17–82.

[FrJa]   M. Fried and M.Jarden, *Field Arithmetic*, Springer-Verlag, (1986)

[FrVo1]   M. Fried and H. Völklein, The inverse Galois problem and rational points on moduli spaces, *Math. Ann.*, **290**, (1991), 771-800.

[FrVo2]   M. Fried and H. Völklein, The embedding problem over a Hilbertian field, *Ann. Math.*, **135**, (1992), 469–481

[GrPoRo] B. Green, F. Pop and P. Roquette, On Rumely's local-global principle, *Jahresbericht der DMV*, **95**, (1995), 43–74

[HaJa]   D. Haran and M. Jarden, Regular split embedding problems over complete valued fields, preprint 1995.

[HaVo]   D. Haran and H. Völklein, Galois groups over complete valued fields, *Israel J. Math.*, to appear.

[Har1]   D. Harbater, Mock covers and Galois extensions, *J. Alg.* **91** (1984), 281–293.

[Har2]   D. Harbater, Galois covering of the arithmetic line, *Lecture Notes in Math.* **1240**, (1987), 165–195.

[Har3]   D. Harbater, Abhyankar's conjecture on Galois groups over curves, *Inv. Math.*, **117** (1994), 1–25

[Har4]   D. Harbater, Fundamental groups and embedding problems in characteristic p, *Cont. Math.* **186**, (1995), 353–369

[Iw]   K. Iwasawa, On solvable extensions of algebraic number fields, *Annals of Math.* **58** (1953), 548–572.

[Ja1]   M. Jarden, The inverse Galois problem over formal power series fields, *Israel J. Math.* **85** (1994),353–369.

[Ja2]   M. Jarden, On free profinite groups of uncountable rank, *Cont. Math.* **186** (1995), 371–383.

[Ja3]   M. Jarden, Totally S-adic extensions of hilbertian fields, preprint 1994.

[Ja4]   M. Jarden, Large normal extensions of Hilbertian fields, *Math. Zeit.*.

[Li]   Q. Liu, Tout groupe fini est groupe de Galois sur $\mathbb{Q}_p(T)$, *Cont. Math.* **186** (1995), 261–265.

[Mat]   B. H. Matzat, *Konstruktive Galoistheorie*, LNM **1284**, Springer, (1987).

[MatMa] B. H. Matzat and G. Malle, *Inverse Galois theory*, (1996).

[Po1]   F. Pop, Half Riemann's existence theorem, Algebra and Number Theory (G. Frey and J. Ritter, eds), de Gruyter Proceedings in Mathematics, (1994)

[Po2]   F. Pop, The geometric case of a conjecture of Shafarevich — $G_{\bar{\kappa}(t)}$ is profinite free —, Heidelberg-Mannheim Preprint Series "Arithmetik", Heft 8, (1993)

[Po3]   F. Pop, Étale Galois covers of affine smooth curves, *Inv. Math.* **120** (1995), 555–578

[Po4]   F. Pop, Embedding problems over large fields, *Annals of Math.* **144**, 1–35, (1996)

[Po5]   F. Pop, Fields of totally Σ-adic numbers, Heidelberg-Mannheim Preprint, (1990)

[Ra]   M. Raynaud, Revêtement de la droite affine en caractéristique p et conjecture d'Abhyankar, *Inv. Math.* **116** (1994), 425–462

[Se]   J.-P. Serre, *Topics in Galois theory*, Jones and Bartlett Publ., Boston, (1992).

[Vo]   H. Völklein, *Groups as Galois groups - an introduction*, Cambridge Univ. Press, (1996).

Univ. Lille, Mathématiques, 59655 Villeneuve d'Ascq Cedex, France.
E-mail: pde@ccr.jussieu.fr, brudesch@ccr.jussieu.fr

The Symplectic Braid Group and Galois Realizations

Karl Strambach and Helmut Völklein*

§0. Introduction

Moduli spaces for covers of the Riemann sphere, as constructed in [FV], have been used for Galois realizations of certain groups $GL_n(q)$ and $PU_n(q)$, see [V1] and [V3]. Here we study what happens when we replace these moduli spaces by certain subvarieties. Our main example is the subvariety obtained by allowing only such branch point sets that are symmetric with respect to the origin. The fundamental group of the space of those branch point sets is the braid group of type C_ℓ (in the sense of Brieskorn [Br]), which we call the symplectic braid group. It replaces the Artin braid group as occurring in [FV], [V1] and [V3]. Proceeding analogously as in [V3], we obtain a general criterion for the realization of groups as Galois groups (including GAL-realizations). This is applied in [MSV] to obtain Galois realizations over the rationals for certain of the simple groups $Sp_n(2^s)$ (in particular, for all groups $Sp_n(2)$).

Notations. The semi-direct product of groups G and H is written as $G \cdot H$ (where G is normal). $Aut(G)$ and $Inn(G)$ denote the group of all automorphisms (resp., all inner automorphisms) of G. Further, S_r is the symmetric group on r letters.

We denote by $\mathbb{Q}$, $\mathbb{C}$, $\mathbb{Q}_{ab}$, $\mathbb{F}_q$ the field of rationals, complexes, the field generated by all roots of unity (in $\mathbb{C}$), and the finite field with q elements, respectively. For a field k we let $\bar{k}$ denote its algebraic closure, and $\mathbf{G}_k = G(\bar{k}/k)$ its absolute Galois group (for perfect k). An extension K of k is called *regular (over k)* if k is algebraically closed in K. Further, $x, x_1, ..., x_r$ are independent transcendentals over k in expressions like $k(x_1, ..., x_r)$.

The algebraic varieties we consider are all defined over subfields of $\mathbb{C}$. We identify a variety $\mathcal{O}$ with its set of complex points. If $\mathcal{O}$ is defined and irreducible over k, we let $k(\mathcal{O})$ be its function field over k. By "covering" we mean an unramified topological covering, in the sense of covering space theory.

* Supported by NSF grant DMS-9306479

§1. Artin's braid group

Let G be a finite group, and let $\mathcal{C} = (C_1, ..., C_r)$ be an r-tuple of conjugacy classes of G. Assume $r \geq 3$, and $C_i \neq \{1\}$ for all i. Let $\mathrm{Ni}(\mathcal{C})$ be the set of r-tuples $(g_1, ..., g_r) \in G^r$ with the following properties: $g_1...g_r = 1$, the group G is generated by $g_1, ..., g_r$, and there is a permutation $\pi \in S_r$ with $g_{\pi(i)} \in C_i$ for all i. We assume that the set $\mathrm{Ni}(\mathcal{C})$ is non-empty.

Let $\mathcal{O}_r$ denote the space of all sets $\mathbf{a} = \{a_1, ..., a_r\}$ of r (distinct) complex numbers. Identifying $\mathbf{a}$ with the polynomial $\prod_{i=1}^r (X - a_i)$, we get a bijection between $\mathcal{O}_r$ and the space of all monic complex polynomials of degree r with non-zero discriminant. Since the discriminant is a polynomial function in the coefficients of the polynomials of degree r, we see that the space $\mathcal{O}_r$ has a natural structure of (affine) algebraic variety defined over $\mathbb{Q}$ (complement of the discriminant locus in $\mathbb{C}^r$). We fix a base point $\mathbf{b} = \{b_1, ..., b_r\}$ in $\mathcal{O}_r$ with $b_1, ..., b_r \in \mathbb{Q}$. (Then $\mathbf{b}$ is a $\mathbb{Q}$-rational point of $\mathcal{O}_r$, but our condition is a little stronger).

For the moment, we view $\mathcal{O}_r$ only as a complex manifold. Let $\mathcal{B}_r$ denote its fundamental group, based at $\mathbf{b}$. This group has generators $Q_1, ..., Q_{r-1}$ (e.g., as given in [V5, Ch. 10.1.6]) that satisfy the defining relations of the *Artin braid group* (and no other relations). This yields the well-known isomorphism between the Artin braid group and the fundamental group of affine r-space minus the discriminant locus.

The braid group $\mathcal{B}_r$ acts on $\mathrm{Ni}(\mathcal{C})$ from the right by the following rule: For $i = 1, ..., r$ we have

$$(1) \qquad (g_1, ..., g_r)^{Q_i} = (g_1, ..., g_{i-1}, g_{i+1}, g_{i+1}^{-1} g_i g_{i+1}, ..., g_r).$$

Let $\mathrm{Aut}_{\mathcal{C}}(G)$ be the group of all automorphisms of G that permute the conjugacy classes $C_1, ..., C_r$. This group acts on the set $\mathrm{Ni}(\mathcal{C})$ as follows: Each $A \in \mathrm{Aut}_{\mathcal{C}}(G)$ sends $(g_1, ..., g_r)$ to $(A(g_1), ..., A(g_r))$. Clearly this action commutes with the action of $\mathcal{B}_r$.

Let $\mathcal{P}$ be an absolutely irreducible non-singular closed subvariety of $\mathcal{O}_r$. Assume $\mathcal{P}$ contains our base point $\mathbf{b}$. Consider the topological fundamental group $\pi_1(\mathcal{P}, \mathbf{b})$ of $\mathcal{P}$. The embedding $\mathcal{P} \to \mathcal{O}_r$ induces a map $\pi_1(\mathcal{P}, \mathbf{b}) \to \mathcal{B}_r$. Let $\mathcal{L} \leq \mathcal{B}_r$ be the image of this map. Let $W \leq S_r$ be the image of $\mathcal{L}$ under the (surjective) homomorphism $\kappa : \mathcal{B}_r \to S_r$ sending Q_i to the transposition $(i, i+1)$.

Let $\mathrm{Ni}^W(\mathcal{C})$ be the subset of $\mathrm{Ni}(\mathcal{C})$ consisting of those $(g_1, ..., g_r)$ for which there is $\pi \in W$ with $g_{\pi(i)} \in C_i$. Let $\mathrm{Aut}_{\mathcal{C}}^W(G)$ be the subgroup of $\mathrm{Aut}_{\mathcal{C}}(G)$ consisting of those α for which there exists $\pi \in W$ such that $\alpha(C_i) = C_{\pi(i)}$. Then $\mathrm{Aut}_{\mathcal{C}}^W(G)$ and $\mathcal{L}$ leave $\mathrm{Ni}^W(\mathcal{C})$ invariant (in their action induced by the action of $\mathrm{Aut}_{\mathcal{C}}(G)$ and $\mathcal{B}_r$, respectively). For each subgroup $\mathbf{A}$ of $\mathrm{Aut}_{\mathcal{C}}^W(G)$, define $\mathrm{Ni}^W(\mathcal{C})^{(\mathbf{A})}$ to be the quotient of $\mathrm{Ni}^W(\mathcal{C})$ by the action of $\mathbf{A}$. Since the action of $\mathcal{L}$ on $\mathrm{Ni}^W(\mathcal{C})$ commutes with the action of $\mathrm{Aut}_{\mathcal{C}}^W(G)$, we get an induced action of $\mathcal{L}$ on $\mathrm{Ni}^W(\mathcal{C})^{(\mathbf{A})}$.

For each integer m let C_i^m be the conjugacy class of g^m with $g \in C_i$. The r-tuple $\mathcal{C}$ is called *W-rational* if for each integer m prime to the order of G there is $\pi \in W$ with $C_i^m = C_{\pi(i)}$ for $i = 1, ..., r$. If $W = S_r$ we call $\mathcal{C}$ rational.

§2. Coverings

§2.1. By covering space theory, there exists an (unramified) covering

$$\Psi : \mathcal{H}(\mathcal{C}) \to \mathcal{O}_r$$

(unique up to equivalence), and an identification of the fiber $\Psi^{-1}(\mathbf{b})$ with the set $\mathrm{Ni}(\mathcal{C})$, such that the natural action of $\mathcal{B}_r = \pi_1(\mathcal{O}_r, \mathbf{b})$ on $\Psi^{-1}(\mathbf{b})$ (see [V5, Ch. 4.1.3]) coincides with the above action on $\mathrm{Ni}(\mathcal{C})$. Since the action of $\mathrm{Aut}_{\mathcal{C}}(G)$ on $\mathrm{Ni}(\mathcal{C})$ commutes with the action of $\mathcal{B}_r$, it extends uniquely to an action of $\mathrm{Aut}_{\mathcal{C}}(G)$ on $\mathcal{H}(\mathcal{C})$ through automorphisms of the covering Ψ. Let ϵ_A denote the automorphism of $\mathcal{H}(\mathcal{C})$ induced by $A \in \mathrm{Aut}_{\mathcal{C}}(G)$.

Here is the basic result that we need from the theory of Hurwitz spaces (see [V4, 3.2, 3.5 and 3.9]).

Theorem. *Set* $k = \mathbb{Q}$ *if the r-tuple $\mathcal{C}$ is rational (see §1), otherwise* $k = \mathbb{Q}_{\mathrm{ab}}$.

(a) The space $\mathcal{H}(\mathcal{C})$ has a structure of non-singular (but not necessarily irreducible) variety defined over k such that Ψ and all ϵ_A become algebraic morphisms defined over k.

(b) The identification of $\Psi^{-1}(\mathbf{b})$ with the set $\mathrm{Ni}(\mathcal{C})$ can be chosen such that the following holds for the action of each $\beta \in \mathbf{G}_k$ on $\Psi^{-1}(\mathbf{b})$: Let m be an integer such that β acts on the $|G|$-th roots of unity as $\zeta \mapsto \zeta^m$. If $\mathbf{p}$ is a point of $\Psi^{-1}(\mathbf{b})$ corresponding to $(g_1, ..., g_r) \in \mathrm{Ni}(\mathcal{C})$ and $\mathbf{p}^\beta$ corresponds to $(h_1, ..., h_r)$ then h_i^m is conjugate to g_i for $i = 1, ..., r$.

Now consider a subvariety $\mathcal{P}$ of $\mathcal{O}_r$ as in §1, and assume $\mathcal{P}$ is defined over k. Let $\tilde{\mathcal{H}}_{\mathcal{P}}(\mathcal{C})$ be the inverse image of $\mathcal{P}$ in $\mathcal{H}(\mathcal{C})$. This is a closed non-singular subvariety of $\mathcal{H}(\mathcal{C})$ defined over k, and Ψ restricts to a morphism

$$\tilde{\mathcal{H}}_{\mathcal{P}}(\mathcal{C}) \to \mathcal{P}$$

defined over k. Topologically, this is a covering. The fiber in $\tilde{\mathcal{H}}_{\mathcal{P}}(\mathcal{C})$ over $\mathbf{b}$ is still identified with $\mathrm{Ni}(\mathcal{C})$. The connected components of $\tilde{\mathcal{H}}_{\mathcal{P}}(\mathcal{C})$ (same as absolutely irreducible components) correspond to the orbits of $\mathcal{L}$ on $\mathrm{Ni}(\mathcal{C})$.

The subset $\mathrm{Ni}^W(\mathcal{C})$ is a union of such orbits (see §1). The union of the corresponding components of $\tilde{\mathcal{H}}_{\mathcal{P}}(\mathcal{C})$ is denoted by $\mathcal{H}_{\mathcal{P}}(\mathcal{C})$.

From now on set $k = \mathbb{Q}$ **if the** r**-tuple** $\mathcal{C}$ **is** W**-rational, otherwise** $k = \mathbb{Q}_{ab}$. Then the space $\mathcal{H}_{\mathcal{P}}(\mathcal{C})$ is defined over k. Indeed, since $\mathbf{b}$ is k-rational, it suffices to show that $\mathcal{H}_{\mathcal{P}}(\mathcal{C}) \cap \Psi^{-1}(\mathbf{b})$ is invariant under $\mathbf{G}_k$; but this follows from Theorem part (b) because the latter set corresponds to $\mathrm{Ni}^W(\mathcal{C})$ by definition.

The restriction of Ψ gives a covering

$$\Psi_{\mathcal{P}} : \ \mathcal{H}_{\mathcal{P}}(\mathcal{C}) \to \mathcal{P}$$

of non-singular varieties defined over k. The space $\mathcal{H}_{\mathcal{P}}(\mathcal{C})$ is invariant under all ϵ_A, $A \in \mathrm{Aut}_{\mathcal{C}}^W(G)$ (since $\mathrm{Ni}^W(\mathcal{C})$ is invariant under $\mathrm{Aut}_{\mathcal{C}}^W(G)$, see §1). For each subgroup $\mathbf{A}$ of $\mathrm{Aut}_{\mathcal{C}}^W(G)$ define $\mathcal{H}_{\mathcal{P}}(\mathcal{C})^{(\mathbf{A})}$ to be the quotient of $\mathcal{H}_{\mathcal{P}}(\mathcal{C})$ by the group of all ϵ_A, $A \in \mathbf{A}$. Then $\mathcal{H}_{\mathcal{P}}(\mathcal{C})^{(\mathbf{A})}$ is a non-singular variety defined over k, and $\Psi_{\mathcal{P}}$ induces a covering

$$\mathcal{H}_{\mathcal{P}}(\mathcal{C})^{(\mathbf{A})} \to \mathcal{P}$$

defined over k. The above identification of the fiber $\Psi_{\mathcal{P}}^{-1}(\mathbf{b})$ with $\mathrm{Ni}^W(\mathcal{C})$ induces an identification of the fiber over $\mathbf{b}$ in $\mathcal{H}_{\mathcal{P}}(\mathcal{C})^{(\mathbf{A})}$ with the set $\mathrm{Ni}^W(\mathcal{C})^{(\mathbf{A})}$.

The absolutely irreducible components (same as connected components in the complex topology) of the space $\mathcal{H}_{\mathcal{P}}(\mathcal{C})^{(\mathbf{A})}$ are in 1-1 correspondence with the orbits of $\mathcal{L}$ on $\mathrm{Ni}^W(\mathcal{C})^{(\mathbf{A})}$.

§2.2. Consider the map

$$\mathcal{O}^{(r)} \overset{\text{def}}{=} \{(a_1, ..., a_r) \in \mathbb{C}^r : \ a_i \neq a_j \text{ for } i \neq j\} \quad \to \quad \mathcal{O}_r$$

where the (ordered) tuple $(a_1, ..., a_r)$ is mapped to the set $\{a_1, ..., a_r\}$. Topologically, the map $\mathcal{O}^{(r)} \to \mathcal{O}_r$ is an unramified Galois covering, whose automorphism group is canonically isomorphic to the symmetric group S_r (permuting $a_1, ..., a_r$). The fiber F over $\mathbf{b}$ contains the point $b := (b_1, ..., b_r)$, and consists of all $b^{\pi} := (b_{\pi(1)}, ..., b_{\pi(r)})$ with $\pi \in S_r$. In the natural permutation representation of $\mathcal{B}_r = \pi_1(\mathcal{O}_r, \mathbf{b})$ on the fiber F of the covering $\mathcal{O}^{(r)} \to \mathcal{O}_r$, each $Q \in \mathcal{B}_r$ sends the point b^{π} to $b^{\pi\kappa(Q)}$. (Recall $\kappa : \mathcal{B}_r \to S_r$ from §1). The kernel $\mathcal{B}^{(r)}$ of κ is called the *pure braid group*. (It is the fundamental group of $\mathcal{O}^{(r)}$).

Let $\tilde{\mathcal{P}}$ be the inverse image of $\mathcal{P}$ in $\mathcal{O}^{(r)}$. The components of $\tilde{\mathcal{P}}$ correspond to the orbits of $\mathcal{L}$ on F, hence correspond to the right cosets of W in S_r. Let $\hat{\mathcal{P}}$ be the component containing the point b. Then the natural map $\hat{\mathcal{P}} \to \mathcal{P}$

is a covering with automorphism group W. In particular, W acts naturally on the function field $k(\hat{\mathcal{P}})$ (and $\hat{\mathcal{P}} \to \mathcal{P}$ is a morphism defined over k).

Proposition. *Let* $\mathbf{A}$ *be a subgroup of* $\mathrm{Aut}_{\mathcal{C}}^{W}(G)$. *Suppose the space* $\mathcal{H}_{\mathcal{P}}(\mathcal{C})^{(\mathbf{A})}$ *has a component* $\mathcal{O}$ *that is defined over* k. *Then* $\mathcal{O}$ *is a rational variety over* k *if the following two conditions hold:*

(i) The group $\mathcal{L}_0 = \mathcal{L} \cap \mathcal{B}^{(r)}$ *fixes one (hence each) element* $\mathbf{g}$ *in the* $\mathcal{L}$-orbit *on* $\mathrm{Ni}^{W}(\mathcal{C})^{(\mathbf{A})}$ *corresponding to* $\mathcal{O}$.

(ii) Let S *be the image of* $\mathrm{Stab}_{\mathcal{L}}(\mathbf{g})$ *in* W, *under the map* $\kappa : \mathcal{L} \to W$. *We require that the fixed field of* S *in* $k(\hat{\mathcal{P}})$ *is of the form* $k(Y_1, ..., Y_s)$ *for independent transcendental elements* $Y_1, ..., Y_s$ *over* k *such that the* k-linear *span of* $Y_1, ..., Y_s$ *is invariant under the normalizer* $N_W(S)$.

More precisely, under conditions (i) and (ii) we have $k(\mathcal{O}) = k(x_1, ..., x_s)$ *for independent transcendentals* $x_1, ..., x_s$ *such that the* k-span *of* $x_1, ..., x_s$ *is fixed by* $\mathrm{Aut}(k(\mathcal{O})/k(\mathcal{P}))$.

Proof. Analogous to that of [V3, Prop. 1.3.2].

Lemma. *Suppose* $\mathbf{A} \le \mathbf{A}' \le \mathrm{Aut}_{\mathcal{C}}^{W}(G)$, *where* $\mathbf{A}$ *is normal in* $\mathbf{A}'$. *Assume* $\mathcal{H}_{\mathcal{P}}(\mathcal{C})^{(\mathbf{A}')}$ *has a component* $\mathcal{O}'$ *defined over* k. *Let* $\mathcal{O}$ *be a component of* $\mathcal{H}_{\mathcal{P}}(\mathcal{C})^{(\mathbf{A})}$ *that lies over* $\mathcal{O}'$.

(i) If $\mathcal{O}$ *is defined over* k, *then* $k(\mathcal{O})$ *is regular over* k, *and* $k(\mathcal{O})/k(\mathcal{O}')$ *is Galois with*

$$G(k(\mathcal{O})/k(\mathcal{O}')) \cong \mathrm{Aut}(\mathcal{O}/\mathcal{O}')$$

where $\mathrm{Aut}(\mathcal{O}/\mathcal{O}')$ *is the group of all topological automorphisms of* $\mathcal{O}$ *over* $\mathcal{O}'$.

(ii) The group $\mathrm{Aut}(\mathcal{O}/\mathcal{O}')$ *embeds naturally into* $\mathbf{A}'/\mathbf{A}$: *It is isomorphic to the stabilizer in* $\mathbf{A}'/\mathbf{A}$ *of the* $\mathcal{L}$-orbit *on* $\mathrm{Ni}^{W}(\mathcal{C})^{(\mathbf{A})}$ *that corresponds to the component* $\mathcal{O}$ *of* $\mathcal{H}_{\mathcal{P}}(\mathcal{C})^{(\mathbf{A})}$.

(iii) If $\mathrm{Aut}(\mathcal{O}/\mathcal{O}')$ *is self-normalizing in* $\mathbf{A}'/\mathbf{A}$, *then* $\mathcal{O}$ *is defined over* k.

Proof. Analogous to that of [V3, Lemma 1.2.3].

Remark. Suppose $\mathbf{A}_0$ is a subgroup of $\mathrm{Aut}(G)$ that fixes each C_i, and acts transitively on the set $\mathcal{E}(\mathcal{C}) = \{(g_1, ..., g_r) \in \mathrm{Ni}(\mathcal{C}) : g_i \in C_i, \ i = 1, ..., r\}$. Then the image in $\mathrm{Ni}^{W}(\mathcal{C})^{(\mathbf{A}_0)}$ of any element $(g_1, ..., g_r)$ of $\mathrm{Ni}^{W}(\mathcal{C})$ depends only on the order in which the conjugacy classes $C_1, ..., C_r$ are represented in $(g_1, ..., g_r)$. It follows that $\mathrm{Ni}^{W}(\mathcal{C})^{(\mathbf{A}_0)}$ is in natural 1-1 correspondence with the set of re-arrangements $C_{\pi(1)}, ..., C_{\pi(r)}$, $\pi \in W$. Thus $\mathcal{L}$ acts transitively on $\mathrm{Ni}^{W}(\mathcal{C})^{(\mathbf{A}_0)}$, and the group $\mathcal{L}_0 = \mathcal{L} \cap \mathcal{B}^{(r)}$ acts trivially. The first fact implies that $\mathcal{H}_{\mathcal{P}}(\mathcal{C})^{(\mathbf{A}_0)}$ is absolutely irreducible, and the second yields condition (i) from the above Proposition.

Let H be a finite group. We say that H *occurs regularly over* k if H is isomorphic to the Galois group of a (Galois) extension $F/k(x_1, ..., x_t)$, regular over k, for some $t \geq 1$. For the notion of GAL-realization see [V5, Ch. 8.3] and [V3, §1.4] (in the latter reference it was called GAT-realization).

Theorem. *Let $\mathcal{P}$ be an absolutely irreducible non-singular closed subvariety of $\mathcal{O}_r$. Let $\mathbf{b} \in \mathcal{P}$ be a base point as in §1. Let $\mathcal{L} \leq \mathcal{B}_r$ be the image of the topological fundamental group $\pi_1(\mathcal{P}, \mathbf{b})$. Let $W \leq S_r$ be the image of $\mathcal{L}$ under the map $\kappa : \mathcal{B}_r \to S_r$.*

Let G be a finite group, and $\mathcal{C} = (C_1, ..., C_r)$ be an r-tuple of conjugacy classes $C_i \neq \{1\}$ of G such that the set $Ni(\mathcal{C})$ is non-empty. Set $k = \mathbb{Q}$ if the r-tuple $\mathcal{C}$ is W-rational, otherwise $k = \mathbb{Q}_{ab}$. Suppose $\mathcal{P}$ is defined over k, and the following holds:

(a) Let S be the stabilizer in W of the arrangement $C_1, ..., C_r$, where $W \leq S_r$ acts on these arrangements by permuting the subscripts. We require that the fixed field of S in $k(\hat{\mathcal{P}})$ (see beginning of §2.2) is of the form $k(Y_1, ..., Y_s)$ for independent transcendental elements $Y_1, ..., Y_s$ over k such that the k-linear span of $Y_1, ..., Y_s$ is invariant under the normalizer $N_W(S)$.

(b) The group

$$\mathbf{A}_0 = \{\alpha \in \operatorname{Aut}(G) : \alpha(C_i) = C_i \text{ for } i = 1, ..., r \}$$

acts transitively on the set $\mathcal{E}(\mathcal{C}) = \{(g_1, ..., g_r) \in Ni(\mathcal{C}) : g_i \in C_i, i = 1, ..., r\}$.

(c) There is a normal subgroup $\mathbf{A}$ of $\mathbf{A}_0$ such that the stabilizer H in $\mathbf{A}_0/\mathbf{A}$ of one (hence each) $\mathcal{L}$-orbit on $Ni^W(\mathcal{C})^{(\mathbf{A})}$ is self-normalizing in $\mathbf{A}_0/\mathbf{A}$.

Then $\mathcal{O}_0 := \mathcal{H}_{\mathcal{P}}(\mathcal{C})^{(\mathbf{A}_0)}$ is rational over k, and all components $\mathcal{O}$ of $\mathcal{H}_{\mathcal{P}}(\mathcal{C})^{(\mathbf{A})}$ are defined over k. Thus $H \cong G(k(\mathcal{O})/k(\mathcal{O}_0))$ occurs regularly over k. The group H has a GAL-realization over k if we have further:

(d) $\mathbf{A}$ and $\mathbf{A}_0$ are normal in some subgroup $\mathbf{A}_1$ of $\operatorname{Aut}_{\mathcal{C}}^W(G)$ such that the centralizer of H in $\mathbf{A}_1/\mathbf{A}$ is trivial, and $[\mathbf{A}_1 : \mathbf{A}_0] = [\operatorname{Aut}(H) : \operatorname{Inn}(H)]$.

The proof is analogous to that of Theorem 1.4 in [V3], using the Proposition, Lemma and Remark above in place of [V3, Lemma 1.2.3 and Cor. 1.3.3].

§3. Varieties associated with the Coxeter group of type C_ℓ

For the rest of the paper we fix an even integer $r = 2\ell \geq 4$.

Let $\mathbb{Z}_2$ be the group of order 2. Embed S_ℓ (resp., $\mathbb{Z}_2^\ell$) into $GL_\ell(\mathbb{C})$ as the group permuting the coordinates of $(a_1, ..., a_\ell) \in \mathbb{C}^\ell$ (resp., as the group changing signs of the coordinates $a_1, ..., a_\ell$). Clearly S_ℓ normalizes $\mathbb{Z}_2^\ell$. The semi-direct product W of these groups is the Coxeter group of type C_ℓ, in its natural reflection representation. Thus by [Br], the space of regular orbits of W (isomorphic to the space $\mathcal{Q}_\ell$ defined below) has its fundamental group isomorphic to the braid group of type C_ℓ. This will be used in §5.

We continue in the set-up of §1. Let $\mathcal{Q}_\ell$ be the set of all monic complex polynomials $g(y)$ of degree ℓ with non-zero discriminant and with $g(0) \neq 0$. Consider the map

$$\mathcal{Q}^{(\ell)} \stackrel{\text{def}}{=} \{(a_1, ..., a_\ell) \in \mathbb{C}^\ell : 0 \neq a_i \neq \pm a_j \text{ for } i \neq j\} \quad \to \quad \mathcal{Q}_\ell$$

where the tuple $(a_1, ..., a_\ell)$ is mapped to the polynomial $g(y) = \prod_{i=1}^{\ell} (y - a_i^2)$. Topologically, the map $\mathcal{Q}^{(\ell)} \to \mathcal{Q}_\ell$ is an unramified Galois covering, with automorphism group $\text{Aut}(\mathcal{Q}^{(\ell)}/\mathcal{Q}_\ell) = W$. Algebraically, it is a finite morphism of affine varieties defined over $\mathbb{Q}$. Hence it induces an embedding of the function field $\mathbb{Q}(\mathcal{Q}_\ell)$ into the function field $\mathbb{Q}(\mathcal{Q}^{(\ell)}) = \mathbb{Q}(T_1, ..., T_\ell)$ (where the T_i are the coordinate functions on $\mathbb{C}^\ell$):

$$\mathbb{Q}(\mathcal{Q}_\ell) = \mathbb{Q}(T_1, ..., T_\ell)^W = \mathbb{Q}(X_1, ..., X_\ell)$$

where the X_i are the elementary symmetric functions in the T_i^2.

Embed $\mathcal{Q}_\ell$ and $\mathcal{Q}^{(\ell)}$ into $\mathcal{O}_r$ and $\mathcal{O}^{(r)}$, respectively, via the map $g(y) \mapsto g(y^2)$ and $(a_1, ..., a_\ell) \mapsto (a_1, ..., a_\ell, -a_\ell, ..., -a_1)$. The image $\mathcal{P}^{(\ell)}$ of $\mathcal{Q}^{(\ell)}$ in $\mathcal{O}^{(r)}$ consists of all $(a_1, ..., a_r) \in \mathcal{O}^{(r)}$ with $a_{r+1-i} = -a_i$, $i = 1, ..., \ell$. It is a closed subvariety of $\mathcal{O}^{(r)}$ defined over $\mathbb{Q}$. Also the image $\mathcal{P}_\ell$ of $\mathcal{Q}_\ell$ in $\mathcal{O}_r$ is a closed subvariety of $\mathcal{O}_r$ defined over $\mathbb{Q}$.

Via these embeddings, the map $\mathcal{Q}^{(\ell)} \to \mathcal{Q}_\ell$ corresponds to the restriction to $\mathcal{P}^{(\ell)}$ of the map $\mathcal{O}^{(r)} \to \mathcal{O}_r$ from 2.2. This way we obtain an unramified covering $\mathcal{P}^{(\ell)} \to \mathcal{P}_\ell$ whose automorphism group is the above W, now embedded into $S_r = \text{Aut}(\mathcal{O}^{(r)}/\mathcal{O}_r)$. From now on view W as a subgroup of S_r in this way; i.e., $W = \mathbb{Z}_2^\ell \cdot S_\ell$ where $\mathbb{Z}_2^\ell$ is generated by all transpositions $(i, r + 1 - i)$, $i = 1, ..., \ell$, and $\pi \in S_\ell$ acts naturally on $\{1, ..., \ell\}$, and $\pi(r + 1 - i) = r + 1 - \pi(i)$, $i = 1, ..., \ell$.

From now on we take $\mathcal{P} = \mathcal{P}_\ell$. We let $\mathcal{L}_r = \mathcal{L}$ denote the image of $\pi_1(\mathcal{P}_\ell, \mathbf{b})$ in $\mathcal{B}_r = \pi_1(\mathcal{O}_r, \mathbf{b})$ (as in §1). We call $\mathcal{L}_r$ the **symplectic braid group** of rank $\ell = r/2$. We define the **pure symplectic braid group** $\mathcal{L}^{(r)} = \mathcal{B}^{(r)} \cap \mathcal{L}_r$. By the remarks at the beginning of this section, $\mathcal{L}_r$ is a homomorphic image of the braid group of type C_ℓ. (Actually, it is an isomorphic image; more precisely, the generators $P_1, ..., P_\ell$ of $\mathcal{L}_r$ given in §5 yield a presentation in terms of the braid relations of type C_ℓ. This is not needed here and will be proved elsewhere).

As in §1 we take the base point $\mathbf{b}$ in $\mathcal{P}_\ell$, say $\mathbf{b} = \prod_{i=1}^{r}(y - b_i)$, with $b_{r+1-i} = -b_i$, $i = 1, ..., \ell$. Then in the notation of 2.2. we have $\hat{\mathcal{P}} = \mathcal{P}^{(\ell)}$.

Then $\mathbb{Q}(\hat{\mathcal{P}}) = \mathbb{Q}(\mathcal{P}^{(\ell)}) = \mathbb{Q}(T_1, ..., T_\ell)$ (where the T_i are the first ℓ coordinate functions on $\mathbb{C}^r$).

Lemma. *Fix $\epsilon = 1$ or $\epsilon = -1$. Assume $C_{r+1-i} = C_i^\epsilon$ for $i = 1, ..., \ell$. Then condition (a) of Theorem 2.2 holds (for the present choice of W).*

Proof. First we consider the case $\epsilon = 1$. There is a partition

$$\{1, ..., \ell\} = M_1 \cup ... \cup M_t$$

with the following properties: For all $i \in M_\mu$ the classes C_i are equal, denote this class by $C(\mu)$. Secondly, for $1 \le \mu < \nu \le t$ we have $C(\mu) \ne C(\nu)$. For $\ell_\mu := |M_\mu|$ we get

$$S = \mathbb{Z}_2^\ell \cdot (S_{\ell_1} \times ... \times S_{\ell_t})$$

Then

$$k(\hat{\mathcal{P}})^S = k(T_1, ..., T_\ell)^S = k(Y_1, ..., Y_\ell)$$

where the Y_i with $i \in M_\mu$ are the elementary symmetric functions in the T_i^2 with $i \in M_\mu$. The normalizer $N_W(S)$ permutes $Y_1, ..., Y_\ell$, which proves the claim in the case $\epsilon = 1$.

It remains to study the case $\epsilon = -1$. Now we can order the classes $C_1, ..., C_r$ (using an element of W) such that $C_i \ne C_j^{-1}$ for $1 \le i < j \le \ell$ and $C_i \ne C_j$. Then there is a partition

$$\{1, ..., \ell\} = M_1 \cup ... \cup M_s \cup M_{s+1} \cup ... \cup M_t$$

with the same properties as above, plus the property that $C(\mu) \ne C(\nu)^{-1}$ for $\mu \ne \nu$. Finally, $C(\mu) = C(\mu)^{-1}$ if and only if $s+1 \le \mu \le t$. Observe that then the classes $C_{\ell+1}, ..., C_r$ are partitioned correspondingly. For $\ell_\mu := |M_\mu|$ we get

$$S = S_{\ell_1} \times ... \times S_{\ell_s} \times (\mathbb{Z}_2^{\ell_{s+1}} \cdot S_{\ell_{s+1}}) \times ... \times (\mathbb{Z}_2^{\ell_t} \cdot S_{\ell_t})$$

viewed naturally as subgroup of $W = \mathbb{Z}_2^\ell \cdot S_\ell$. Thus

$$k(\hat{\mathcal{P}})^S = k(T_1, ..., T_\ell)^S = k(Y_1, ..., Y_\ell)$$

where the Y_i with $i \in M_\mu$ are the elementary symmetric functions in the T_i (resp., T_i^2) with $i \in M_\mu$, for $1 \le \mu \le s$ (resp., $s+1 \le \mu \le t$). The normalizer $N_W(S)$ permutes $Y_1, ..., Y_\ell$ up to sign. Hence the claim. □

§4. Choosing the group G

Now we apply the above results with the following particular choice of G and $\mathcal{C}$.

§4.1.   Let $n = r - 2 = (2\ell - 2) \geq 2$. Let q be a power of the prime p, and let $\mathbb{F}_q$ be the finite field with q elements. Let $\zeta_1, ..., \zeta_r$ be elements of the multiplicative group $\mathbb{F}_q^*$, with $\zeta_1 \cdots \zeta_r = 1$, $\mathbb{F}_q = \mathbb{F}_p(\zeta_1, ..., \zeta_r)$, and $\zeta_i \neq 1$ for all i. Let Z be the subgroup of $\mathbb{F}_q^*$ generated by $\zeta_1, ..., \zeta_r$, and set $\zeta = (\zeta_1, ..., \zeta_r)$. Let V be the elementary abelian group $\mathbb{F}_q^n$, and set

$$G \overset{\text{def}}{=} V \cdot Z,$$

the semi-direct product of V and Z (where Z acts on V via scalar multiplication). We write the elements of G as pairs $[v, z]$ with $v \in V$, $z \in Z$. For $i = 1, ..., r$, let $C(\zeta_i)$ be the conjugacy class of G consisting of all $[v, \zeta_i]$, $v \in V$. Set

$$\mathcal{C} = \mathcal{C}_\zeta \overset{\text{def}}{=} (C(\zeta_1), ..., C(\zeta_r))$$

The group $G = V \cdot Z$ embeds naturally as a normal subgroup into

$$\mathbf{A}_0 \overset{\text{def}}{=} V \cdot \mathrm{GL}_n(q)$$

and the centralizer of G in $\mathbf{A}_0$ is trivial. Via conjugation action on G we view $\mathbf{A}_0$ as a group of automorphisms of G. It fixes all conjugacy classes $C(\zeta_i)$. Let M be a subgroup of $\mathbb{F}_q^*$, view it as a scalar subgroup of $\mathrm{GL}_n(q)$, and set

$$\mathbf{A} \overset{\text{def}}{=} V \cdot M.$$

Consider the natural action on $\mathrm{Ni}(\mathcal{C}_\zeta)^{(\mathbf{A})}$ of the group

(2) $$\mathbf{A}_0/\mathbf{A} = \mathrm{GL}_n(q)/M$$

§4.2. Let $\mathcal{B}_r(\zeta)$ be the group of all $Q \in \mathcal{B}_r$ such that for the permutation $\pi = \kappa(Q)$ we have $(\zeta_{\pi(1)}, ..., \zeta_{\pi(r)}) = (\zeta_1, ..., \zeta_r)$ (equivalently, $(C_{\pi(1)}, ..., C_{\pi(r)}) = (C_1, ..., C_r)$). Set $\mathcal{L}_r(\zeta) = \mathcal{B}_r(\zeta) \cap \mathcal{L}_r$. The group $\mathcal{B}_r(\zeta)$ (resp., $\mathcal{L}_r(\zeta)$) contains the pure braid group $\mathcal{B}^{(r)} = \ker(\kappa)$ (resp., the pure symplectic braid group $\mathcal{L}^{(r)} = \mathcal{B}^{(r)} \cap \mathcal{L}_r$).

In [V3, §2] (and [V5, Ch. 9.4.1]) the homomorphism $\Phi_\zeta : \mathcal{B}_r(\zeta) \to \mathrm{GL}_n(q)$ is defined. It describes the braiding action on $\mathrm{Ni}(\mathcal{C}_\zeta)$. More precisely, the image Δ_ζ of this homomorphism is a subgroup of $\mathrm{GL}_n(q)$ with the following property: For each $M \leq \mathbb{F}_q^*$ the group $\Delta_\zeta M/M$ is the stabilizer in $\mathbf{A}_0/\mathbf{A} = \mathrm{GL}_n(q)/M$ (see (2)) of some $\mathcal{B}_r$-orbit in $\mathrm{Ni}(\mathcal{C}_\zeta)^{(\mathbf{A})}$. (The stabilizers of the other $\mathcal{B}_r$-orbits are conjugates because $\mathbf{A}_0/\mathbf{A}$ acts transitively on these orbits). See [V3, Lemma 2.2].

Similarly, the orbits of the symplectic braid group $\mathcal{L}_r \leq \mathcal{B}_r$ on $\mathrm{Ni}(\mathcal{C}_\zeta)$ are determined by the image of $\mathcal{L}_r(\zeta)$ under Φ_ζ. Denote this image by Λ_ζ. Then

$$\Lambda_\zeta \leq \Delta_\zeta \leq \mathrm{GL}_n(q)$$

Analogously to [V3, Lemma 2.2] we conclude that $\Lambda_\zeta M/M$ is the stabilizer in $\mathbf{A}_0/\mathbf{A}$ of an $\mathcal{L}_r$-orbit in $\mathrm{Ni}^W(\mathcal{C}_\zeta)^{(\mathbf{A})}$.

Define ζ to be W-rational if and only if $\mathcal{C}_\zeta$ is W-rational. Now we apply Theorem 2.2. Take $\mathcal{P} = \mathcal{P}_\ell$ as in section 3. Then $\mathcal{L} = \mathcal{L}_r$ and W are as in section 3. Let G and $\mathcal{C} = \mathcal{C}_\zeta$ as chosen in the present section. By [V3, Lemma 2.2(c)] condition (b) of the Theorem holds. By the preceding paragraph condition (c) of the Theorem holds if the group $H = \Lambda_\zeta M/M$ is self-normalizing in $\mathbf{A}_0/\mathbf{A} = \mathrm{GL}_n(q)/M$. If ζ satisfies $\zeta_{r-i+1} = \zeta_i^\epsilon$ with fixed $\epsilon = \pm 1$, then also condition (a) of the Theorem holds (by the Lemma in §3). Thus we get

Corollary. *Assume ζ is W-rational. (This holds e.g. if the tuple $(\zeta_1, ..., \zeta_\ell)$ is rational, see §1). Fix $\epsilon = 1$ or $\epsilon = -1$. Assume $\zeta_{r-i+1} = \zeta_i^\epsilon$ for $i = 1, ..., \ell$. Let M be a subgroup of $\mathbb{F}_q^*$. If $H \stackrel{\mathrm{def}}{=} \Lambda_\zeta M/M$ is self-normalizing in $\mathrm{GL}_n(q)/M$, then H occurs regularly over $\mathbb{Q}$. If additionally the centralizer of H in $\Gamma L_n(q)/M$ is trivial and $[\mathrm{Aut}(H) : \mathrm{Inn}(H)] = m$, where $q = p^m$, then H has a GAL-realization over $\mathbb{Q}$.*

Here $\Gamma L_n(q)$ is the group of semi-linear transformations of the vector space $\mathbb{F}_q^n$. If we drop the condition of W-rationality then $\Lambda_\zeta M/M$ occurs regularly over $\mathbb{Q}_{ab}$; and it has a GAL-realization over $\mathbb{Q}_{ab}$ if $\mathrm{Aut}(\mathbb{F}_q)$ permutes $\zeta_1, ..., \zeta_\ell$.

In the next section we give generators of the group $\mathcal{L}_r$. Using these generators, and the basic pairing between Φ_ζ and $\Phi_{\zeta^{-1}}$ from [V2], it is possible to compute the group Λ_ζ explicitly, see [MSV]. The case $\epsilon = -1$ is the case of main interest, since then Λ_ζ is mostly the symplectic group $\mathrm{Sp}_n(q)$ or $\mathrm{Sp}_n(\sqrt{q})$. If q is even then $\mathrm{PSp}_n(q)$ is self-normalizing in $\mathrm{PGL}_n(q)$, and the Corollary applies with $M = \mathbb{F}_q^*$. This yields GAL-realizations over $\mathbb{Q}$ for the simple groups $\mathrm{Sp}_n(2^s)$ with $n \geq 2^{s+1}$, $s \neq 2$ (see [MSV]). GAL-realizations for certain of the groups $\mathrm{PGL}_n(q)$ and $\mathrm{PU}_n(q)$ were found in [V3] (using the Artin braid group).

§5. Generators of the symplectic braid group

First we specify the base point b. Let $b_1 = -\ell$, $b_2 = -\ell+1$,..., $b_\ell = -1$ and $b_{r-i+1} = -b_i$ for $i = 1, ..., \ell$.

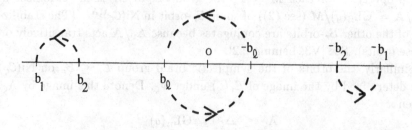

Generators $Q_1, ..., Q_{r-1}$ of the Artin braid group $\mathcal{B}_r$ can be specified as follows (see [V5, Ch. 10.1.6]): Fix some $i = 1, ..., r - 1$. For $j = 1, ..., r$, $i \neq j \neq i+1$, let $a_j(t)$ be the constant path b_j. Let $a_i(t)$ (resp., $a_{i+1}(t)$) be a path in the complex plane with initial point b_i (resp., b_{i+1}) and endpoint b_{i+1} (resp., b_i), travelling counter-clockwise on the respective circle with center on the real line. Let $\tilde{Q}_i(t) = (a_1(t), ..., a_r(t))$, a path in $\mathcal{O}^{(r)}$. Note that $\tilde{Q}_{r-i} = -\tilde{Q}_i$. Define $Q_i \in \mathcal{B}_r = \pi_1(\mathcal{O}_r, \mathbf{b})$ to be represented by the image of $\tilde{Q}_i$ in $\mathcal{O}_r$.

Define $P_1, ..., P_\ell \in \mathcal{B}_r$ by

$$P_i = Q_i \, Q_{r-i}, \quad i = 1, ..., \ell - 1$$

and

$$P_\ell = Q_\ell$$

These elements are naturally represented by paths in the subvariety $\mathcal{P}_\ell$ of $\mathcal{O}_r$. Hence $P_1, ..., P_\ell$ lie in the symplectic braid group $\mathcal{L}_r = \pi_1(\mathcal{P}_\ell, \mathbf{b})$. We claim:

Proposition. *The elements $P_1, ..., P_\ell$ generate $\mathcal{L}_r$.*

Proof. We show that via the homeomorphism $\mathcal{Q}_\ell \to \mathcal{P}_\ell$ the elements P_i get identified with the generators of $\pi_1(\mathcal{Q}_\ell)$ given by Brieskorn [Br].

For a moment view W again as acting on $\mathbb{C}^\ell$, in the natural representation of the Coxeter group of type C_ℓ. A Coxeter (generating) system of W consists of the reflections $\sigma_1, ..., \sigma_\ell$, where σ_i for $i < \ell$ acts on the points $(a_1, ..., a_\ell) \in \mathbb{C}^\ell$ as transposition $(i, i + 1)$ on the coordinates (i.e., interchanging a_i and a_{i+1}); and σ_ℓ maps $(a_1, ..., a_\ell)$ to $(a_1, ..., a_{\ell-1}, -a_\ell)$. Take $c = \{b_1, ..., b_\ell\}$ as base point in $\mathcal{Q}^{(\ell)} \subset \mathbb{C}^\ell$.

Restricting the action of W to $\mathbb{R}^\ell \subset \mathbb{C}^\ell$, we obtain the natural reflection representation of W in Euclidean space. The fixed point hyperplanes of the reflections in W decompose $\mathbb{R}^\ell$ into chambers. The chamber containing c consists of all $(a_1, ..., a_\ell)$ with $a_1 < ... < a_\ell < 0$. The hyperplanes associated with the σ_i are the walls of this chamber (i.e., form the boundary of this chamber). For each $i = 1, ..., \ell$ let L_i be the complex line joining c and $\sigma_i(c)$. Let $c^{(i)}$ be the intersection point of L_i and the fixed point hyperplane of σ_i. Let P_i^* be a path in L_i going from c to $\sigma_i(c)$ on a semi-circle in counter-clockwise direction around $c^{(i)}$.

Let $\mathbf{c}$ be the image of c in $\mathcal{Q}_\ell$ under the natural map $\mathcal{Q}^{(\ell)} \to \mathcal{Q}_\ell \cong \mathcal{Q}^{(\ell)}/W$ (see §3). Let $P_i' \in \pi_1(\mathcal{Q}_\ell, \mathbf{c})$ be the image of P_i^*. Then $P_1', ..., P_\ell'$ generate $\pi_1(\mathcal{Q}_\ell, \mathbf{c})$ by [Br]. It is easy to see that the P_i' correspond to the above elements $P_i \in \pi_1(\mathcal{P}_\ell, \mathbf{b})$ (under the homeomorphism $\mathcal{Q}_\ell \to \mathcal{P}_\ell$). We illustrate the case of P_1. We have $\sigma_1(c) = (b_2, b_1, b_3, ..., b_\ell)$, hence the line L_1 consists of all

$$((b_1 + b_2)/2 + \tau(b_1 - b_2)/2, \ (b_1 + b_2)/2 + \tau(b_2 - b_1)/2, \ b_3, \ ..., b_\ell), \quad \tau \in \mathbb{C}$$

and we obtain the path P_1^* by taking $\tau = e^{it}$, $0 \le t \le \pi$. Clearly, the image of P_1^* under the map $\mathcal{Q}^{(\ell)} \to \mathcal{P}^{(\ell)}$, $(a_1, ..., a_\ell) \mapsto (a_1, ..., a_\ell, -a_\ell, ..., -a_1)$ is the path $\tilde{Q}_1 \tilde{Q}_{r-1}$. Hence P_1' corresponds to P_1.

References

[Br]   E. Brieskorn, Die Fundamentalgruppe des Raumes der regulären Orbits einer endlichen komplexen Spiegelungsgruppe, *Inv. Math.* **12** (1971), 57-61.

[FV]   M.D. Fried and H. Völklein, The inverse Galois problem and rational points on moduli spaces, *Math. Ann.* **290** (1991), 771-800.

[MSV] K. Magaard, K. Strambach and H. Völklein, Finite quotients of the pure symplectic braid group, in preparation.

[V1]   H. Völklein, $GL_n(q)$ as Galois group over the rationals, *Math. Ann.* **293** (1992), 163-176.

[V2]   H. Völklein, Braid group action through $GL_n(q)$ and $U_n(q)$, and Galois realizations, *Israel J. Math.* **82** (1993), 405-427.

[V3]   H. Völklein, Braid group action, embedding problems and the groups $PGL(n,q), PU(n,q^2)$, *Forum Math.* **6** (1994), 513-535.

[V4]   H. Völklein, Moduli spaces for covers of the Riemann sphere, *Israel J. Math.* **85** (1994), 407-430.

[V5]   H. Völklein, *Groups as Galois Groups - an Introduction*, Cambridge University Press, 1996.

Karl Strambach, Universität Erlangen
Helmut Völklein, University of Florida and Universität Erlangen

Applying Modular Towers to the Inverse Galois Problem

Michael D. Fried and Yaacov Kopeliovich

Let G be a finite (possibly simple) group, and let p be a prime dividing the order of G. The characteristic finite quotients ${}^{k}_{p}\tilde{G}$ of the universal p-Frattini cover of G are strikingly similar groups. It takes an effort to distinguish them for finding $\mathbb{Q}$ regular realizations for the Inverse Galois problem. This paper starts a program to show one can't realize *all* these groups as Galois groups of extensions $L/\mathbb{Q}(x)$ with at most r (fixed) branch points. Let $\mathbf{C}$ be an r-tuple of p-regular conjugacy classes of G. To compare realizations of these groups we use a sequence of varieties—a *Modular Tower*—attached to $(G, p, \mathbf{C})$. The notation for this sequence is $\mathcal{H}({}^{k}_{p}\tilde{G}, \mathbf{C})$, $k = 0, 1, \ldots$: $\mathcal{H}({}^{k}_{p}\tilde{G}, \mathbf{C})$ is the kth *level* of the Modular Tower. Crucial properties of level k translate to properties of the characteristic modular representation of ${}^{k}_{p}\tilde{G}$. Properties of these representations support the following statement.

Conjecture. *For each r there exists k_r so that for $k > k_r$, $\mathbb{Q}$ regular realization of ${}^{k}_{p}\tilde{G}$ requires more than r branch points.*

For $r = 4$ this reduces to showing two pieces of geometric information.

(a)   There is a uniform (with k) bound on the number of absolutely irreducible components at the kth level.

(b)   For k large $\mathcal{H}({}^{k}_{p}\tilde{G}, \mathbf{C})$ has no *obstructed* components.

The main example of this paper and [FrK] is $G = A_5$, $p = 2$ and $\mathbf{C} = \mathbf{C}_{3^r}$ with $r = 4$ repetitions of the conjugacy class of 3-cycles. It allows full explanation and illustration of the significance of obstructed components.

§0. Introduction to the main problem.

[MT] introduced Modular Towers. This paper concentrates on exactly one application of these to the Inverse Galois Problem. Fix an indeterminate x

Support from NSF #DMS-9622928 for both authors; first author support from Alexander von Humboldt Foundation and Institut für Experimentelle Mathematik, July 1996. Correspondence with John Thompson, after an Oct. 1996 conference at UF, induced us to write §4 reducing our Main Conjecture to a diophantine statement on points on Modular Towers. Conversations with Bob Guralnick improved our confidence with the group theory parts of the paper. Finally, Elena Black's conference talk exclamation, "Do you know how hard it is to work with one group at a time?" inspired the philosophy of §1.B: Fix conjugacy classes and change the group.

as a uniformizing parameter for the projective x line $\mathbb{P}^1 = \mathbb{P}^1_x$. For any field K let $\bar{K}$ be an algebraic closure of K. The characteristic of K is 0, and K is usually a subfield of the complex numbers $\mathbb{C}$. The symmetric group of degree n is S_n and its alternating subgroup is A_n.

The framework of this paper has appeared nowhere else. It includes an introduction to more comprehensive results of [Fr1] and [FrK]. These will quote this paper. Modular Towers join *Hurwitz space* constructions and the *universal Frattini cover* of a finite group. App.I reviews relevant definitions from [MT]. One crucial point: Levels of a Modular Tower aren't (usually) homogeneous spaces. Yet, modular representation theory provides a replacement for the semisimple Lie group theory often attached to homogeneous spaces.

§0.A. Basic Notation.

Recall what is a K *regular realization* of a finite group G: G is the Galois group of an extension $L/K(x)$ with $L \cap \bar{K} = K$.

The main combinatorial data for a regular realization is a set of conjugacy classes $\mathbf{C}$ from G. A place x' of $\mathbb{C}(x)$ denotes the specialization of the elements of $\mathbb{C}(x)$ associated to $x \mapsto x'$. Consider any place $\mathfrak{p}$ of $\mathbb{C} \cdot L$ over x'. Let $e = e(\mathfrak{p}/x')$ be the ramification index of $\mathfrak{p}$ over x'. If $e > 1$ for some $\mathfrak{p}$ over x', call x' a *branch point* of the extension $\mathbb{C} \cdot L/\mathbb{C}(x)$. Then, the completion $L_{\mathfrak{p}}$ of L at $\mathfrak{p}$ embeds in the Laurent field $\mathbb{C}\{\{(x - x')^{\frac{1}{e}}\}\}$. Restricting the automorphism $(x - x')^{\frac{1}{e}} \mapsto \exp^{2\pi i/e}(x - x')^{\frac{1}{e}}$ to L gives $g_{\mathfrak{p}} \in G$. Referencing only x', and not $\mathfrak{p}$, defines $g_{\mathfrak{p}}$ up to conjugacy in G.

This attaches a unique conjugacy class C of G to x'. Thus, the complete set $\mathfrak{x} = \{x_1, \ldots, x_r\}$ of branch points of the extension produces a collection $\mathbf{C} = \{C_1, \ldots, C_r\}$ of conjugacy classes in G. A description of branch cycles for $L/K(x)$ consists of an r-tuple $\mathbf{g} = (g_1, \ldots, g_r) \in G^r$ with these properties:

(0.0a) $\mathbf{g} \in \mathbf{C}$, $g_1 \cdots g_r = \Pi(\mathbf{g}) = 1$; and

(0.0b) $\langle \mathbf{g} \rangle = G$.

Here $\mathbf{g} \in \mathbf{C}$ means, in some order, entries of $\mathbf{g}$ are in the classes $\mathbf{C}$.

For any regular extension $L/\mathbb{Q}(x)$ such an r-tuple satisfying (0.0) always exists. It isn't, however, unique. The *Nielsen class* $\mathrm{Ni}(G, \mathbf{C})$ of a cover is the complete collection of elements $\mathbf{g}$ satisfying the conditions (0.0). This codifies the relation between different covers and all descriptions of branch cycles (see App.I).

§0.B. The Main Conjectures.

Let G be a finite group and p a prime dividing $|G|$. Also, assume G is centerless (has no center). The pair (G, p)

produces an infinite sequence of *characteristic Frattini covers*:

(0.1) $\cdots \to {}_p^k\tilde{G} \to {}_p^{k-1}\tilde{G} \to \cdots \to {}_p^0\tilde{G} = G.$

§1.B reminds of the basics. The characteristic homomorphism $\phi_k : {}_p^k\tilde{G} \to G$ has p-group kernel for each nonnegative integer k. Let ${}_p\mathcal{C}(G)$ be those conjugacy classes of G whose elements have order prime to p. These are the *p-regular conjugacy classes* of G. Thus, $\mathbf{C} \in {}_p\mathcal{C}(G)$ lifts uniquely to a conjugacy class in ${}_p^k\tilde{G}$ of elements with the same order. Also, refer to this lifted conjugacy class as C: $\mathbf{C} \in {}_p\mathcal{C}({}_p^k\tilde{G}) = {}_p\mathcal{C}(G)$. Denote the cardinality of the conjugacy classes in **C** by $r = r(\mathbf{C})$.

Main Conjecture 0.1. *Let r_0 be any positive integer.*

(†.a) $\exists\ k_a = k_a(G, p, r_0)$ *with this property. For $k > k_a$, $\mathbb{Q}$ regular realization of ${}_p^k\tilde{G}$ with r branch cycles in ${}_p\mathcal{C}(G)$ requires $r > r_0$.*

(†.b) $\exists\ k_b = k_b(G, p, r_0)$ *with this property. For $k > k_b$, $\mathbb{Q}$ regular realization of ${}_p^k\tilde{G}$ requires $r > r_0$ branch cycles.*

It simplifies notation to consider this over $\mathbb{Q}$. So, we restrict often to that case. Still, for K finitely generated over $\mathbb{Q}$ replacing $\mathbb{Q}$ this should hold with k_a and k_b depending on K.

Let r be an integer and **C** a collection of r conjugacy classes (possibly with repetitions) from ${}_p\mathcal{C}(G)$. To this data [MT] canonically attaches a sequence of reduced Hurwitz spaces (App.II):

(0.2) $\cdots \mathcal{H}({}_p^{k+1}\tilde{G}, \mathbf{C})^{\mathrm{rd}} \to \mathcal{H}({}_p^k\tilde{G}, \mathbf{C})^{\mathrm{rd}} \to \cdots \to \mathcal{H}({}_p^0\tilde{G}, \mathbf{C})^{\mathrm{rd}}.$

The kth level is the manifold $\mathcal{H}({}_p^k\tilde{G}, \mathbf{C})^{\mathrm{rd}}$ and ${}_p^0\tilde{G} = G$. Special Case: $G = D_p$ is the dihedral group of order $2p$, p is an odd prime and **C** is $r = 4$ repetitions of the conjugacy class of involutions. Then, (0.2) is the classical sequence of modular curves:

$$\cdots \to Y_1(p^{k+2}) \to Y_1(p^{k+1}) \to \cdots \to Y_1(p).$$

This case agrees with the conclusion of the Main Conjecture. For this case, however, it is easy to show the levels have only one (unobstructed) component. [Fr3, §7] gives the reduction of (†.b) to (†.a) for $G = D_p$ and for any value of r_0.

For D_p and $r_0 = 4$ (or 5) the results are very strong; they translate to known results of Frey, Kamienny, Mazur and Merrill. They even give a statement uniform in p ([DFr, §5] or [Fr3, §7]). Conjecture 0.1 asks for less in this case: that $\mathbb{Q}$ regular D_{p^k} realization requires at least 6 branch points if $k > k_b$. (Also, k_b here may depend on p.)

The goal is to generalize the following argument. Fix p and let K be any finitely generated extensions over $\mathbb{Q}$. The genus (of the single component) of $Y_1(p^k)$ goes up (quickly) with k. Thus, Faltings' Theorem implies the K points $Y_1(p^k)(K)$ on $Y_1(p^k)$ are finite, excluding finitely many values of p^k. The conjecture thus comes to eliminating a possible projective system $\eta_k \in Y_1(p^{k+1})$ of K rational points. [Fr3, §7] handles this when $G = D_p$ and $r_0 = 4$ (or 5).

For an arbitrary finite group G, each step above encounters problems. [FrK] reveals the main difficulties. This case has $G = A_n$ and $\mathbf{C} = \mathbf{C}_{3^r}$, r repeats of the conjugacy class of 3-cycles. The Nielsen class $\mathrm{Ni}_{n,r}$ of r 3-cycles in A_n appears often. Use the notation $\mathcal{H}_{n,r}$ for the corresponding (inner) Hurwitz space (App.I). Though most groups produce difficulties that don't appear for the dihedral group, conjecturally Modular Towers do mimic properties of modular curve towers. Even, however, with $G = D_p$ the conjecture is open if $r_0 > 5$.

Here are three accomplishments of this paper.

(0.3a) For any G it shows that the (†.b) version of Main Conjecture 0.1 reduces to showing (†.a): For k large, $\mathcal{H}(^k_p\tilde{G}, \mathbf{C})^{\mathrm{rd}}$ has no $\mathbb{Q}$ points.

(0.3b) For $G = A_5$, $p = 2$ and $\mathbf{C} = \mathbf{C}_{3^r}$ it describes progress on bounding components of the kth level, $k \geq 0$, in the Modular Tower.

(0.3c) It shows how *obstructed components* (§0.C) affect the Main Conjecture. Further, A_5 examples support their disappearance for k large.

§0.C. **Obstructed components of a Modular Tower.** The projective limit of sequence (0.1) is the *universal p-Frattini cover* $_p\tilde{G}$ of G. §1.C explains how similar are all the groups $^k_p\tilde{G}$.

Subtheme 0.2 Replacing G with $_p\tilde{G}$ produces conclusions on regular realizations of all $_p\tilde{G}$ characteristic quotients.

For fixed r, computations suggest there is a uniform bound on absolutely irreducible components of $\mathcal{H}(^k_p\tilde{G}, \mathbf{C})^{\mathrm{rd}}$. When $r=4$, reduced Hurwitz spaces are curves. Bounding the number of components in the kth level is necessary to assure Faltings' Theorem applies. That is, the genus of absolutely irreducible components at level k must exceed 1 for k large. Still, that doesn't preclude rational points at arbitrary high levels on the most mysterious of components, those we call *obstructed*.

Suppose no points of $\mathcal{H}(^{k+1}_p\tilde{G}, \mathbf{C})^{\mathrm{rd}}$ lie above a component $\mathcal{H}'$ of $\mathcal{H}(^k_p\tilde{G}, \mathbf{C})^{\mathrm{rd}}$. We say $\mathcal{H}'$ is obstructed. [MT, §III.D] gave a *big invariant* detecting obstructed components. Obstruction Lemma 3.2 reformulates this invariant using modular representations.

There is good news and bad news in the appearance of obstructed components. The former occurs if there is exactly one component at level k and it is obstructed. Then, (0.3a) has a positive answer. The reduced Hurwitz space at level $k+1$ is empty, so it has no rational points. [Fr1] gives examples of exactly that (see Thm. 3.1 in §3.A).

Many cases, however, have at least one unobstructed component. General results describing components of a level of a Modular Tower require conjugacy classes in C to appear with high multiplicity. The rest of this subsection illustrates this.

Definition 0.3 Consider a collection C of conjugacy classes from a finite group G. Let s_0 be a positive integer. Then, C has *multiplicity* at least s_0 if *each* conjugacy class in C appears at least s_0 times.

Suppose C has suitably (explicitly) high multiplicity compared to k. Then, an effective version of a Conway and Parker result precisely bounds the components at level k. The following is a special case of a result from [FrK]. Here $M\binom{k}{p}\tilde{G}$ is the *Schur multiplier* of $\frac{k}{p}\tilde{G}$.

Theorem 0.4 *Fix a value of k. Assume G is perfect and centerless, $p \mid |G|$, all classes in C are from $_pC(G)$ and C has multiplicity at least s_0. Decompose $\mathcal{H}(\frac{k}{p}\tilde{G}, C)^{\mathrm{rd}}$ into absolutely irreducible (dimension $r-3$) components $\cup_{i=0}^{t_k} \mathcal{H}'_i$. Then, there is an explicit $s_0(G,k) = s_0$ so the following hold.*

(0.4a) $t_k + 1 = |M\binom{k}{p}\tilde{G})|$.

(0.4b) $\mathcal{H}'_0$ is unobstructed.

(0.4c) If C is a rational union (see §1.A), $\mathcal{H}'_0$ has field of definition $\mathbb{Q}$.

(0.4d) $\mathcal{H}'_i$ is obstructed, $i = 1, \ldots, t_k$.

§0.D. Disappearance of obstructed components for k large. Suppose G has a nontrivial Schur multiplier (as does A_n). Then, so does $\frac{k}{p}\tilde{G}$ for all k (Schur Multipliers Result 3.3). Thus, Theorem 0.4 implies obstructed components appear in $\mathcal{H}_{n,r}^{\mathrm{rd}}$ for r large (notation from §0.B). The crucial investigation must consider minimal values $r_0 = r_0(n,k)$ of r that produce obstructed components in $\mathcal{H}(\frac{k}{2}\tilde{G}, C_{3^r})^{\mathrm{rd}}$. By contrast, a particular Modular Tower considers k large compared to a fixed value of $r = r(C)$. There are serious diophantine troubles for the Main Conjecture should new obstructed components appear at each level.

For example, suppose there is an s_0 giving the following.

(0.5a) Theorem 0.4 holds with s_0 independent of k.

(0.5b) There is a bound on $|M(^k_p\tilde{G})|$ independent of k.

This would produce a bound in (0.3b), at least for r large. That wouldn't, however, contribute to the Main Conjecture. Even the strongest diophantine conjectures can't eliminate rational points on obstructed components at arbitrary high levels of a Modular Tower. Worse yet, (0.5b) doesn't hold.

So, it is a surprise that the case $G = A_5$, $p = 2$ and $\mathbf{C} = \mathbf{C}_{3^r}$ supports the following conjecture. Recall: A projective nonsingular variety V is of *general type* if it's canonical bundle is ample. If V is a curve this means V has genus at least two. Using the phrase *general type* on $\mathcal{H}(^k_p\tilde{G}, \mathbf{C})^{\mathrm{rd}}$ means apply it to some nonsingular compactification of its components.

Conjecture 0.5. *Assume G is centerless, $p\,|\,|G|$ and $\mathbf{C}$ are conjugacy classes from $_p\mathcal{C}(G)$. Then, there are two constants $B = B(G,r)$ and $k_0 = k_0(G,r)$ so that for $k \geq k_0$, the following hold.*

(0.6a) $\mathcal{H}(^k_p\tilde{G}, \mathbf{C})^{\mathrm{rd}}$ *has at most B absolutely irreducible components.*

(0.6b) *Each component of $\mathcal{H}(^k_p\tilde{G}, \mathbf{C})^{\mathrm{rd}}$ is of general type.*

(0.6c) *No components of $\mathcal{H}(^k_p\tilde{G}, \mathbf{C})^{\mathrm{rd}}$ are obstructed.*

The classical connection has $G = D_p$, p is an odd prime, and $\mathbf{C}$ is $r = 2r'$ repetitions of the conjugacy class of involutions for some integer r'. Braid group action on Nielsen classes here is easy to calculate. As in App.I, orbits of the braid group B_r on Nielsen classes $\mathrm{Ni}(G, \mathbf{C})$ correspond to the components of $\mathcal{H}(G, \mathbf{C})$. [Fr4, §3] shows there is one orbit. So (0.6a) and (0.6c) hold. When $r' = 2$, §0.B notes (0.6b) holds. This seems to be unknown for higher values of r'.

Beyond dihedral groups, evidence for (0.6c) is from the case $G = A_5$. Lemma 3.2 uses the action of the characteristic quotient $^k_p\tilde{G}$ on the characteristic kernel $\ker_k / \ker_{k+1}$ for all values of k. Consider this contrast with k_0 fixed and r large. Then, Theorem 0.4 says $\mathcal{H}(^{k_0}_2\tilde{A}_5, \mathbf{C}_{3^r})^{\mathrm{rd}}$ has at least two absolutely irreducible components; one unobstructed and all others obstructed. For $r = 4$, however, we expect—a result of [FrK] for k small—no obstructed components for any value of k. There should be a minimal value $r' = r'(k_0)$ of r for which there are no obstructed components in $\mathcal{H}(^k_2\tilde{A}_5, \mathbf{C}_{3^{r'}})^{\mathrm{rd}}$ for $k \geq k_0$. In support of (0.6c), [FrK] makes an explicit guess for $r' = r'(k_0)$.

§1. Precise versions of the main conjecture.

Here is the data for a Modular Tower: a finite group G, a prime $p\,|\,|G|$ and a collection of r conjugacy classes $\mathbf{C}$ from $_p\mathcal{C}(G)$. Recall: D_r is the *discriminant locus* in projective r-space. We concentrate on $\mathbb{Q}$ realizations of a centerless group G. [FrV] shows this is equivalent to finding a $\mathbb{Q}$ point on an *inner* Hurwitz space $\mathcal{H}(G, \mathbf{C}) = \mathcal{H}(G, \mathbf{C})^{in}$ for some $\mathbf{C}$. See §0.A and App.I for *Nielsen classes* and how $(G, \mathbf{C})$ canonically produces the moduli space $\mathcal{H}(G, \mathbf{C})$ with a cover $\Phi_{G,\mathbf{C}} = \Phi : \mathcal{H}(G, \mathbf{C}) \to \mathbb{P}^r \setminus D_r$. Suppose K is a field of definition of $\mathcal{H}(G, \mathbf{C})$ and $\mathfrak{p} \in \mathcal{H}(G, \mathbf{C})$. Denote the field generated by coordinates of $\mathfrak{p}$ over K as $K(\mathfrak{p})$. In this paper points are *geometric* points. Also, having coordinates for $\mathfrak{p}$ means $\mathcal{H}(G, \mathbf{C})$ is a quasi-projective (even affine) algebraic set ([Fr2], [FrV] or [V]).

§**1.A. Notation for $\mathbb{Q}$ moduli spaces.** Let $N_{\mathbf{C}} = N$ be the least common multiple of the orders of elements in the collection $\mathbf{C}$. The group $\hat{\mathbb{Z}}^*$ is the projective limit of the invertible elements of $\mathbb{Z}/N$ over all positive integers N. Consider a field K (a subfield of $\mathbb{C}$). A Galois extension $L/K(x)$ is a K regular $(G, \mathbf{C})$ *realization* if the following hold.

(1.1a) $G(L/\mathbb{Q}(x)) = G$ and $L \cap \bar{K} = K$, and

(1.1b) points of the x line ramified in L have associated *branch cycles* in $\mathbf{C}$.

Covers associated to such extensions have r branch points. According to the *branch cycle argument* ([Fr2, before Thm. 5.1], [Fr3], [V, p. 34]), there is a first criterion for such a realization. The minimal field containing all roots of 1 is $\mathbb{Q}^{cyc}$, the cyclotomic closure of $\mathbb{Q}$. The formulation uses the K-cyclotomic group:

$$H_K = \{\sigma \in G(\mathbb{Q}^{cyc}/\mathbb{Q}) \cong \hat{\mathbb{Z}}^* \mid \sigma \text{ fixes elements of } K \cap \mathbb{Q}^{cyc}\}.$$

Each $n \in H_K$ defines an invertible integer modulo $N_{\mathbf{C}}$. So $\mathbf{C}^n$, all elements of $\mathbf{C}$ to the power n, makes sense.

Definition 1.1 Call $\mathbf{C} = (\mathbf{C}_1, \dots, \mathbf{C}_r)$ a K-*rational union* if $\mathbf{C}^n = \mathbf{C}$ for each $n \in H_K$. Then, $\mathbf{C}$ is a rational union if $\mathbf{C}^n = \mathbf{C}$ for each $n \in \hat{\mathbb{Z}}^*$.

[MT, p. 161] explains the meaning of this phrase: $\mathcal{H}(G, \mathbf{C})$, as a *moduli space* has field of definition K. Roughly: For any point $\mathfrak{p} \in \mathcal{H}(G, \mathbf{C})$ the least field of definition of a Galois cover in the Nielsen class representing $\mathfrak{p}$ is $K(\mathfrak{p})$. The K-rational condition in the next result is necessary for $\mathcal{H}(G, \mathbf{C})$ to be a moduli space over K. Again, the elementary branch cycle argument gives this. That K-rationality suffices is tougher—see [FrV] or [V, §10.3.2].

K-**Branch Cycle Argument 1.2** [Fr2, §5] Assume G is a centerless group. The moduli space $\mathcal{H}(G, \mathbf{C})$ (with its morphism Φ) has field of definition K if and only if $\mathbf{C}$ is a K-rational union. If $\mathcal{H}(G, \mathbf{C})(K)$ is nonempty, then $\mathbf{C}$ is a K-rational union and $\mathcal{H}(G, \mathbf{C})$ contains an absolutely irreducible component over K. Also, K points on $\mathcal{H}(G, \mathbf{C})$ produce K regular $(G, \mathbf{C})$ realizations.

For simplicity, the remainder of the paper concentrates on $K = \mathbb{Q}$. In particular, assume the conjugacy classes $\mathbf{C}$ from G is a rational union.

§1.B. Fix conjugacy classes, change the group. Our Main Conjecture divides into cases: $(p, N_{\mathbf{C}}) = 1$ and $p \mid N_{\mathbf{C}}$. §2 and §3 consider the case $(p, N_{\mathbf{C}}) = 1$. The Main Conjecture reduces to this case according to Theorem 4.4. Analysis of components of $\mathcal{H}(_p^k\tilde{G}, \mathbf{C})$ as k varies requires information about $_p\tilde{G}$. To help, we expand here on [FrJ, Chap. 20] and [MT, Part II]. This discussion will continue in [FrK].

The universal p-Frattini cover $\phi_G : {}_p\tilde{G} \to G$ of G is *versal* for embedding problems with p-group kernel. The meaning of versal: Given $\psi : A \to G \to 1$ exact, there is a map $\psi' : {}_p\tilde{G} \to A$ giving the obvious commutative diagram. Versal, unlike universal, does not mean ψ' is unique. If G is perfect, the universal p-central extension of G is an example of a natural quotient of $_p\tilde{G}$. It is, however, a very small quotient. We explain this.

Let $\ker_0 \to {}_p\tilde{G} \to G$ be the natural short exact sequence. Then, $\ker_0$ is a pro-free p-group of finite rank. Define $\ker_k$ inductively: It is the closed subgroup of $_p\tilde{G}$ that $[\ker_{k-1}, \ker_{k-1}]\ker_{k-1}^p$ generates. Then, $_p^k\tilde{G} = {}_p\tilde{G}/\ker_k$. Follow [FrJ, Chap. 20] in calling the minimal number of elements that generate it its *rank*.

The hypothesis $(p, N_{\mathbf{C}}) = 1$ allows a special choice of lift of entries of $\mathbf{C}$ to corresponding conjugacy classes of $_p\tilde{G}/\ker_k = {}_p^k\tilde{G}$. Choose lifting representatives with the same orders as their images in G. To emphasize this choice of conjugacy classes in $_p^k\tilde{G}$, keep the notation $\mathbf{C}$ for these *lifted* classes. Further, as conjugacy classes in $_p^k\tilde{G}$, $\mathbf{C}$ is also a rational union, inheriting this property from G.

The Frattini cover property appears here. Suppose elements of $_p^k\tilde{G}$ are entries of $\mathfrak{g}' \in \mathbf{C}$. Also, assume $\mathfrak{g}'$ lifts entries of $\mathfrak{g} \in \mathrm{Ni}(G, \mathbf{C})$. Then, $\langle \mathfrak{g}' \rangle = {}_p^k\tilde{G}$. From the character theory viewpoint—including generalizations of *rigidity*—$(_p^k\tilde{G}, \mathbf{C})$ realizations are much like $(G, \mathbf{C})$ realizations.

Reminder Statement 1.3. [MT, Lemma 3.6] Suppose G is centerless and perfect. Then, so is $_p^k\tilde{G}$, $k \geq 0$. In particular, a $\mathbb{Q}$ point on $\mathcal{H}(_p^k\tilde{G}, \mathbf{C}) = \mathcal{H}_k$ is equivalent to giving a $\mathbb{Q}$ regular $(_p^k\tilde{G}, \mathbf{C})$ realization [FrV]. Such a point automatically gives $\mathbb{Q}$ regular $(_p^j\tilde{G}, \mathbf{C})$ realizations, from the images of the corresponding point in $\mathcal{H}_j$, $0 \leq j \leq k$.

One goal of this paper is to show progress on the following problem.

Modular Tower Conjecture 1.4. *Suppose G is a centerless group and r is a positive integer. With $k = k(G, r)$ suitably large the following hold.*

(1.2a) *For any prime p dividing $|G|$ and conjugacy classes* **C** *supported in* $_p\mathcal{C}(G)$, $\mathcal{H}(^k_p\tilde{G}, \mathbf{C})^{\mathrm{rd}}(\mathbb{Q})$ *is empty.*

(1.2b) *More generally, there are no* $^k_p\tilde{G}$ *realizations with at most r branch points.*

Theorem 4.4 shows (1.2a) (for some value of $k(G, r)$) implies (1.2b) (for a possibly larger value of $k(G, r)$). [MT] concentrated on the standard Hurwitz spaces (as in [FrV]) arising from representations of the fundamental group of $\mathbb{P}^r \setminus D_r$ acting on Nielsen classes. App.II explains reduced Hurwitz spaces as a $\mathrm{PSL}_2(\mathbb{C})$ quotient of standard Hurwitz spaces.

§1.C. Braid action and $\mathrm{PSL}_2(\mathbb{C})$ quotients. Let G be a finite group with conjugacy classes **C**. Consider one $\mathbb{Q}$ regular $(G, \mathbf{C})$ realization of a group by a cover $\phi : X \to \mathbb{P}^1_x$. This produces infinitely many $\mathbb{Q}$ regular realizations; compose ϕ with any $\alpha \in \mathrm{PSL}_2(\mathbb{Q})$. Thus, to measure $(^k_p\tilde{G}, \mathbf{C})$ realizations requires counting $\mathbb{Q}$ points on reduced Hurwitz spaces $\{\mathcal{H}(^k_p\tilde{G}, \mathbf{C})^{\mathrm{rd}} = \mathcal{H}^{\mathrm{rd}}_k\}^\infty_{k=0}$. These spaces have natural maps

$$\cdots \to \mathcal{H}^{\mathrm{rd}}_{k+1} \to \mathcal{H}^{\mathrm{rd}}_k \to \cdots \to \mathcal{H}^{\mathrm{rd}}_0 \to J_r.$$

App.II gives the definition of J_r. Since **C** is a rational union, these spaces and maps have field of definition $\mathbb{Q}$. Investigating this tower requires information on absolutely irreducible components of $\mathcal{H}^{\mathrm{rd}}_k$ for each level k. We explain more of the group theory behind these diophantine problems.

Statement 1.3 says a $\mathbb{Q}$ regular $(^{k+1}_p\tilde{G}, \mathbf{C})$ realization produces a $\mathbb{Q}$ $(^k_p\tilde{G}, \mathbf{C})$ realization. It also suggests these groups are similar, for example, when G is perfect and centerless. Lifting Lemma 4.1 adds to this. It characterizes $_p\tilde{G}$ quotients giving a cover $\psi : H \to G$ as having the following properties.

(1.3a) The kernel of ψ has p-group kernel.

(1.3b) For $\mathbf{g}' = \{g'_1, \ldots, g'_s\} \in H$ elements of order prime to p, $\langle \mathbf{g}' \rangle = H$ if and only if $\langle \psi(\mathbf{g}') \rangle = G$.

Here is a more subtle similarity between H and G satisfying (1.3). Brauer's Theorem [MT, §II.B] says there as many simple $\bar{\mathbb{F}}_p$ modules as there are p-regular conjugacy classes. Therefore, the two groups have the same simple $\bar{\mathbb{F}}_p$ modules. §3 shows the significance of an appearance of the module **1** (for $^k_p\tilde{G}$) in the Loewy display of the $\mathbb{F}_p$ module $\ker_k / \ker_{k+1}$. It is the whole story of obstructed components.

Fix r_0 and a field K. Theorem 4.4 says the following are equivalent.

(1.4a) There are K regular $^k_p\tilde{G}$ realizations with at most r_0 branch points for each $k \geq 0$.

(1.4b) For some $r \leq r_0$ there is an r-tuple **C** of p-regular conjugacy classes of G and for each $k \geq 0$, there are K regular $(^k_p\tilde{G}, \mathbf{C})$ realizations.

The Main Conjecture reduces to showing (1.4b) is impossible when K is a number field for each r-tuple $\mathbf{C}$ of p-regular conjugacy classes of G. Proving the Main Conjecture comes to considering rational points on reduced Hurwitz spaces ($\mathrm{PSL}_2(\mathbb{C})$ quotients), curves when $r_0 = 4$ (App.II).

Suppose you show (0.6a) (for large k). Faltings' Theorem [Fa] applies if (0.6b) holds: the genus of the components at level k goes up with k. [FrK] gives evidence for this rise in genus from our special case.

§1.D. Projective systems of rational points. Suppose the program of §1.C works. Then, there are only finitely many $\mathbb{Q}$ points on some level k_0 of Modular Towers satisfying the hypotheses. It is, however, a formidable task to eliminate the possibility of a nontrivial set of $\mathbb{Q}$ points at every level. This is where hypothesis (0.6c) enters. It assumes there are no obstructed components at level k_0 or beyond. Then, having points at each level above k_0 produces a *projective system* of rational points $\{\mathfrak{p}_k \in \mathcal{H}(_p^k \tilde{G}, \mathbf{C})^{\mathrm{rd}}\}_{k=0}^\infty$. That is, the canonical map $_p^{k+1}\tilde{G} \to {}_p^k\tilde{G}$ inducing $\mathcal{H}(_p^{k+1}\tilde{G}, \mathbf{C})^{\mathrm{rd}} \to \mathcal{H}(_p^k\tilde{G}, \mathbf{C})^{\mathrm{rd}}$ gives $\mathfrak{p}_{k+1} \mapsto \mathfrak{p}_k$. Precluding projective systems is the last leg for $r = r_0 = 4$. To do so for any value of r is a separate problem independent of other diophantine considerations.

Projective System Conjecture 1.5. *Assume the setup for the Modular Tower above associated to $(G, \mathbf{C})$, with r arbitrary and K finitely generated over a number field. Then, there exists no projective system of K rational points on the Modular Tower.*

§2. Construction of universal Frattini covers.

[FrK] expands the Normalizer Observation of [MT, Remark 2.10]. It helps decipher the whole universal p-Frattini cover from representation observations on the original group G. Thus, representation facts about G translate to a presentation of the universal p-Frattini cover of G, and so of its infinite string of characteristic quotients. This section presents two preliminary results in this direction.

§2.A. Relation of $_p\tilde{G}$ to $_p\tilde{H}$ with $H \leq G$. Let $P = P_p = P_{G,p}$ be a p-Sylow subgroup of G. The notation $\ker_0(G)$ allows simultaneously distinguishing the characteristic kernels of the universal p-Frattini covers for G and for a subgroup H. Use $\phi_G : {}_p\tilde{G} \to G$ for the canonical map. Also, denote $\ker_i(G)/\ker_{i+1}(G)$ by $M_i(G)$. For any subgroup H of G, restricting H to $M_i(G)$ gives an H module, $M_i(G)_H$. The *rank* of any profinite group G is the smallest number of elements that topologically generate it. Finally, $\tilde{P}_{G,p} = \phi_G^{-1}(P_{G,p})$ is the p-Sylow subgroup of $_p\tilde{G}$.

Example 2.1. *The p-Sylow Frattini hull.* Let G be any finite group. Suppose P_p has rank u. The free pro-p group on u generators is ${}_p\tilde{F}_u$. Let $N_G(P_p)$ be the normalizer of P_p in G. Then, the universal p-Frattini cover of $N_G(P_p)$ is ${}_p\tilde{F}_u \times^s N_G(P_p)/P_p$ [Ri, Th. 3.2]. This a special case of Principle 2.3 (below) by extending the action of $N = N_G(P_p)/P_p$ to ${}_p\tilde{F}_u$. Since ${}_p\tilde{F}_u$ is pro-free, there is a subgroup B of $\text{Aut}({}_p\tilde{F}_u)$ mapping surjectively to N.

The essential point is that the kernel of the map $B \to N$ is a pro-p-group. (Prove this by inductively considering the characteristic Frattini quotients of ${}_p\tilde{F}_u$.) Apply Schur-Zassenhaus to split off a copy of N. Thus produce the action of N on ${}_p\tilde{F}_u$ [FrJ, Lemma 20.45]. Important point: Action of N is unique only up to conjugation by an automorphism of ${}_p\tilde{F}_u$.   □

Let $\ker_0^*$ be the kernel of ${}_p\tilde{F}_u \to P_p$ as in Ex. 2.1. For Ex. 2.1, the action of $N_G(P_p)$ on $\ker_0^* / \ker_1^*$ extends to G.

Example 2.2. *The p-Sylow Frattini hull can be all of ${}_p\tilde{G}$.* Take $G = A_5$, $p = 2$ and $N_G(P_p) = A_4 \le A_5$. Then, we can identify $\ker_0^*$ as the kernel of the universal 2-Frattini of A_5 [MT, Prop. 2.9]. When, however, $p = 3$ or 5 (and $G = A_5$) $\ker_0^* / \ker_1^*$ is cyclic, and not an A_5 module. In particular, $\ker_0^*$ for these values of p isn't the kernel of ${}_p\tilde{G} \to G$.   □

[FrK] generalizes the observation of Ex. 2.2. Principle 2.3 characterizes the universal p-Frattini cover ${}_p\tilde{G}$ of G to relate it to that of its subgroups.

Subgroup Frattini Principle 2.3. Let G be any profinite group and p any prime dividing $|G|$. Then ${}_p\tilde{G}$ is the smallest group that covers G and has its p-Sylow subgroup pro-free.

Suppose $H \le G$ are profinite groups. Then, there is a surjective pro-p group homomorphism $\rho_{G,H} : \ker_0(G) \to \ker_0(H)$. Further, $\rho_{G,H}$ is unique up composition on the left with an automorphism of $\ker_0(G)$. This induces a surjective H module morphism $\rho_{G,H}^* : M_0(G) \to M_0(H)$. Finally, if G is a finite group, the following are equivalent:

(2.1a) $\rho_{G,H}^*$ is an isomorphism.

(2.1b) $\dim_{\mathbb{F}_p}(M_0(G)) = \dim_{\mathbb{F}_p}(M_0(H))$.

(2.1c) $\phi_G^{-1}(H) = {}_p\tilde{H}$.

Proof. First consider the opening characterization of ${}_p\tilde{G}$. For G any profinite group, a simple property determines the universal Frattini cover of $\tilde{G}$. It is the minimal projective cover of G [FrJ, Prop. 20.33]. A profinite group is projective if and only if each p-Sylow subgroup of it is pro-free [FrJ, Prop. 20.47]. The kernel of ϕ_G is a nilpotent group [FrJ, Prop. 20.44]. So $\tilde{G}$ is the fiber product over G of closed subgroups ${}_p\tilde{G}$ for each prime $p \mid |G|$.

Suppose ${}_p\phi^* : {}_p\tilde{G}^* \to G$ is any profinite group cover of G for which ${}_p\tilde{G}^*$

has a pro-free p-Sylow subgroup. Consider the fiber product over G of the groups $_p\tilde{G}^*$ and $_{p'}\tilde{G}$ for all primes p', $p' \neq p$, dividing $|G|$. Call this fiber product $\tilde{G}^*$. From the Sylow subgroup characterization of projective groups, $\tilde{G}^*$ is projective. Also, it is a minimal projective cover of G if and only if $_p\tilde{G}^*$ is a minimal cover of G having a pro-free p-Sylow. This concludes the opening characterization of $_p\tilde{G}$.

The remainder of the proof concentrates on universal p-Frattini covers. All notation (say, for $\ker_i$) refers to the fixed prime p. Now consider $H \leq G$ profinite groups. Then, $\tilde{H}^* = {}_p\phi_G^{-1}(H)$ is a closed subgroup of $_p\tilde{G}$. The p-Sylow subgroup of $\tilde{H}^*$ is a closed subgroup of the p-Sylow of $_p\tilde{G}$. Thus, it is projective and therefore pro-free.

Apply the characterization above for the universal p-Frattini cover of H. Conclude there is a surjective map $\beta : \tilde{H}^* \to {}_p\tilde{H}$ commuting with the natural surjective maps to H. The image of β is a closed subgroup of $_p\tilde{H}$ mapping surjectively to H. Since $_p\tilde{H}$ is a Frattini cover of H, β is surjective. So, it induces the surjective map $\rho_{G,H} : \ker_0(G) \to \ker_0(H)$ in the statement. The rest follows because the 1st characteristic Frattini quotient determines the rank of a pro-free p-group. □

§2.B. Decomposing Frattini cover kernels.
The following is a special case of [Be, Exer. 1, p. 11]. We fill in details, isolating it from its generalities.

Indecomposability Lemma 2.4. *The $_p\tilde{G}/\ker_k(G)$ module $M_k(G)$ is indecomposable.*

Proof. Let M be any G module. Define $\Omega(M)$ to be the kernel of a surjective G module homomorphism $\psi : P \to M$ with P a projective module, minimal for direct sum decomposition. Suppose P' is another module with these properties. Then, projectivity of the two modules gives maps $\alpha : P \to P'$ and $\beta : P' \to P$, each commuting with the surjective maps from P and P' to M. Define $f = \alpha \circ \beta$ and the composition of f, n times, by f_n. Similarly, for g_n with g the composition $\beta \circ \alpha$. If n is large, then $P = \ker(g_n) \oplus \mathrm{Im}(g_n)$. Note: $\ker(g_n)$, being a summand of a projective module, is also projective.

Suppose P and P' aren't isomorphic. Then $\ker(g_n)$ goes to 0 under the map from P to M. This contradicts the minimality of P, and proves that $\Omega(M)$ is well defined. Further, from Schanuel's Lemma [Be, Lemma 1.5.3], up to projective summands $\Omega(M)$ is well defined. Similarly, there is the operator $\Omega^{-1}(M)$: the cokernel of the embedding of M in a minimal injective module.

In the category of finite dimensional modules over group rings, projectives and injectives are the same [Be, Prop. 1.6.2]. Thus, P above is both the minimal projective covering M and an injective module containing $\Omega(M)$.

That is, up to addition of a projective module, $\Omega^{-1}(\Omega(M)) = M$. So, M is indecomposable, if and only if $\Omega(M)$ is. Apply Ω twice to the indecomposable module $M = \mathbf{1}_{k \tilde{G} \atop p}$. From [MT, Proj. Indecomp. Lemma 2.3], $\ker_k / \ker_{k+1} = \Omega^2(M)$. So, $\Omega^2(M)$ is indecomposable.   □

Consider any cover $\phi : H \to G$ with abelian kernel M. Then G acts on M: $g \in G$ maps $m \in M$ to $\bar{g}m\bar{g}^{-1}$ where $\bar{g}$ is any lift of g to H.

Remark 2.5. *Frattini kernels may decompose.* Let M_1 and M_2 be nonisomorphic simple modules appearing as respective kernels for nonsplit extensions G_1 and G_2 of G. Consider the fiber product $H = G_1 \times_G G_2$ and the natural map $\phi_{1,2} : H \to G$. We show $\phi_{1,2}$ is a Frattini cover.

Suppose H_1 is a subgroup of H mapping surjectively to G. As $G_1 \to G$ and $G_2 \to G$ are Frattini covers, H_1 factors surjectively through them both. Thus, by the Jordan-Hölder Theorem, the composition series of the kernel of $H_1 \to G$ includes M_1 and M_2. It must therefore be $M_1 \oplus M_2$.

Two series of simple groups agree when $n = 8$: $A_8 \cong \mathrm{SL}(4, \mathbb{Z}/2)$. Let M_4 be the standard 4 dimensional representation of $\mathrm{SL}(4, \mathbb{Z}/2)$. [Be] shows $H^2(A_8, M_4)$ has dimension 1. This gives a Frattini extension of A_8 not factoring through the universal central extension of A_8. It is an example of the above.   □

§3. Progress on the case A_5 and $\mathbf{C} = \mathbf{C}_{3^r}$.

The main example of [FrK] is the A_5 Modular Tower associated with $r = 4$, $p = 2$, and all conjugacy classes those of the 3-cycle in A_5: $\mathbf{C} = \mathbf{C}_{3^4}$. Denote the corresponding reduced Modular Tower (of inner spaces) by

$$(3.1) \qquad \cdots \to \mathcal{H}(\textstyle\frac{k+1}{2}\tilde{A}_5, \mathbf{C})^{\mathrm{rd}} \to \mathcal{H}(\textstyle\frac{k}{2}\tilde{A}_5, \mathbf{C})^{\mathrm{rd}} \to \cdots \to \mathcal{H}(A_5, \mathbf{C})^{\mathrm{rd}}.$$

This example reveals major phenomena that don't occur for modular curves.

§3.A. Illustration of obstructed components.
Obstructed components produce both encouraging and discouraging diophantine results. First we illustrate the former. Actual examples go precisely as the outline in §0.C. The proof of the next result explains the variety $\mathcal{H}(G, \mathbf{C})^{\mathrm{abs}}$.

Theorem 3.1. ([Fr1]) *Assume $n \geq 5$ is odd. Then, $\mathcal{H}(A_n, \mathbf{C}_{3^{n-1}})^{\mathrm{rd}}(\mathbb{Q})$ is dense in $\mathcal{H}(A_n, \mathbf{C}_{3^{n-1}})^{\mathrm{rd}}$. Now assume $n \geq 6$ is even. Then, there are no $\mathbb{Q}$ regular $(\frac{1}{2}\tilde{A}_n, , \mathbf{C}_{3^{n-1}})$ realizations.*

Further, there is a $\mathbb{Q}$ unramified cover $\mathcal{H}(G, \mathbf{C})^{in} \to \mathcal{H}(G, \mathbf{C})^{\mathrm{abs}}$, Galois with group $\mathbb{Z}/2$ with the following properties.

(3.2a) *Each $\mathfrak{p} \in \mathcal{H}(G, \mathbf{C})^{\mathrm{abs}}$ corresponds to a degree n cover $\phi_{\mathfrak{p}} : X \to \mathbb{P}^1_x$ whose Galois closure represents $\mathfrak{p}' \in \mathcal{H}(G, \mathbf{C})^{in}$ lying over $\mathfrak{p}$.*

(3.2b) *For* $\mathfrak{p} \in \mathcal{H}(G, \mathbf{C})^{\mathrm{abs}}$, $\phi_{\mathfrak{p}} : X \to \mathbb{P}^1_x$ *has (minimal) field of definition* $\mathbb{Q}(\mathfrak{p})$. *Its Galois closure is defined over* $\mathbb{Q}(\mathfrak{p}')$ *with* $\mathfrak{p}'$ *lying over* $\mathfrak{p}$.

(3.2c) $\mathfrak{p} \in \mathcal{H}(G, \mathbf{C})^{\mathrm{abs}}$ *corresponds to a* $\mathbb{Q}(\mathfrak{p})$ *regular* $(A_n, \mathbf{C}_{3^{n-1}})$ *realization if and only if* $\mathbb{Q}(\mathfrak{p}') = \mathbb{Q}(\mathfrak{p})$.

For all $n \geq 5$, *the set* $\{\mathfrak{p} \in \mathcal{H}(G, \mathbf{C})^{\mathrm{abs}}(\bar{\mathbb{Q}})$ *with* $\mathbb{Q}(\mathfrak{p}') \neq \mathbb{Q}(\mathfrak{p})\}$ *is dense. For* $n = 5$, *even* $\{\mathfrak{p} \in \mathcal{H}(G, \mathbf{C})^{\mathrm{abs}}(\mathbb{Q})$ *with* $\mathbb{Q}(\mathfrak{p}') \neq \mathbb{Q}(\mathfrak{p})\}$ *is dense.*

Proof. This example compares the group theoretic Hurwitz space ideas, with explicit equation calculation. Hurwitz space computations are explicit group theory computations. It's more than a matter of taste, for we are after *properties* of the spaces, not their equations. Other examples of this are in [Fr3] and [Fr4]. The remaining proof has four parts, starting from statements relating inner Hurwitz spaces to absolute Hurwitz spaces. This lets us focus on the effect of Mestre's calculation for n odd, distinguishing this from the more mysterious case when n is even. Note: If $\mathbb{Q}(\mathfrak{p}') \neq \mathbb{Q}(\mathfrak{p})$ (notation as in (3.2c)), then the Galois closure of $\phi_{\mathfrak{p}} : X \to \mathbb{P}^1_x$ in (3.2b) has group S_n (see below).

Part 1: Absolute Nielsen classes. For $k = 0$, a special case of [Fr1] shows there is one B_{n-1} orbit on $\mathrm{Ni}(A_n, \mathbf{C}_{3^{n-1}})$. Another Hurwitz space is handy for this problem. It is closest to Mestre's computations [Me] below.

For any $G \leq S_n$, consider the normalizer $N_{S_n}(G)$ of G in S_n. Let $N(\mathbf{C}) = N_{S_n}(G, \mathbf{C})$ be the subgroup of $N_{S_n}(G)$ whose elements conjugate the entries of $\mathbf{C}$ among themselves. We also use $G(1)$, the stabilizer of 1 in this degree n representation. Form the natural quotient $\mathrm{Ni}(G, \mathbf{C})/N(\mathbf{C}) = \mathrm{Ni}(G, \mathbf{C})^{\mathrm{abs}}$. Then, B_{n-1}, and its Hurwitz monodromy group $H_{n-1} = \pi_1(\mathbb{P}^{n-1} \setminus D_{n-1})$ quotient, act on $\mathrm{Ni}(G, \mathbf{C})^{\mathrm{abs}}$. Covering space theory produces a sequence of unramified covers

(3.3) $\mathcal{H}(G, \mathbf{C}) = \mathcal{H}(G, \mathbf{C})^{in} \to \mathcal{H}(G, \mathbf{C})^{\mathrm{abs}} \to \mathbb{P}^{n-1} \setminus D_{n-1}$.

As in [Fr2], [FrV] and [V], interpret $\mathcal{H}(G, \mathbf{C})^{\mathrm{abs}}$ as a moduli space of equivalence classes of degree n covers. For $\mathfrak{p} \in \mathcal{H}(G, \mathbf{C})^{\mathrm{abs}}$, let $\phi_{\mathfrak{p}} : X_{\mathfrak{p}} \to \mathbb{P}^1_x$ be a representing cover. Cover equivalence here is (II.1a) (in App.II). This cover has the following properties.

(3.4a) The Galois closure $\hat{X} \to \mathbb{P}^1_x$ of $\phi_{\mathfrak{p}}$ represents $\mathfrak{p}' \in \mathcal{H}(G, \mathbf{C})^{in}$ over $\mathfrak{p}$.

(3.4b) X is the quotient of $\hat{X}$ by $G(1)$.

Suppose $\mathcal{H}^{in}_*$ is a connected component of $\mathcal{H}^{in}$ and $\mathcal{H}^{\mathrm{abs}}_*$ is the image of $\mathcal{H}^{in}_*$ in $\mathcal{H}^{\mathrm{abs}}$. Then, $\mathcal{H}^{in}_*$ corresponds to an orbit O^{in}_1 of B_r on $\mathrm{Ni}(G, \mathbf{C})$. The natural image O^{abs}_1 of O^{in}_1 in $\mathrm{Ni}(G, \mathbf{C})^{\mathrm{abs}}$ corresponds to $\mathcal{H}^{\mathrm{abs}}_*$. The following set is actually a group:

$$\mathcal{G}(\mathbf{C}) = \{h \in N(\mathbf{C}) \mid \exists\, Q \in B_{n-1} \text{ and } \mathfrak{g} \in O^{in}_1 \text{ with } (\mathfrak{g})Q = h\mathfrak{g}h^{-1}\}.$$

The cover $\mathcal{H}(G,\mathbf{C})^{in} \to \mathcal{H}(G,\mathbf{C})^{abs}$ on the left of (3.3) is Galois with group $\mathcal{G}(\mathbf{C})/G$. Note: [MT, Lemma 3.3] says $G \leq \mathcal{G}(\mathbf{C})$.

Part 2: Using $\mathcal{G}(\mathbf{C}_{3^{n-1}})/G \equiv \mathbb{Z}/2$. With $A_n \leq S_n$ the standard representation of the alternating group, $N(\mathbf{C}) = S_n$. Thus, $\mathcal{G}(\mathbf{C})$ is a subgroup of $S_n/A_n = \mathbb{Z}/2$. Also, $\mathcal{H}(G,\mathbf{C})^{abs}$ is the moduli space of degree n covers with $n-1$ 3-cycles as branch cycles. From the Riemann-Hurwitz formula, this is a moduli space of genus 0 covers. Further, as in [Fr2], since the subgroup $G(1) = A_n(1)$ is self normalizing in A_n, $\mathcal{H}(G,\mathbf{C})^{abs}$ is a *fine* moduli space. In particular, $\mathfrak{p} \in \mathcal{H}(G,\mathbf{C})^{abs}(K)$ corresponds to a K cover $\phi_{\mathfrak{p}} : X_{\mathfrak{p}} \to \mathbb{P}_x^1$.

[Fr1] says B_{n-1} is transitive on $\mathrm{Ni}(A_n,\mathbf{C}_{3^{n-1}})$. Thus, $\mathcal{G}(\mathbf{C}_{3^{n-1}}) = \mathcal{G} = \mathbb{Z}/2$. An example gives the sense of why this holds. Take $n = 5$. From the transitivity result, $\mathcal{G}/G = \mathbb{Z}/2$ if and only if there exists $\mathfrak{g} \in \mathrm{Ni}(A_n,\mathbf{C}_{3^{n-1}})$ and $Q \in B_4$ with $(\mathfrak{g})Q = (4\,5)\mathfrak{g}(4\,5)$. Take

$$\mathfrak{g} = ((1\,2\,3),(1\,2\,3)^{-1},(1\,4\,5),(1\,4\,5)^{-1}).$$

Here applying Q_3 (App.I) has the same affect on $\mathfrak{g}$ as conjugating by $(4\,5)$.

Apply Hilbert's irreducibility theorem to the sequence of covers in (3.3). This says a dense set of points $\mathfrak{x} \in \mathbb{P}^{n-1}(\mathbb{Q})$ have above them $\mathfrak{p}' \in \mathcal{H}(G,\mathbf{C})^{in}$ lying over $\mathfrak{p} \in \mathcal{H}(G,\mathbf{C})^{abs}$ for which $\mathbb{Q}(\mathfrak{p}') \neq \mathbb{Q}(\mathfrak{p})$. Excluding the statement about $\mathbb{Q}$ points, this completes the proof.

Part 3: Interpreting Mestre's calculation. Continue the discussion at the beginning of Part 2. Here is the first place where there is a distinction between n even and odd. If $\phi_{\mathfrak{p}}$ has odd degree, then $X_{\mathfrak{p}}$ has a rational point over $\mathbb{Q}(\mathfrak{p})$. In particular, $\phi_{\mathfrak{p}}$ is $\mathbb{Q}(\mathfrak{p})$ equivalent (in the sense of (II.1a) in App.II) to a cover $\mathbb{P}_y^1 \to \mathbb{P}_x^1$. [DFr1, p. 115] and [DFr2, p. 115] discuss whether this means a space like $\mathcal{H}_{n,n-1}^{abs}$ is a *family of rational functions*. That would mean, there is a natural map $\Psi^{abs} : \mathcal{H}_{n,n-1}^{abs} \times \mathbb{P}_y^1 \to \mathcal{H}_{n,n-1}^{abs} \times \mathbb{P}_x^1$. Also, for any $\mathfrak{p} \in \mathcal{H}_{n,n-1}^{abs}$, $\Psi_{\mathfrak{p}}$ is a rational function cover representing $\mathfrak{p}$. We don't think there is such a map Ψ^{abs}, though we haven't excluded it.

Still, as in [DFr1, 2], if you add the branch points of the cover to the moduli space, you nearly have coordinates for a family of rational functions. When n is odd that is exactly what [Me] does. In fact, he gives a dense set of rational points in $\mathcal{H}_{n,n-1}^{in}$. Here is how it goes [Se, p. 100–101].

Let $P(y) = \prod_{i=1}^{n}(y - a_i)$ with the a_i algebraically independent indeterminates over $\mathbb{Q}$. If $Q(y)$ is a polynomial of degree $n-1$, then Q/P maps ∞ to 0. We look to choose those polynomials Q so the derivative of $Q(y)/P(y)$ is a square $(R(y)/P(y))^2$. If Q and R are suitably generic, that means the cover from Q/P has 3-cycles as branch cycles. It is then in the Nielsen class of our interest. Here are the expressions for Q and R:

$$(3.5a) \qquad Q(y)/P(y) = \sum_{i=1}^{n} -c_i^2/(y-a_i), \qquad R/P = \sum_{i=1}^{n} c_i/(y-a_i)$$

where the c_i s satisfy

$$(3.5b) \qquad \sum_{\substack{j=1 \\ j \neq i}} c_j/(a_i - a_j) = 0 \text{ for all } i.$$

This is the second place using n odd. The matrix with $i \times j$ entry $1/(a_i - a_j)$ for $i \neq j$ and 0 for $i = j$ is skew-symmetric and $n \times n$. Thus, for n odd its determinant is 0. So, there is a line of solutions for the vector of c_i s.

[Fr1] says $\mathcal{H}(A_n, \mathbf{C}_{3^{n-1}})^{in}$ is an absolutely irreducible $\mathbb{Q}$ variety. So, Mestre's collection of rational functions shows $\mathcal{H}^{abs}$ is a unirational variety ([Me] when n is odd; [Se, p. 100-101] asserts a variant works for even n, too). So it gives a $\mathbb{Q}$ dense subset of $\mathcal{H}^{abs}$. These points are the image of $\mathbb{Q}$ points of $\mathcal{H}^{in}$. This is a dense set of $\mathfrak{p} \in \mathcal{H}(G, \mathbf{C})^{abs}(\mathbb{Q})$ for which $\mathbb{Q}(\mathfrak{p}') = \mathbb{Q}(\mathfrak{p})$. When $n = 5$, [Fr5, Thm. 5.9] shows there is a dense set of $\mathfrak{p} \in \mathcal{H}(G, \mathbf{C})^{abs}(\mathbb{Q})$ for which $\mathbb{Q}(\mathfrak{p}') \neq \mathbb{Q}(\mathfrak{p})$.

Part 4: For n even, $\mathcal{H}(A_n, \mathbf{C}_{3^{n-1}})$ is obstructed. There is nothing in the Nielsen class $\text{Ni}(\frac{1}{2}\tilde{A}_n, \mathbf{C}_{3^{n-1}})$ when n is even. It goes like this. Let $\hat{A}_n$ be the universal central exponent 2 extension of A_n. Then, $\frac{1}{2}\tilde{A}_n \to A_n$ factors through $\hat{A}_n \to A_n$. Suppose $\frac{1}{2}\tilde{\mathfrak{g}} \in \text{Ni}(\frac{1}{2}\tilde{A}_n, \mathbf{C})$. The canonical map $\frac{1}{2}\tilde{A}_n \to \hat{A}_n$ sends $\frac{1}{2}\tilde{\mathfrak{g}}$ to $\hat{\mathfrak{g}} \in \hat{A}_n^r$ with $\Pi(\hat{\mathfrak{g}}) = 1$. Let $\mathfrak{g}$ be the image of $\hat{\mathfrak{g}}$ in $\text{Ni}(A_n, \mathbf{C}_{3^{n-1}})$. Again use transitivity of B_{n-1} on $\text{Ni}(A_n, \mathbf{C}_{3^{n-1}})$. So, if one element of $\text{Ni}(A_n, \mathbf{C}_{3^{n-1}})$ lifts to an element of $\text{Ni}(\hat{A}_n, \mathbf{C}_{3^{n-1}})$, then any element does. On the other hand, [MT, Ex. III.12] explicitly gives $\mathfrak{g} \in \text{Ni}(A_n, \mathbf{C}_{3^{n-1}})$ with no lift to $\text{Ni}(\hat{A}_n, \mathbf{C}_{3^{n-1}})$. Thus, level one of the Modular Tower for $(A_n, \mathbf{C}_{3^{n-1}})$ is an empty variety. $\qquad \square$

§3.B. Invariants for obstructed components.

Obstructed components and pure group theory establish Modular Tower Conjecture 1.4 for $(A_n, \mathbf{C}_{3^r})$ with even $n \geq 6$ and $r = n - 1$. [Fr1], however, shows a different outcome for the cases (n, r), $r \geq n$ that Theorem 3.1 doesn't cover. There are unobstructed components in all levels of the Modular Tower.

Lemma 3.2 is a completely general result applying to all levels of any Modular Tower. As in §0.A, with $\mathfrak{g} = (g_1, \ldots, g_r) \in G^r$ denote the product $g_1 \cdots g_r$ by $\Pi(\mathfrak{g})$. Let G be a finite group. Suppose $\phi : H \to G$ is a central extension. Then, any $g \in G$ of order prime to $|\ker(\phi)|$, has a *unique* lift $\hat{g}$ to an element of H of the same order as G. Assume $\ker(\phi)$ has order prime to the orders of all entries of $\mathfrak{g} = (g_1, \ldots, g_r) \in G^r$. For $\Pi(\mathfrak{g}) = 1$ let $s(\mathfrak{g})$ be $\Pi(\hat{g}_1 \cdots \hat{g}_r)$. In Lemma 3.2, the exact sequence has $\mathbf{1}$ as kernel. Here $\mathbf{1}$ is the identity module for the action of $^k_p\tilde{G}$.

Obstruction Lemma 3.2. *Suppose O is a B_r orbit in $\mathfrak{g} \in \mathrm{Ni}(^k_p\tilde{G}, \mathbf{C})$. Let*

$$S(O) = \{\Pi(\hat{\mathfrak{g}}) \mid \hat{\mathfrak{g}} \in \,^{k+1}_p\tilde{G}, \ \hat{\mathfrak{g}} \in \mathbf{C} \text{ and } \hat{\mathfrak{g}} \bmod \ker_k = \mathfrak{g}\}.$$

This is a union of conjugacy classes in $^{k+1}_p\tilde{G}$. Then, $\mathbf{1} \notin S(O)$ exactly if there exists a sequence of covers

$$^{k+1}_p\tilde{G} \to H_2 \to H_1 \to \,^k_p\tilde{G}$$

with the following properties.
(3.6a) The kernel of $H_2 \to H_1$ is $\mathbf{1}$.
(3.6b) There exists $\mathfrak{g}^* \in \mathrm{Ni}(H_1, \mathbf{C})$, $\mathfrak{g}^* \bmod \ker_k = \mathfrak{g}$ and $s(\mathfrak{g}^*) \neq 0$.

Suppose $\mathcal{H}_O$ is the level k component of $\mathcal{H}_k$ corresponding to O. The previous condition holds exactly when $\mathcal{H}_O$ is obstructed.

Proof. Use induction on the Loewy layers of $\ker_k / \ker_{k+1}$. Replace $\mathbf{1}$ in the kernel in $H_2 \to H_1$ by an irreducible module $A \neq \mathbf{1}$. Consider the set S' of $\Pi(\mathfrak{g}')$ as $\mathfrak{g}'$ runs over all allowable lifts of r-tuples $\mathfrak{g} \in H_2^r$ with $\Pi(\mathfrak{g}) = 1$. Then $S' = A$. Here is the argument, using that the set S' is a braid invariant set. That is, let $\tilde{\mathfrak{g}}^0$ be one lift. In the orbit, you can braid $\mathfrak{g}^0$ to something whose lift has the same product, but in the braid g_i^0 appears on the right side. Now form a lift replacing g_i^0 by $ag_i^0a^{-1} = g_i^0 a^{g_i^0} a^{-1}$. So, this lift gives $\tilde{\mathfrak{g}}^0 a^{g_i^0} a^{-1}$. You can do this for any i and any $a \in A$. If $A \neq \mathbf{1}$ is irreducible, the possible products of lifts gives $\tilde{\mathfrak{g}}^0 a$ with $a \in A$ arbitrary. So, the corresponding Nielsen class is empty only if $A = \mathbf{1}$. □

Schur Multipliers Result 3.3. *Assume $n \geq 4$. For each $k \geq 0, ^k_2\tilde{A}_n$ has a nontrivial Schur multiplier. Generally, suppose $^k_p\tilde{G}$ has a nontrivial Schur multiplier. That is, there is a sequence $^{k+1}_p\tilde{G} \to H_1 \to \,^k_p\tilde{G}$ with $\mathbf{1} \to H_1 \to \,^k_p\tilde{G}$ short exact as in Obstruction Lemma 3.2. Then, $[\ker_k, H_1]H_1^p$ generates a proper closed subgroup H_2 of $\ker_{k+1}$. Also, $^{k+1}_p\tilde{G}$ has trivial action on $M = \ker_{k+1}/H_2$, producing a nontrivial Schur multiplier for $^{k+1}_p\tilde{G}$.*

Proof. As is well-known [MT, §II.C], the exponent 2-part of the Schur multiplier of A_n ($n \geq 4$) is $\mathbb{Z}/2$. Thus, the statement on the alternating groups follows from the general inductive statement. Since $\ker_k$ is a pro-free group, H_2 is a proper subgroup of $\ker_{k+1}$. The action of $^{k+1}_p\tilde{G}$ on M is trivial if it is trivial on generators of M. One type of generator is v^p with $v \in \ker_k$. Conjugating v by $g \in \,^{k+1}_p\tilde{G}$ gives vh for some $h \in H_1$. Modulo

$H_1^p[\ker_k, H_1]$, set $h^p = 1$ for $h \in H_1$ and $vh = hv$ for $v \in \ker_k$ and $h \in H_1$. Thus, the following hold.

(3.7a) $gv^p g^{-1} = (vh)^p = v^p \bmod H_1^p[\ker_k, H_1]$.

(3.7b) $g[v, v']g^{-1} = [vh, v'h'] = [v, v'] \bmod H_1^p[\ker_k, H_1]$.   □

§4. When $p|N_{\mathbf{C}}$.

Let G be a finite group and let $\mathbf{C}$ be a collection of conjugacy classes in G. Theorem 4.4 assumes Modular Tower Conjecture 1.4 in the form (1.2a) holds. From this it concludes Conjecture 1.4 is true. The core of the proof starts by assuming $\mathbf{C}$ contains at least one conjugacy class that isn't p-regular. It then shows, for k large, lifts of $\mathbf{C}$ to conjugacy classes in $_p^k\tilde{G}$ can't be a rational union. The gist of this is it suffices to establish our major conjectures when $(p, N_{\mathbf{C}}) = 1$.

§4.A. Lifting elements of order p.

Continue notation from §2.A: $\phi = \phi_G : {}_p\tilde{G} \to G$ is the canonical map having kernel $\ker_0(G, p) = \ker_0(G)$.

Lifting Lemma 4.1. *The following are equivalent.*

(4.1a) $\mathbf{C}$ *consists of p-regular conjugacy classes.*

(4.1b) *For each $k \geq 1$, classes from $\mathbf{C}$ lift uniquely to classes $_p^k\tilde{\mathbf{C}}$ of the same order in $_p^k\tilde{G}$.*

(4.1c) (4.1b) *holds for $k = 1$.*

Further, let $\mathfrak{g}$ any set of generators of G, with each having order prime to p. Let $\alpha : H \to G$ be a cover of G with p-group as kernel. Then, α is a Frattini cover if and only if lifts of $\mathfrak{g}$ to elements $\tilde{\mathfrak{g}}$ of the same order in H implies $\langle\tilde{\mathfrak{g}}\rangle = H$.

Proof. Let g be in a conjugacy class from $\mathbf{C}$. Apply Schur-Zassenhaus [MT, Intro. to Part III] to the sequence

$$\ker_0(G, p) \to \phi^{-1}(\langle g\rangle) \to \langle g\rangle.$$

This shows (4.1a) implies (4.1b).

For the converse, assume p divides the order of g. Suppose g lifts to $^k\tilde{g} \in {}_p^k\tilde{G}$ of the same order, for each k. Then, so does g^a with a the order of g divided by p. The result follows if we show lifts of g to higher characteristic quotients must increase their order when g has order exactly p. A p-Sylow $\tilde{P}_G$ of $_p\tilde{G}$ is a pro-free p-group. The projective limit $\lim_{\infty\leftarrow k} {}^k\tilde{g}$ is an element of order p in the pro-free group $\tilde{P}_G$. Nontrivial projective profinite groups have no elements of finite order [FrJ, Cor. 20.14]. So, this is impossible.

Equivalence of (4.1a) and (4.1b) with (4.1c) follows by showing g, of order p, lifts to no element of order p in $\frac{1}{p}\tilde{G}$. Assume $P = P_G$ is a p-Sylow of G containing g. Let $\tilde{F}_P$ be the pro-free p-group of the same rank as P. Choose a surjective map $\alpha : \tilde{F}_P \to P$. Then, $\tilde{F}_P$ is the universal p-Frattini cover of P [FrJ, Chap. 20]. Denote its kernel by $\ker_0(P)$. As $\tilde{P}_G$ is also free and covers P, there is a surjective homomorphism $\psi : \tilde{P}_G \to \tilde{F}_P$ commuting with the respective maps ϕ and α of $\tilde{P}_G$ and $\tilde{F}_P$ to P. Hypothesis (4.1c) says g lifts to g' of order p in $\tilde{P}_G/\langle[\ker_0(G), \ker_0(G)]\ker_0(G)^p\rangle$.

Since $\ker_0(G)$ maps surjectively to $\ker_0(P)$, $\ker_0(G)^p$ maps onto $\ker_0(P)^p$ and $[\ker_0(G), \ker_0(G)]$ maps onto $[\ker_0(P), \ker_0(P)]$. Thus, $\ker_0(G)/\ker_1(G)$ maps surjectively to $\ker_0(P)/\ker_1(P)$. The image of g' in $\ker_0(P)/\ker_1(P)$ has order p and it is also a lift of g. Thus, a lift of g to something of order p in $\frac{1}{p}\tilde{G}$ implies a lift of it has order p in $\ker_0(P)/\ker_1(P)$. Assume $g' \in \tilde{F}_P/\ker_1(P)$ of order p maps to g. Here, apply a similar argument. Consider the pullback $\alpha^{-1}(\langle g\rangle)$ of $\langle g\rangle$ in $\tilde{F}_P$. This must map surjectively to $\mathbb{Z}_p$, the universal p-Frattini cover of $\langle g\rangle$. Therefore, an element of order p in $\mathbb{Z}/p^2$ would map to the generator of $\mathbb{Z}/p$. This contradiction concludes the first part of the lemma.

Now consider the last condition in the lemma. If $H \to G$ is a Frattini cover the condition holds by definition. Suppose, however, it holds and $H \to G$ is not a Frattini cover. Then, a proper subgroup H_1 of H maps surjectively to G. Apply Schur-Zassenhaus to H_1. This lifts the entries of $\mathfrak{g}$ to elements $\mathfrak{g}_1$ of H_1 of the same respective orders. By hypothesis, $H_1 \geq \langle\mathfrak{g}_1\rangle = H$. This contradicts H_1 being a proper subgroup of H. □

§4.B. Irrational characters.

Consider the universal p-Frattini cover of G. For $g \in G$ let $\tilde{g} \in {}_p\tilde{G}$ be a lift of g. The p'-order of g is the prime to p part of the order of g. This subsection discusses the values of the irreducible characters of $\frac{k}{p}\tilde{G}$ on the image $^k\tilde{g}$ of $\tilde{g}$ in $\frac{k}{p}\tilde{G}$. The p'-order of $\tilde{g}$ equals that of g. A relevant example might have $A_n = G$, $n \geq 5$, $p = 3$, and the conjugacy classes those of 3-cycles. Then, given a 3-cycle, choose lifts to 3-power orders in $\frac{k}{3}\tilde{A}_n$.

Lift Question 4.2. The values of all irreducible G characters at g generate a field $\mathbb{Q}_g$. Similarly, let $\mathbb{Q}_{\tilde{g}}$ be the direct limit of the fields generated by the values of $^k\tilde{g}$ at irreducible characters of $\frac{k}{p}\tilde{G}$. Suppose G is perfect and p divides the (supernatural) order of ^kg. Is it possible that $\mathbb{Q}_g = \mathbb{Q}_{\tilde{g}}$?

If yes, it could be all characteristic ${}_p\tilde{G}$ quotients are groups of regular Galois extensions of $\mathbb{Q}(x)$ having a bounded number of branch points, yet infinite inertia groups. The answer, however, is "No!" Call $\tilde{g} \in {}_p\tilde{G}$ rational if its image in $\frac{k}{p}\tilde{G}$ defines a rational conjugacy class for each $k \geq 0$.

Lemma 4.3. *If the order of g is prime to p, then $\mathbb{Q}_g = \mathbb{Q}_{\tilde{g}}$. Now suppose g (of any order) defines a rational conjugacy class in G. Then, Lift Question 4.2 has a yes answer if and only if there is a rational lift $\tilde{g} \in {}_p\tilde{G}$ of g.*

Proof. Suppose g and $\tilde{g}$ have p'-order. Then, characters of ${}_p^k\tilde{G}$ restricted to $\langle \tilde{g} \rangle$, give sums of characters of the cyclic group $\langle g \rangle$. Thus, $\mathbb{Q}_{k\tilde{g}} = \mathbb{Q}_g$ for each $k \geq 0$.

Assume g defines a rational conjugacy class in G. Suppose $\tilde{g}$ is a lift of g with $\mathbb{Q} = \mathbb{Q}_{\tilde{g}}$. This means ${}^k\tilde{g}$ has a rational value in each irreducible representation of ${}_p^k\tilde{G}$. Assume u is prime to the order of ${}^k\tilde{g}$. Then, ${}^k\tilde{g}^u$ also takes the same values under each irreducible representation of ${}_p\tilde{G}$. Values of a conjugacy class on irreducible characters determine the conjugacy class. Thus, ${}^k\tilde{g}^u$ is conjugate to ${}^k\tilde{g}$ in ${}_p^k\tilde{G}$, so ${}^k\tilde{g}$ determines a rational conjugacy class in ${}_p^k\tilde{G}$. □

§4.C. Conjecture 1.4 reduces to Modular Tower property (1.2a).

Now consider g having order a power of p. Let ${}_p^k\tilde{P}$ be the p-Sylow of ${}_p^k\tilde{G}$. The zeroth characteristic quotient is G; denote ${}_p^0\tilde{P}$ by $P = P_p$. Lemma 4.1 shows any lift of g to ${}_p^1\tilde{P}$ has larger order than does g.

Theorem 4.4. *Only p' elements of ${}_p\tilde{G}$ can be rational. In particular, let r_0 be any positive integer and let K be a number field. There is an integer $k_0 = k_0(r_0, K, G)$ with the following property. Suppose $k > k_0$ and there is a K regular realization of ${}_p^k\tilde{G}$. Then, one of the following holds:*

(4.2a) *there is a regular realization corresponding to a point of $\mathcal{H}({}_p^{k_0}\tilde{G}, \mathbf{C})^{\mathrm{rd}}(K)$ for some $\mathbf{C}$ with at most r_0 entries from ${}_p\mathcal{C}(G)$; or*

(4.2b) *the regular realization has more than r_0 branch points.*

Proof. Suppose $g \in {}_p\tilde{G}$ is a rational element and g is not a p'-element. For any integer v, g^v is also a rational element of ${}_p\tilde{G}$. Let v be the order of the image of g in G. Then, from Lemma 4.1 as restated above, g^v is a nontrivial p-power element in $\ker_0$. Replace g by g^v. Find the minimal integer n with $g \in \ker_n \setminus \ker_{n+1}$. In particular, $g = g_1$ has nontrivial image in $\ker_n / \ker_{n+1}$. Since $\ker_n$ is a pro-free pro-p group, any collection of representatives of the nontrivial cosets of $\ker_n / \ker_{n+1}$ give topological generators of $\ker_n$. Compliment g_1 with elements $g_2, \ldots, g_u$ that freely generate $\ker_n$. Now, we show g_1 is not a rational element of ${}_p\tilde{G}$.

For each p'-integer m, our rationality assumption says there is an $h_m \in {}_p\tilde{G}$ for which $g^m = h_m g h_m^{-1}$. Since ${}_p\tilde{G}/\ker_m$ is a finite group, there are infinitely many p'-powers g^m of g with corresponding h_m in $\ker_n$. Let m' be a nontrivial choice of such an integer.

This is our setup. Let $B = \langle g \rangle$ and $D = \langle g_2, \ldots, g_u \rangle$. Then, the groups B and D freely generate $\ker_n$. Further, there exists $h_{m'} \in \ker_n$ with $h_{m'} B h_{m'}^{-1} = B$. Now apply [HR]. This says, for each $h' \in \ker_n$, either $h' \in B$ (and $B^g = B$) or $h' B (h')^{-1} \cap B = 1$. Taking $h' = h_{m'}$ violates both these conclusions. Therefore, g is not a rational element.

To finish the proof, consider the conclusion on regular realizations. Suppose $^k_p \mathbf{C}$ is any set of r conjugacy classes (not necessarily p-regular) of $^k_p \tilde{G}$. A K regular $(^k_p \tilde{G}, {}^k_p \mathbf{C})$ realization corresponds to a point of $\mathcal{H}(^k_p \tilde{G}, {}^k_p \mathbf{C})^{\mathrm{rd}}(K)$ (end of App.I). Further, this produces a K regular $(^j_p \tilde{G}, {}^j_p \mathbf{C})$ realization with $^j_p \mathbf{C}$ the conjugacy class image of $^k_p \mathbf{C}$ in $^j_p \tilde{G}$, $0 \le j \le k$. There are only finitely many choices of p-regular conjugacy classes $\mathbf{C}$ with at most r_0 entries. So, if (4.2a) holds for each k, there exists some $\mathbf{C}$ (independent of k) consisting of p-regular classes. Suppose there are $^k_p \tilde{G}$ realizations for every k with at most r_0 branch points, and (4.2a) does not hold with $k \ge k_0'$. Then, for large k, the following hold.

(4.3a) There exists $^k_p \mathbf{C} = (^k \mathbf{C}_1, \ldots, {}^k \mathbf{C}_r)$ and a point of $\mathcal{H}(^k_p \tilde{G}, {}^k_p \mathbf{C})^{\mathrm{rd}}(K)$
giving a $^k_p \tilde{G}$ realization with $r \le r_0$ branch points.

(4.3b) At least one entry of $^k_p \mathbf{C}$ mod $\ker_{k_0'}$ is not p-regular.

Reorder the entries of $^k_p \mathbf{C}$ to assume $\mathbf{C}_1$ is not p-regular. The branch cycle argument (§1.A) says $^k_p \mathbf{C}$ must be a K-rational union. Let $g \in \mathbf{C}_1$. This puts a bound of r_0 on $\mathcal{G}_k = \{g^n \mid (n, N_{^k_p\mathbf{C}}) = 1\}/^k_p \tilde{G}$. For suitably large k, this contradicts the first part of the proof.   □

Remark 4.5. *Effective k_0 in Theorem 4.4.* The proof of Theorem 4.4 assumes existence of k_0' with no p-regular realizations of $^k_p \tilde{G}$ for $k > k_0'$. Assuming an explicit such k_0', it is possible to produce an explicit k_0. The proof above returns this to the following. Let $a \in {}_p \tilde{F}_u$ have nontrivial image in the first Frattini quotient of $_p \tilde{F}_u$. Then, we must give an explicit lower bound c_k on the number of prime to p powers of a in $_p \tilde{F}_u / \ker_k$ conjugate to a, where $\lim_{k \mapsto \infty} c_k \mapsto \infty$. The argument of [HR] can give such a bound.

§Appendix I. Nielsen classes and Modular Towers.

This is a quick review of fundamental definitions from [MT]. Excluding Theorem 3.1, Hurwitz spaces in this paper are the $\mathcal{H}(G, \mathbf{C})^{in}$ inner spaces of [FrV] and [MT]. These parametrize Galois covers $X \to \mathbb{P}^1_x$ whose branch cycles fall in the Nielsen class $\mathrm{Ni}(G, \mathbf{C})$ and have a fixed isomorphism of the automorphism group of $X \to \mathbb{P}^1_x$ with G. Let $^k_p \tilde{\mathbf{C}}$ be any conjugacy classes in $^k_p \tilde{G}$—not necessarily p-regular classes as in §0.B—mapping to $\mathbf{C}$ by the canonical quotient with $\ker_0$.

Suppose $\mathfrak{g} \in \mathbf{C}$ with $\langle \mathfrak{g} \rangle = G$ lifts to $\tilde{\mathfrak{g}} \in {}^k_p\mathbf{C}$. The Frattini covering property means $\langle \tilde{\mathfrak{g}} \rangle = {}^k_p\tilde{G}$ is automatic. So, after the first level, $\Pi(\tilde{\mathfrak{g}}) = 1$ is the significant formula for defining the Nielsen class (§0.A). Specifically:

$$\mathrm{Ni}({}^k_p\tilde{G}, {}^k_p\tilde{\mathbf{C}}) = \{ \tilde{\mathfrak{g}} \in {}^k_p\tilde{\mathbf{C}} \mid \tilde{\mathfrak{g}} \bmod \ker_0 \in \mathrm{Ni}(G, \mathbf{C}) \text{ and } \Pi(\tilde{\mathfrak{g}}) = 1 \}$$

is the kth level Nielsen class.

Consider the free group on generators Q_i, $i = 1, \ldots, r-1$, with these relations:

(I.1a) $Q_i Q_{i+1} Q_i = Q_{i+1} Q_i Q_{i+1}$, $i = 1, \ldots, r-2$;

(I.1b) $Q_i Q_j = Q_j Q_i$, $|i - j| > 1$; and

(I.1c) $Q_1 Q_2 \cdots Q_{r-1} Q_{r-1} \cdots Q_1 = 1$.

Conditions (I.1a) and (I.1b) define the *Artin braid group* B_r. Add (I.1c) to get the *Hurwitz monodromy group* H_r of degree r, a quotient of B_r. The Q_is in B_r act on $\mathfrak{g} \in \mathrm{Ni}(G, \mathbf{C})$:

(I.1d) $(\mathfrak{g})Q_i = (g_1, \ldots, g_{i-1}, g_i g_{i+1} g_i^{-1}, g_i, g_{i+2}, \ldots, g_r)$, $i = 1, \ldots, r-1$.

Mod out by inner automorphisms of G to induce an action by H_r. Irreducible components of $\mathcal{H}(G, \mathbf{C})$ correspond to orbits of this action.

Braid group action on $\mathrm{Ni}({}^k_p\tilde{G}, {}^k_p\tilde{\mathbf{C}})$ extends that on the level 0 Nielsen class. This produces the corresponding sequence of moduli spaces

(I.2) $\cdots \to \mathcal{H}({}^{k+1}_p\tilde{G}, {}^{k+1}_p\tilde{\mathbf{C}}) \to \mathcal{H}({}^k_p\tilde{G}, {}^k_p\tilde{\mathbf{C}}) \to \cdots \to \mathcal{H}(G, \mathbf{C}).$

Suppose ${}^k_p\tilde{G}$ has no center for $k \geq 0$. ([FrK] shows this holds if G has no center and no $\mathbb{Z}/p$ quotient; in particular, if G is a centerless perfect group [MT, Lemma 3.6].) Let K be a field of characteristic prime to $|G|$. Then, as in [MT, Part III], a K point $\mathfrak{p} \in \mathcal{H}({}^k_p\tilde{G}, {}^k_p\tilde{\mathbf{C}})$ gives a sequence of K covers

(I.3) ${}^kX_{\mathfrak{p}} \to {}^{k-1}X_{\mathfrak{p}} \to \cdots \to {}^0X_{\mathfrak{p}} \to \mathbb{P}^1.$

Further, ${}^jX_{\mathfrak{p}} \to \mathbb{P}^1$ gives a K regular $({}^j_p\tilde{G}, {}^j_p\tilde{\mathbf{C}})$ realization, $j = 0, \ldots, k$. When $\mathbf{C} = {}^k_p\tilde{\mathbf{C}}$ consists of p-regular conjugacy classes, then (I.2) is what we call a *Modular Tower* as in §0.B.

§Appendix II. Equivalence of covers of the sphere

There are two natural equivalences of covers of the sphere.

(II.1a) $\phi_i : X_i \to \mathbb{P}^1$, $i = 1, 2$, are equivalent if there exists $\alpha : X_1 \to X_2$
with $\phi_2 \circ \alpha = \phi_1$.
(II.1b) As in (II.1a), except there is $\beta : \mathbb{P}^1 \to \mathbb{P}^1$ with $\phi_2 \circ \alpha = \beta \circ \phi_1$.

§App.II.A. Action of $\mathrm{SL}_2(\mathbb{C})$ on $\mathbb{P}^r \setminus D_r$.

Below, $\mathcal{H} = \mathcal{H}(G, \mathbf{C})$ refers to a space of covers in a given Nielsen class up to equivalence (II.1a) .

The group S_r acts on the space $(\mathbb{P}^1)^r$ by permutation of its coordinates. This gives a natural map $\Psi_r : (\mathbb{P}^1)^r \to \mathbb{P}^r$. Consider $\mathfrak{x} = (x_1, \ldots, x_r)$ with none of the coordinates equal ∞. The point whose coordinates are the coefficients of the polynomial $\prod_{i=1}^{r}(z - x_i)$ in z represents the image of $\mathfrak{x}$ under Ψ_r. (If $x_i = \infty$, replace the factor $z - x_i$ by 1.) Also, Ψ_r takes the fat diagonal Δ_r to D_r. This interprets $U^r = \mathbb{P}^r \setminus D_r$ as the space of r distinct unordered points in $\mathbb{P}^1$. Thus, $\Psi_r : U^r \to U_r$ is an unramified Galois cover with group S_r.

Consider $\mathrm{PSL}_2(\mathbb{C})$ as linear fractional transformations acting diagonally on the r copies of $\mathbb{P}^1_x$. For $\alpha \in \mathrm{PSL}_2(\mathbb{C})$ and $\mathfrak{x} \in U^r$, $\mathfrak{x} \mapsto (\alpha(x_1), \ldots, \alpha(x_r))$. The action of $\mathrm{PSL}_2(\mathbb{C})$ is on the left, commuting with the coordinate permutation action of S_r. The quotient $\mathrm{PSL}_2(\mathbb{C}) \setminus U^r = \Lambda_r$ generalizes the λ line minus the points $0, 1, \infty$. Further, $\mathrm{PSL}_2(\mathbb{C}) \setminus U_r = J_r$ generalizes the j line minus the point at ∞ from the theory of modular curves. It has complex dimension $r - 3$. The case $r = 4$ is crucial to us, so we reassure the reader by displaying these identifications.

Given $\mathfrak{x} = \{x_1, x_2, x_3, x_4\}$ up to equivalence (II.1a) there is a unique degree 2 cover $X_{\mathfrak{x}} \to \mathbb{P}^1$ ramified exactly at $\mathfrak{x}$. Thus, X is a genus 1 curve; its j invariant determines its isomorphism class. Further, $X_{\mathfrak{x}}$ is equivalent to $X_{\mathfrak{x}'}$ if and only if there exists $\alpha \in \mathrm{PSL}_2(\mathbb{C})$ with $\alpha(\mathfrak{x}) = \mathfrak{x}'$. This identifies J_4 and the j line minus ∞. The natural λ line (ramified) cover of the j line is Galois with group S_3 [R, I.59]. Don't confuse this copy of S_3 with an S_3 inside the coordinate permutation action of S_4.

§App.II.B. Extending $\mathrm{PSL}_2(\mathbb{C})$ action to $\mathcal{H}(G, \mathbf{C})$.

Suppose $\mathfrak{p} \in \mathcal{H}$ has a representative cover $\phi_{\mathfrak{p}} : X_{\mathfrak{p}} \to \mathbb{P}^1$. Extend the action of $\alpha \in \mathrm{PSL}_2(\mathbb{C})$ by composing $\phi_{\mathfrak{p}}$ with α to give $\alpha \circ \phi_{\mathfrak{p}} : X_{\mathfrak{p}} \to \mathbb{P}^1$, a new G cover in the Nielsen class. Thus, $\mathrm{PSL}_2(\mathbb{C})$ action extends to $\mathcal{H}(^k_p \tilde{G}, \mathbf{C})$; denote its quotient by $\mathcal{H}(^k_p \tilde{G}, \mathbf{C})^{\mathrm{rd}}$. Since $\mathcal{H}(^k_p \tilde{G}, \mathbf{C})$ is an affine algebraic set, so is $\mathcal{H}(^k_p \tilde{G}, \mathbf{C})^{\mathrm{rd}}$ [MFo, Thm. 1.1]. These reduced spaces generalize the spaces of modular curves $Y_1(p^{k+1})$ [MT, Intro.].

So, the spaces $\mathcal{H}^{\mathrm{rd}}$ are moduli spaces for covers up to equivalence (II.1b): *reduced* Hurwitz spaces. Many have asked: "Which equivalence is more important?" My answer: Classical geometers often like equivalence (II.1b). (Also, it is usual to use the pullback of $\mathcal{H}^{\mathrm{rd}}$ over Λ_r. When, however, there are repetitions in the conjugacy classes **C**, this pullback is inappropriate, often wiping out the significant arithmetic information.) Their justification is the complex dimension of $\mathcal{H}^{\mathrm{rd}}$ is 3 less than that of $\mathcal{H}$.

We like that, too. For example, the reduced spaces are curves when $r = 4$, allowing use of Faltings' Theorem. Further, any rational point on $\mathcal{H}$ automatically produces infinitely many others in its $\mathrm{PSL}_2(\mathbb{Q})$ orbit. Allowing this would defy the finiteness results this paper conjectures. Still, it is equivalence (II.1a) that supports the Modular Tower construction. From that construction we compatibly reduce all levels of the tower by the $\mathrm{PSL}_2(\mathbb{C})$ action. Conclusion: We require both equivalences for Modular Towers.

References

[Be]   D. J. Benson, *Representations and cohomology, I: Basic representation theory of finite groups and associative algebras*, Cambridge studies in Advanced Mathematics **30**, Camb. Univ. Press, 1991.

[Be2]   D. J. Benson, The Loewy structures for the projective indecomposable modules for A_8 and A_9 in characteristic 2, *Comm. in Alg.* **11** (1983), 1395–1451.

[DFr1] P. Dèbes and M. Fried, Arithmetic variation of fibers in families: Hurwitz monodromy criteria for rational points, *J. Crelle* **409** (1990), 106–137.

[DFr2] P. Dèbes and M. Fried, Integral specialization of rational function families, preprint. Jan. 1997, 20 pgs.

[Fa]   G. Faltings, Diophantine approximation on abelian varieties *Annals of Math.* **133** (1991), 549–576

[Fr1]   M. Fried, Alternating groups and lifting invariants, Preprint as of 07/01/96.

[Fr2]   M. Fried, Fields of Definition of Function Fields and Hurwitz Families and Groups as Galois Groups, *Comm. in Algebra* **5** (1977), 17–82.

[Fr3]   M. Fried, Enhanced review of J.-P. Serre's Topics in Galois Theory, with examples illustrating braid ridigity, Proceedings AMS-NSF, Summer Conference Cont. Math series **186**, *Recent Developments in the Inverse Galois Problem* (1995), 15–32.

[Fr4]   M. Fried, Global construction of general exceptional Covers: with motivation for applications to encoding Applications and Algorithms, *Cont. Math.* **168**, G.L. Mullen and P.J. Shiue, editors, 1994, 69–100.

[Fr5]   M. Fried, Arithmetic of 3 and 4 branch point covers: a bridge provided by noncongruence subgroups of $SL_2(\mathbb{Z})$, in *Progress in Mathematics* **81**, Birkhauser, 1990, 77–117.

[FrJ]   M. Fried and M. Jarden, *Field Arithmetic*, Ergebnisse der Mathematik III, **11**, Springer Verlag, Heidelberg, 1986.

[FrK]   M. Fried and Y. Kopeliovich, A_5 Modular Towers, 30 page preprint, 1996.

[FrV]   M. Fried and H. Völklein, The inverse Galois problem and rational points on moduli spaces, *Math. Annalen* **290** (1991), 771–800.

[MT]   M. Fried, Modular Towers: Generalizing the relation between dihedral groups and modular curves, in Proceedings AMS-NSF Summer Conference, Cont. Math series **186**, *Recent Developments in the Inverse Galois Problem*, 1995, 111-171.

[HR]   W. Herfort and L. Ribes, Torsion elements and centralizers in free products of profinite groups, *J. Crelle* **358** (1985), 155-161.

[Me]   J-F. Mestre, Extensions régulières de $\mathbb{Q}(T)$ de groupe de Galois $\tilde{A}_n$, *J. Alg.* **131** (1990), 483-495.

[MFo]   D. Mumford and J. Fogarty, Geometric Invariant Theory, *Ergeb. der Math. und ihrer Grenzgebiete* **34**, Springer Verlag, 2nd enlarged edition, 1982.

[Ri]   L. Ribes, Frattini covers of profinite groups, *Archiv der Math.* **44** (1985), 390–396.

[R]   A. Robert, *Elliptic curves*, Lecture Notes in Mathematics **326**, Springer-Verlag, Heidelberg-New York, 1973.

[Se]   J.-P. Serre, *Topics in Galois Theory*, Bartlett and Jones Publishers, 1992.

[V]   H. Völklein, *Groups as Galois Groups*, Cambridge Studies in Advanced Mathematics **53**, Cambridge Univ. Press, 1996.

University of California at Irvine, Irvine CA 92717
mfried@math.uci.edu

Part III. Galois actions and mapping class groups

Galois group $G_{\mathbb{Q}}$, Singularity E_7, and Moduli M_3

Makoto Matsumoto

Monodromy of iterated integrals and non-abelian unipotent periods

Zdzisław Wojtkowiak

Part III. Galois actions and mapping class groups

Galois group $G_{\mathbb{Q}}$, Singularity E_7, and Moduli $\mathcal{M}_3$

Makoto Matsumoto

An Artin group is a generalization, defined for any graph, of the usual braid groups. The Artin group of E_7 Dynkin diagram has a natural surjection to the mapping class group of genus three curves, defined by sending the canonical generators to a set of Dehn twists, called Humphries generators.

We first show that this group homomorphism comes from a geometric morphism between moduli spaces by taking π_1.

Next, we describe the notion of tangential morphisms. Let U be an affine variety and $D_1, \ldots, D_l$ be normal crossing divisors which are principal. The tangential morphism from $D_1 \cap \cdots \cap D_l$ to $U - (D_1 \cup \cdots \cup D_l)$ is defined when we specify the generator of the defining ideal of D_i for each i. This yields a morphism between the algebraic fundamental groups. A tangential morphism from a point $\operatorname{Spec} k$ coincides with the notion of k-rational tangential base point.

Using this, we describe the action of the absolute Galois group on the profinite completion of Artin groups of type A_n, B_n, C_n, D_n, and E_7. (The first type gives the Braid group investigated before.) Description uses only the Belyi's coordinates, i.e. the action of the Galois group on the algebraic fundamental group of $\mathbb{P}^1$ minus $0, 1, \infty$.

As a corollary, we give an explicit description of the Galois action on the profinite completion of the mapping class group of genus 3 curve.

Contents

The author was supported by Japan Association for Mathematical Sciences for attending the Luminy conference.

§3. Galois action on Artin groups

  3.1. Motivation

  3.2. Tangential base point on $V(E_7)$

  3.3. $\mathbb{P}^1_{01\infty}$ and $\mathbb{P}^1_{0\pm1\infty}$

  3.4. Galois action on each generator

  3.5. The result on E_7 and M_3

  3.6. A_n, B_n, C_n, D_n cases

§0. Introduction

This manuscript consists of three parts. The first two parts are logically independent, and the last part uses the results of the first two.

The first part of §1 is purely geometric, and is independent of the Galois group.

We recall Brieskorn's result [5], [6] on the deformation of E_7-singularity, which says that (1) the smooth locus of the deformation space has the E_7 Artin group as the fundamental group, and (2) on the smooth locus, each fiber is a genus three curve. By (2), we have a classifying map from the smooth deformation space to the moduli stack of genus three curves. Our first result is that by taking π_1 of this morphism, we have a group homomorphism from the E_7 Artin group to the mapping class group of genus three curve, which maps each generator of the Artin group to the Humphries' Dehn-twist generators of the mapping class group. This is an interesting coincidence, since the two sets of generators were invented independently.

Then the algebraic meaning of the Humphries generators are clear, so we move to the investigation of the Galois action. As a preparation, in the second part of §2, we give the notion of tangential morphisms. This notion is a slight generalization of the tangential base point introduced by Deligne [7], developed and effectively used by Ihara, Nakamura, and the author [15] [17] [22] [18] [27] (Ihara and Nakamura even dealt with the formal scheme version in [18]). Note that there are also log-scheme theoretic interpretations, which fit to motivic fundamental groups [10] [30].)

In the third part §3, we give a construction of a tangential base point for the deformation space of the E_7 singularity. (The construction mimics that for A_n in [17].) We give seven tangential morphisms from $\mathbb{P}^1$ minus four points to the deformation space, then the Galois action on the Artin group is written in terms of $f_\sigma, \chi(\sigma)$, which appears for $\mathbb{P}^1$ minus three points. As a corollary, the Galois action on profinite mapping class group with genus three is described. We also treat some other type of Artin groups, A_n, B_n, C_n, D_n.

The motivation of this research is as follows. Let V be a geometrically connected variety over the rational number field $\mathbb{Q}$, and x be a $\mathbb{Q}$-rational point. Then, we have a group action of the absolute Galois group $G_{\mathbb{Q}} = \mathrm{Gal}(\overline{\mathbb{Q}}/\mathbb{Q})$ of $\mathbb{Q}$ on the profinite completion of the fundamental group of V as a complex variety [11]: $\rho_{V,x} : G_{\mathbb{Q}} \to \mathrm{Aut}\widehat{\pi_1}(V(\mathbb{C}),x) \cong \mathrm{Aut}\pi_1^{alg}(\overline{V},x)$ (here $\overline{V}$ denotes the coefficient extension $V \otimes \overline{\mathbb{Q}}$).

Grothendieck conjectured[1] that V can be recovered from $\rho_{V,x}$, for any "anabelian" variety V. Also he asserted the importance [12] of $\rho_{V,x}$ with V moduli spaces of (g,n)-curves, where g is the genus and n the number of punctures.

If $V = \mathbb{P}^1_{01\infty}$ (i.e., $\mathbb{P}^1$ minus three points $0, 1, \infty$), then $\rho_{V,x}$ is injective (Belyĭ [1]) (actually is injective for any affine curve V with nonabelian fundamental group [22]). Since $\mathbb{P}^1_{01\infty}$ is the moduli space of four pointed $\mathbb{P}^1$, and since it is the simplest, we have a conjecture that $\rho_{V,x}$ can be "described naturally" in terms of $\rho_{\mathbb{P}^1_{01\infty},b}$ under good choices of b on $\mathbb{P}^1_{01\infty}$ and x on V, for V moduli of some good objects[2].

This is the case for V the configuration space of n points on the affine plane. In this case, the fundamental group of V is the n-string Braid group B_n. We can choose canonical base point x, b, which are actually on the boundary of the spaces (i.e., tangential base point of Deligne [7]), such that the image of the each canonical generator of B_n by $\sigma \in G_{\mathbb{Q}}$ can be written down in terms of those of $\pi_1^{top}(\mathbb{P}^1_{01\infty}(\mathbb{C}),b)$ simply by substitutions (note that $\pi_1^{top}(\mathbb{P}^1_{01\infty}(\mathbb{C}),b)$ is a free group of two generators x, y, that $\sigma(x), \sigma(y)$ is an element of the profinite completion, and substituting x, y in $\sigma(x), \sigma(y)$ to some elements in another profinite group makes sense). This was proved by Ihara and the author [17], where the insight comes from Drinfeld's result on the action of the Grothendieck-Teichmüller group on the pro-unipotent completed braid groups [9].

Here we will describe some generalization of this, i.e., for V deformation space of rational singularity of surfaces (the above case corresponds to the singularity of A_{n-1}-type). Then, using a tight relation (see Theorem 1.3) between the deformation of E_7 singularity and the moduli stack M_3 of genus-three curves (see also Looijenga [21]), we extend the result to $V = M_3$ (see Theorem 3.1).

---

[1] Note that this conjecture for curves has been settled. The genus zero case was solved by H. Nakamura (see his integrated paper [25]). Recently, A. Tamagawa [33] settled this for affine curves (also finite field case), and S. Mochizuki [23][24] for the proper case (he proved a stronger p-adic version).

[2] After the author's talk on the genus three case in Luminy, H. Nakamura [27] succeeded in describing the Galois action on mapping class groups with any genus g by $\mathbb{P}^1 - \{0, 1, \infty\}$. In the talk at Luminy, the author depended on $\mathbb{P}^1 - \{0, \pm 1, \infty\}$ rather than the desirable $\mathbb{P}^1 - \{0, 1, \infty\}$. Now the result of this manuscript involves only $\mathbb{P}^1 - \{0, 1, \infty\}$, and this improvement comes from the stimulation by the Nakamura's success. The author is thankful to him for informing him of his beautiful result.

We shall exchange the order of the explanation: explain first about E_7 and M_3, and next about Galois actions. This is because the former is purely geometric and independent of the mess of the Galois actions, and it seems better for the readability.

§1. E_7 and M_3

1.1. Fundamental group of moduli stack

Let $M_{g,n}$ denote the moduli stack of the genus g n-ordered pointed curves [8] over $\mathbb{Q}$. The coarse moduli scheme of $M_{g,n}$, equipped with $\mathbb{C}$-topology, is analytically isomorphic to the quotient of the Teichmüller space by the mapping class group. Then, this analytic space has a trivial fundamental group. However, there is Oda's theory of the fundamental group of analytic stacks [28]. Roughly speaking, considering the fundamental group of the stacks corresponds to considering the orbifold, and $\pi_1^{an}(M_{g,n} \otimes_{\mathbb{Q}} \mathbb{C})$ is proved to be isomorphic to the mapping class group $\Gamma_{g,n}$ of the genus g oriented surface fixing n distinguished points. Moreover, the monodromy representation of the universal fiber $M_{g,n+1} \to M_{g,n}$, i.e., the outer monodromy action of the fundamental group of the base space on that of a fiber, coincides with the canonical morphism $\Gamma_{g,n} \to \mathrm{Out}\pi_1^{top}(\Sigma_{g,n})$, where $\Sigma_{g,n}$ is the n-punctured genus g oriented surface.

For $n = 0$, we simply write $M_g = M_{g,0}$, $\Gamma_g = \Gamma_{g,0}$, and $\Sigma_g = \Sigma_{g,0}$.

1.2. Humphries generators and Artin group of E_7-type

It is known that the mapping class group Γ_g is generated by the following $2g + 1$ Dehn twists [13], which are called the Humphries generators.

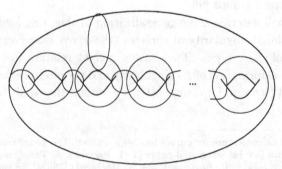

The relation of these Dehn-twist generators was determined by Wajnryb. There are four types of relations (for the explicit description, see Wajnryb [34] and Birman's survey article [3]).

(i) Braid relations: $DD' = D'D$, if D and D' have no intersection. $DD'D = DD'D$ if D and D' have one intersection point (the number of intersections is at most one among the above curves).

(ii) Monocular (or chain) relation, which is supported by the five leftmost twists.

(iii) Lantern relation.

(iv) Commutativity with the hyperbolic involution.

The relation (i) shows similarity with the braid group. We write an intersection diagram Γ of these $2g + 1$ twists. Thus, Γ has $2g + 1$ vertices corresponding to the $2g + 1$ curves, and $2g$ edges corresponding to the intersection point.

The group generated by $2g + 1$ generators with relation (i) is called *Artin group of type* Γ, and denoted by $A(\Gamma)$. In general, let G be any graph each of whose edge e is labeled with non negative integer m_e. Then, the corresponding Artin group $A(G)$ is defined as the group with the vertices as generators and the relations

$$\underbrace{aba\cdots}_{m_e+2} = \underbrace{bab\cdots}_{m_e+2}$$

for each pair a, b of vertices. Here, e is the edge between a and b, m_e is the number assigned for this edge, and if there is no edge then we put $m_e = 0$, i.e., a and b commute.

In this terminology, the braid group of n strings is nothing but the Artin group of straight line with $n - 1$ vertices.

Brieskorn [5] [6] proved the following profound theorem. If the graph is a Dynkin diagram of some semisimple Lie algebra over $\mathbb{C}$, then the associated Artin group is realized as the fundamental group of a rational variety, denoted here by $V(G)$ with G the type of Lie algebra. This $V(G)$ has a moduli interpretation. It is the smooth fiber locus of the miniversal deformation space of the rational singularity corresponding to the Dynkin diagram. Actually, the Artin group is defined to describe the fundamental group of these varieties.

The braid group appears for A_{n-1} singularities, to which the corresponding rational variety is the configuration space of n points on a plane.

Now, we consider the case $g = 3$. Then, the intersection diagram of Humphries generators coincides with E_7 Dynkin diagram. So, there is a surjection from the Artin group of E_7 type to the mapping class group of genus 3:

$$A(E_7) \twoheadrightarrow \Gamma_3,$$

by mapping canonical generators to canonical generators. Now, a natural question arises: does this morphism come from an algebraic map?

The answer is yes (see Theorem 1.3 for the precise statement): there exists a natural family of two pointed genus three curves over $V(E_7)$ such that the classifying map to M_3 obtained by forgetting two points induces $A(E_7) \twoheadrightarrow \Gamma_3$ by taking the fundamental groups.

The rational variety $V(E_7)$ can be described explicitly by using the Weyl group, and the Galois action on $\pi_1^{alg}(\overline{V(E_7)}, x)$ can be described in terms of that on $\pi_1^{alg}(\overline{\mathbb{P}^1_{01\infty}}, b)$, for some appropriate tangential base points x, b.

Thus, by the functoriality of π_1^{alg}, we can describe the Galois action on Γ_3. This will be explained in §3.

1.3. Variety $V(E_7)$

$V(E_7)$ is a miniversal deformation space of E_7 singularity. Brieskorn solved conjecture of Grothendieck on the explicit construction by Lie group [5] of this deformation space. Here we need this construction. For the reference on the following facts, see Slodowy [31].

Let Φ be a root system of a semisimple Lie algebra, i.e., one of A_n, B_n, C_n, D_n, E_6, E_7, E_8, F_4, G_2. This is a finite subset of a vector space U over $\mathbb{Q}$. The set of reflections generates a finite group, called the Weyl group of type Φ. For each root $\alpha \in \Phi$, let H_α be the reflection hyperplane (i.e., the fixed divisor by the reflection along α) of the complexified vector space $U_{\mathbb{C}} := U \otimes_{\mathbb{Q}} \mathbb{C}$. Let us define $V(\Phi)$ as the quotient of $U_{\mathbb{C}} - \cup_\alpha H_\alpha$ by the action of the Weyl group. This variety $V(\Phi)$ is known to be a smooth rational variety defined over $\mathbb{Q}$ by invariant theory: the quotient of the affine space by a finite group G is an affine space if and only if G is a reflection group, i.e. generated by (possibly nonorthogonal) reflections (see for example Theorem 4.2.5 in [32]).

Theorem 1.1. (Brieskorn [5] [6]) *The fundamental group of $V(\Phi)$ is isomorphic to the Artin group of type Φ. This isomorphism is given as follows. Take the fundamental Weyl chamber C in $U_{\mathbb{R}}$ as the base point of $V(\Phi)$ (since the chamber is simply connected, this makes sense). For each fundamental root α, take a path in $U_{\mathbb{C}} - H_\alpha$ from this chamber to the reflection image of C with respect to α, which goes around the divisor H_α counterclockwise. Then, taking the quotient by the Weyl group, we obtain a closed path in $V(\Phi)$. This is the corresponding generator of $\pi_1^{top}(V(\Phi), C)$ to the generator associated with α in the Artin group $A(\Phi)$.*

The simplest example is A_n type, which yields a usual braid group. In this case, U is the set of vectors $(x_1, x_2, \ldots, x_{n+1})$ with $\sum_i x_i = 0$. If we denote by e_i the i-th unit vector $(0, \ldots, 0, 1, 0, \ldots, 0)$, then $\Phi = \{e_i - e_j | 1 \leq i, j \leq n, i \neq j\}$, and the set of fundamental roots is $e_1 - e_2, e_2 - e_3, \ldots, e_n - e_{n+1}$. The fundamental Weyl chamber is the set of real vectors whose inner product with any fundamental root is positive, i.e., $(x_1, \ldots, x_{n+1})$ with $x_1 > x_2 > \cdots > x_{n+1}$. The reflection associated with $e_i - e_j$ is the exchange of x_i and x_j, and the Weyl group is the symmetric group S_{n+1}. The reflection hyperplane is the locus with $x_i = x_j$. Thus, $V(A_n)$ is $(U -$ the locus with $x_i = x_j$ for some $i \neq j)$ divided by S_{n+1}, i.e., the configuration space of unordered $n + 1$ points. Its fundamental group is the braid group, i.e., $A(A_n)$. It is easy to see that the path described above corresponds to the standard generator of the braid group.

In §3, we describe the Galois action on the profinite fundamental group of some of these varieties. Each $V(\Phi)$ has a natural meaning as a moduli of singularities as follows. We don't use the universality in this manuscript, so we let the reader consult the reference for the definition of universality of these moduli spaces. (For the case of B_n, C_n, F_4, G_2, see [31].)

Theorem 1.2. [5] *Let Φ be a root system of A, C, E type. Then, the germ of U divided by the Weyl group near the origin is a miniversal deformation space of the rational singularity of Φ type. That is, there is a family of curves on U, such that the fiber at the origin is a curve of Φ-type singularities, satisfying the universal property for the deformation. The quotient of $\cup_\alpha H_\alpha$ by the Weyl group is the locus with singular fibers.*

By the universality of $M_{g,n}$, we have a classification morphism from $V(\Phi)$ to $M_{g,n}$ for suitable g, n. If $\Phi = A_{n-1}$, then it is known that the deformation of the singularity is given by the equation $y^2 = x^n + a_1 x^{n-1} + a_2 x^{n-2} + \cdots + a_n$, i.e., the hyperelliptic curves. Here $(a_1, \ldots, a_n)$ is the parameter of $\mathbf{A}^n / S_n$. Thus, if $(b_1, \ldots, b_n)$ is the parameter of $\mathbf{A}^n$, then a's are the fundamental symmetric polynomials of b's. After compactification of the fibers, we have $V(A_n) \to M_g$ with $n = 2g + 2$. The image is the hyperelliptic locus of M_g, and it can be easily checked that the generators of $\pi_1(V(A_n))$ are mapped to the Dehn twists along the following circles.

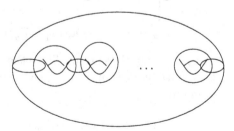

This group homomorphism is surjective if and only if $g \leq 2$. For $g = 3$, we can consider an E_7 analogue.

Theorem 1.3. *The deformation of singularities of E_7 type gives a family of two pointed curves with genus three. By forgetting the two points, we have a classification morphism $V(E_7) \to M_3$. Then, the corresponding group homomorphism $\pi_1(V(E_7)) \to \Gamma_3$ is given by mapping the generators of the Artin group to the Humphries generators of Γ_3.*

Proof. The E_7 singularity is given by the equation $y(y^2 - x^3) = 0$. It is known that a smooth deformation of this singularity becomes a two-pointed curve with genus three. See for example [31], or, consider a small deformation of the one punctured cusp $y^2 - x^3 = 0$ to an elliptic curve. This yields $y(y^2 - x^3 + \alpha) = 0$, which is the three points normal intersection of one-punctured projective line and one-punctured elliptic curve. By deforming the three points, we obtain two pointed genus three curves. By forgetting the punctures, we have a smooth family $F \to V(E_7)$ of genus three curves. Now we have the classifying morphism $V(E_7) \to M_3$. We shall show that the π_1 gives the canonical surjection $A(E_7) \to \Gamma_3$.

Since $F \to V(E_7)$ is a smooth family of genus 3 curves, by the universality of $M_{g,1} \to M_g$, we have the following pullback diagram:

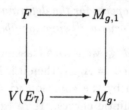

This yields a compatible monodromy map

$$\pi_1^{top}(V(E_7)) \longrightarrow \mathrm{Out}(\pi_1^{top}(\Sigma_g))$$

$$\Gamma_g \longrightarrow \mathrm{Out}(\pi_1^{top}(\Sigma_g)),$$

where Γ_g is the genus g mapping class group. Since the lower map is injective (Dehn-Nielsen's Theorem), it is enough to show that the image of each generator by the monodromy

$$\pi_1(V(E_7)) \to \mathrm{Out}(\pi_1(\Sigma_g))$$

coincides with prescribed Dehn twists.

First, in the miniversal deformation of singularities, it is known that the codimension one component of the singular locus corresponds to one-pinched Riemann surface, and the corresponding monodromy is a Dehn twist. Thus, we may assume each generator of the Artin group is mapped to a Dehn twist. Moreover, this Dehn twist is not trivial, even if we pass to the monodromy in the first cohomology [31].

Next, we use Ishida's result [19]: Let C, C' be two Dehn twists in Γ_g, $g \geq 2$, along the curves with same notation. We choose the curves from their homotopy classes so that the number of the intersections is the minimum. Then, Ishida's result is:

(i) C and C' commute if and only if $C \cap C' = \emptyset$,

(ii) C and C' satisfy the braid relation $CC'C = C'CC$ if and only if $C \cup C'$ =one point,

(iii) C and C' satisfy no relation, i.e., generate a free group, otherwise.

Now we shall complete the proof. We put the names C_0, C_1, C_2, C_3, C_4, C_5, C_6 for the generators of $A(E_7)$ and give the same names to the closed curves in Σ_g, such that the corresponding Dehn twists are the images of C_i's in Γ_3.

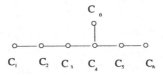

Then, up to homotopy equivalence and diffeomorphisms, there are only finite possibilities of C_1 in Σ_g: the one which does not separate Σ_g, or, the one which separates Σ_g into two pieces with genus g_1, g_2, $g_1 + g_2 = g$. If C_1 separates the surface, then, any other curves intersects with C_1 at even number of points. Then, by Ishida's theorem, there is no Dehn twist satisfying the braid relation with C_1; in other words, we can't take C_2, and this leads to a contradiction. Thus, we may assume C_1 is mapped to the curve shown in the figure below.

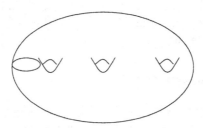

Next, we consider the image of C_2. The tubular neighborhood of C_1, C_2, respectively, is a tube. Since C_1 and C_2 intersect at exactly one point, the tubular neighborhood of $C_1 \cup C_2$ in Σ_g is the two tubes glued along a square neighborhood of the crossing point. This is homeomorphic to a torus minus a closed disk. Then, the whole surface is a union of this torus and the complement surface, glued at the boundary homeomorphic to a circle. If we redraw the picture along this decomposition, we may assume C_2 is as follows, up to a homeomorphism.

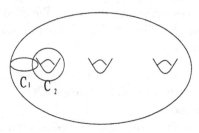

We shall consider C_3. C_3 intersects with C_2 at one point, and does not intersect with C_1. Thus, the tubular neighborhood of $C_1 \cup C_2 \cup C_3$ is isomorphic to a genus one surface with two closed boundaries. There are two kinds of possibilities: these two closed boundaries are connected in the complement of the tubular neighborhood in Σ_g, or not connected. If not connected, then we can't choose C_4, which intersects at one point in C_3 and has no intersection with the others. Thus, the two boundaries are connected in the complement, so we may assume that C_3 is just as expected. Similarly, we can prove that C_i, $i = 1, 2, 3, 4, 5, 6$ are like the following picture.

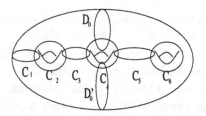

Now, C_0 intersects with only C_4. If we cut C_1, C_2, C_3, C_5, C_6, then we obtain a tube with several disks removed. Since C_0 does not intersect with these curves, we conclude that the only possibility is D_0 or its mirror image D_0'. This completes the proof.   ◇

§2. Tangential morphisms

First we will give a rough intuitive explanation on the notion of tangential morphisms, which includes as a zero dimensional case tangential base points.

Let $V^* = \operatorname{Spec} A$ be an affine variety over a field $k \subset \mathbb{C}$, and let $t_1, \ldots, t_l$ be regular functions on V^*. Let $D_1, \ldots, D_l$ denote the zero divisor of $t_1, \ldots, t_l$, respectively, and assume that these are normal crossing. Put $Z := D_1 \cap D_2 \cap \cdots \cap D_l$ and $V := V^* - D_1 \cup D_2 \cup \cdots \cup D_l$. Then, a tangential morphism $Z \to V$ is a morphism from Z to V, obtained by "infinitesimally small perturbation" of $Z \hookrightarrow V^*$ "in the direction" specified by $t_1, \ldots, t_l$.

We regard Z, V^* as complex analytic spaces via $k \subset \mathbb{C}$. Then Z is the locus in V^* defined by $t_1 = t_2 = \cdots = t_l = 0$. We perturb Z to obtain Z', which is defined by $t_1 = t_2 = \cdots = t_l = \varepsilon$, with infinitesimally small positive real number ε. Since D_i's are normal crossing, Z' is homeomorphic to Z, and we have $Z \cong Z' \to V \subset V^*$ and $\pi^{top}(Z) \cong \pi^{top}(Z') \to \pi_1^{top}(V)$. Less trivial is the algebraic structure, which is explained in the next section.

In the case that Z is a point $\operatorname{Spec} k$, we have a section $\pi_1^{alg}(\operatorname{Spec} k, \operatorname{Spec} \bar{k})$ $\to \pi_1^{alg}(\overline{V}, \overline{Z})$, which is nothing but a k-rational tangential base point.

In §2, we need only the characteristic zero case. However, the results in 2.1-2.3, using Galois categories are valid for positive characteristics. We treat these cases in parallel for future reference.

2.1. Tangential morphisms.

Let k be a field of characteristic $p \geq 0$. We denote by $\mathbb{N}'$ the set of natural numbers coprime with p. Thus, if k has the characteristic zero, then $\mathbb{N} = \mathbb{N}'$.

Let V be a geometrically connected variety over k. We denote by V_{fet} the category of finite etale covers of V. We mainly deal with the case $V = V^* - D$, where V^* is a variety and D a divisor of V^*. If $p > 0$, we consider V_{fet}^{*D}, the category of finite covers of V^* that are etale outside D and tamely ramified at D. When it cannot cause confusion, we denote V_{fet}^{*D} simply by V_{fet}. V_{fet} is a Galois category. For fiber functors F, F' : $V_{fet} \to \{\text{Finite Sets}\}$, we denote $\pi_1(V; F, F')$ the set of isomorphisms from the functor F to F'. This constitutes a groupoid. If $F = F'$, this is nothing but the algebraic fundamental group.

We shall deal with the following case.

Definition 2.1. Let V be a Zariski open subscheme of a geometrically connected variety V^* over k. Let $U = \operatorname{Spec} A$ be an affine open subscheme of V^*. We assume that there are regular functions $t_1, t_2, \ldots, t_l$ on U, such that (1) $t_i = 0$ $(1 \leq i \leq l)$ give normal crossing divisors (each denoted by D_i) on U and (2) $V \cap U$ is the complement of these divisors in U. Let Z be the intersection of D_i, $i = 1, 2, \ldots, l$.

The above datum $(V, V^*, U, t_1, \ldots, t_l, Z)$ is called a tangential morphism from Z to V, denoted by $\tau : Z \to V$. (I.e., $\tau : D_1 \cap \cdots \cap D_l \to U - (D_1 \cup \cdots \cup D_l) \subset V$.)

This tangential morphism is called k-rational. Let L be an algebraic extension field of k. A tangential morphism to $V \otimes L$ is called an L-rational tangential morphism.

If Z is a point $\operatorname{Spec} k$, then $\tau : Z \rightharpoonup V$ is called *k-rational tangential (base) point*.

A tangential morphism induces an exact functor from V_{fet} to Z_{fet}, and hence a morphism between fundamental groupoids, as follows.

Let $\hat{U}$ be the completion of U along Z, i.e. $\operatorname{Spec} \hat{A} := \operatorname{Spec} \lim_{\leftarrow n \in \mathbb{N}} A/I_Z^n$, where $I_Z = (t_1, \ldots, t_l)$ is the ideal defining Z.

Let us denote by $Z\{t\} := Z\{t_1, \ldots, t_l\}$ the spectrum of

$$A_{\{t_1,\ldots,t_l\}} := \bigcup_{N \in \mathbb{N}'} \hat{A}[[t_1^{1/N}, \ldots, t_l^{1/N}]],$$

where $t_i^{1/N}$ denotes a compatible system of N-th roots of $t_i \in A$. Z is identified with the closed subscheme of $Z\{t\}$ defined by $t_1 = \cdots = t_l = 0$. When Z is $\operatorname{Spec} k$, by the regularity of V, $Z\{t\} = \operatorname{Spec} k\{t_1, \ldots, t_l\}$, which is the ring of Puiseux series over k.

There is a natural morphism $Z\{t\} \to V^*$ obtained as $Z\{t\} \to \hat{U} \to V^*$. The image of this map is contained in any Zariski open neighborhood of Z in V^*.

Let W be a finite etale cover of V. Then we take the normal closure of V^* with respect to W, and call it W^*. In a neighborhood of Z in V^*, the possibly ramified locus is $t_1 \cdots t_l = 0$. We pullback W^* along $Z\{t\} \to V^*$, and denote the fiber product by $W^*|_{Z\{t\}}$. Take the normal closure of $Z\{t\}$ with respect to the generic fiber of the fiber product, and denote it $N(W|_{Z\{t_1,\ldots,t_l\}})$. Then this is finite etale over $Z\{t_1, \ldots, t_l\}$. This follows from Abhyankar's Lemma for $Z\{t_1, \ldots, t_l\} \to \hat{U}$, since the ramification is preserved by completion. Then we obtain an etale cover of Z by base change, which is denoted by $\tau_{fet}^*(W)$.

We have constructed a functor $\tau_{fet}^* : V_{fet} \to Z_{fet}$. This functor preserves the fiber product, since the generic fiber of $W|_{Z\{t_1,\ldots,t_l\}}$ does not change after taking normalization. By uniqueness of normalization, taking fiber products commutes with normalization. This functor maps direct sum to direct sum and epimorphisms to epimorphism, for the same reason. Thus, by SGA1 [11], τ_{fet}^* is an exact functor between Galois categories, which yields a morphism between groupoids $\pi_1(\tau) : \pi_1(Z; F, F') \to \pi_1(V; \tau(F), \tau(F'))$, where F, F' are any fiber functors of Z_{fet}, and $\tau(F)$ is the fiber functor $F \circ \tau_{fet}^* : V_{fet} \to \{\text{Finite Sets}\}$. Now we proved:

Proposition 2.1. *For a tangential morphism $\tau : Z \rightharpoonup V$, we have an associated exact functor $\tau_{fet}^* : V_{fet} \to Z_{fet}$, which induces a homomorphism of groupoids $\pi_1(\tau) : \pi_1(Z; F, F') \to \pi_1(V; \tau(F), \tau(F'))$.*

Assume $Z = \operatorname{Spec} k$. In this case, τ is called a *k-rational tangential point*. This yields a group homomorphism $\pi_1(\operatorname{Spec} k, x) \to \pi_1(V, \tau(x))$, where $x : \operatorname{Spec} \Omega \to \operatorname{Spec} k$ is any geometric point. This is a section to the structural morphism $\pi_1(V, \tau(x)) \to \pi_1(\operatorname{Spec} k, x)$. It is easy to check this at Galois category level. This is nothing but a tangential base point considered in [7] [15].

This yields an action of $G_k = \pi_1(\operatorname{Spec} k, x)$ on $\pi_1(V; \tau(x), F)$ for any fiber functor F from left by $\sigma \in G_k : \gamma \in \pi_1(V; \tau(x), F) \mapsto \gamma \circ \pi_1(\tau)(\sigma^{-1})$, and similarly an action from left on $\pi_1(V; F, \tau(x))$ by $\gamma \mapsto \pi_1(\tau)(\sigma)\gamma$.

Let τ' be another k-rational tangential point. Then, it is proved that

$$\pi_1(\overline{V}; \bar{\tau}(x), \bar{\tau}'(x)) \subset \pi_1(V; \tau(x), \tau'(x))$$

is the inverse image of the unit element by $\pi_1(V; \tau(x), \tau'(x)) \to \pi_1(\operatorname{Spec} k, x)$. Here, $\bar{\tau}$ is the $\bar{k}$-rational base point, which is obtained by extending the coefficient in the construction of τ_{fet}^*.

By $\gamma \mapsto \pi_1(\tau')(\sigma)\gamma\pi_1(\tau)(\sigma)^{-1}$, $\pi_1(\operatorname{Spec} k, x)$ acts on $\pi_1(\overline{V}; \tau(x), \tau'(x))$. If $\tau = \tau'$, then this is an automorphism of $\pi_1(\overline{V}, \tau(x))$. This representation $\operatorname{Gal}(\bar{k}/k) \to \operatorname{Aut} \pi_1(\overline{V}, \tau(x))$ is what we want to study.

2.2. Homotopy between tangential morphisms.

Let $F, G : \mathcal{C} \to \mathcal{D}$ be two exact functors from a Galois category to another. We call an isomorphic natural transformation from F to G a *homotopy* from F to G. If the target category $\mathcal{D}$ is {Finite sets}, then F and G are fiber functors, and a homotopy between fiber functors is nothing but an element of $\pi_1(\mathcal{C}; F, G)$. This is an analogy to that a path is a homotopy between two points.

Let $\tau, \tau' : Z \to V$ be two tangential morphisms, which comes from $Z\{t\} = Z\{t_1, \ldots, t_l\} \to V^*$, $Z\{s\} = Z\{s_1, \ldots, s_l\} \to V^*$, respectively.

Definition 2.2. An infinitesimal homotopy from τ to τ' is an isomorphism $Z\{t\} \xrightarrow{\sim} Z\{s\}$, commuting with the morphisms to V^*, and mapping the divisors (t_i) to (s_i) for each i.

Proposition 2.2. *An infinitesimal homotopy yields a homotopy from τ_{fet}^* to $\tau_{fet}'^*$.*

Proof. Let W be a finite etale cover of V. The normalizations on $Z\{t\}$ and on $Z\{s\}$ are canonically isomorphic, since they coincide outside the normal crossing divisor. Thus, we have a functorial isomorphism $\tau_{fet}^*(W) \to \tau_{fet}'^*(W)$, which yields a homotopy. $\diamondsuit$

This proposition says that the tangential morphisms essentially depend only on the choice of the basis of normal cotangent bundle of Z in V^*. For example, $t_i \mapsto t_i(1 - t_j)^{-1}$ does not change the morphism. This proposition is useful when we compare two distinct tangential base points which are mapped each other by an automorphism of V.

Another interesting example is $\mathbb{Q}\{t\} \to \mathbb{P}^1_{01\infty}$ and $\mathbb{Q}\{-t\} \to \mathbb{P}^1_{01\infty}$. These two are not infinitesimally homotopic. If we add $e^{2\pi i/N}$ for all N, then we have an infinitesimal homotopy $\mathbb{Q}^{ab}\{t\} \to \mathbb{Q}^{ab}\{-t\}$ by $t^{1/N} \to e^{\pi i/N}e^{1/N}$. This expresses the fact that a small half circle from $0\vec{1}$ to $1\vec{0}$ yields a path, on which $G_{\mathbb{Q}^{ab}}$ acts trivially. This path realizes the infinitesimal homotopy. By taking the conjugate by this path, we may identify the action of $G_{\mathbb{Q}^{ab}}$ on these two $\mathbb{Q}^{ab}$-rational tangential base points. If we change the base field to $\mathbb{Q}^{ab}$, then this is true for any $\mathbb{Q}^{ab}\{\zeta t\}$ with ζ a root of unity. We can regard this as $\operatorname{Spec}\mathbb{Q}^{ab}$ homotopically moves along an infinitesimally small circle around zero. A similar thing can be said about $\mathbb{Q}\{t\}$ and $\mathbb{Q}\{at\}$, with a a rational number. There exists a homotopy if we "restrict" to $\cup_{N\in\mathbb{N}}\mathbb{Q}[a^{1/N}]$.

2.3. Composition of tangential morphisms. Let us consider the diagram

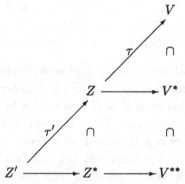

Here, τ and τ' are tangential morphisms which are composable in the following sense. The vertical inclusions are open immersions, and the horizontal arrows are immersions. There is an open affine subscheme U of V^{**}, on which there are functions $t_1, \ldots, t_l, s_1, \ldots, s_m$.

(1) $(V, V^{**}, U, t_1, \ldots, t_l, s_1, \ldots, s_m, Z')$ gives a tangential morphism $Z' \dashrightarrow V$, which is symbolically denoted by $\tau\tau'$. Thus, Z' is defined by $t_1 = \cdots = t_l = s_1 = \cdots = s_m = 0$ in U, and $V \cap U$ is the complement of $t_1 \ldots t_l s_1 \ldots s_m = 0$ in U.

(2) $(V, V^*, U \cap V^*, t_1, \ldots, t_l, Z)$ gives a tangential morphism $\tau : Z \dashrightarrow V$. (3) Z^* is the closed subscheme of U defined by $t_1 = \ldots = t_l = 0$. $(Z, Z^*, Z^*, s_1, \ldots, s_m, Z')$ gives a tangential morphism $\tau' : Z' \dashrightarrow Z$.

Proposition 2.3. *There is a canonical homotopy between two functors*

$$\tau'^*_{fet} \circ \tau^*_{fet}, \ (\tau\tau')^*_{fet} : V_{fet} \to Z'_{fet}.$$

Proof. Let W be a finite etale cover of V. τ^*_{fet} is defined by the following diagram

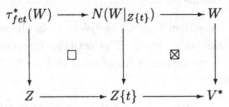

where $\square$ denotes the fiber product, and $\boxtimes$ denotes: to take the fiber product and take normalization of the base scheme at the generic fiber (denoted by $N()$). It is easy to see that this double-square diagram extends to

$$
\begin{array}{ccccc}
N(\tau^*_{fet}(W)) & \longrightarrow & N(W|_{Z^*\{t\}}) & \longrightarrow & W \\
\downarrow & & \downarrow & & \downarrow \\
& \square & & \boxtimes & \\
\downarrow & & \downarrow & & \downarrow \\
Z^* & \longrightarrow & Z^*\{t\} & \longrightarrow & V^{**}
\end{array}
$$

By uniqueness of normalization, we have a unique morphism $N(W|_{Z\{t\}}) \to N(W|_{Z^*\{t\}})$ which is a pullback along $Z\{t\} \to Z^*\{t\}$. Then, by pulling back along $Z \to Z\{t\}$, we have $\tau^*_{fet}(W)$. $N(\tau^*_{fet}(W))$ is the pullback of $N(W|_{Z^*\{t\}})$ if restricted to Z, and hence on all Z^*. Next we consider

$$
\begin{array}{ccccc}
\tau'^*_{fet} \circ \tau^*_{fet}(W) & \longrightarrow & N(\tau_{fet}(W)|_{Z'\{t\}}) & \longrightarrow & N(\tau^*_{fet}(W)) \\
\downarrow & & \downarrow & & \downarrow \\
& \square & & \boxtimes & \\
\downarrow & & \downarrow & & \downarrow \\
Z' & \longrightarrow & Z'\{t\} & \longrightarrow & Z^*
\end{array}
$$

and

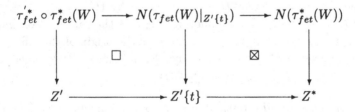

There exists a canonical morphism from the upper diagram to the lower one extending the pullback $Z^* \to Z^*\{t\}$ at the right end. This is because when restricted to the locus $t \neq 0$ of $Z'\{t\}$, $\boxtimes$ is just a pullback, and the normalization is unique. Now we have a canonical morphism $\tau'^*_{fet} \circ \tau^*_{fet}(W) \to (\tau\tau')^*_{fet}(W)$ which is a pullback of the identity $Z' \to Z'$, hence an isomorphism. $\diamond$

2.4. GAGA. In this section, we assume k to be a subfield of $\mathbb{C}$ and denote by $\bar{k}$ the algebraic closure in $\mathbb{C}$.

2.4.1. Tangential morphism without completion. In the construction of τ_{fet}^* in 2.1, the completion of U along Z is actually unnecessary. For a finite etale cover W of V, we obtain a finite etale cover $\tau_{fet}^*(W)$ of Z as in the following diagram:

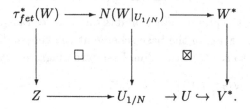

$U_{1/N}$ is defined by

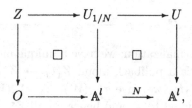

where the right down arrow $U \to \mathbb{A}^l$ is given by $(t_1, \ldots, t_l)$, $\mathbb{A}^l \xrightarrow{N} \mathbb{A}^l$ is given by $t_i \mapsto s_i^N$ with N a common multiple of the ramification index of each divisor corresponding to $t_i = 0$, and O is the origin of the affine space. This gives the same τ_{fet}^*, since by Abhyankar's Lemma $N(W)$ is etale over $U_{1/N}$ and $Z \to U_{1/N}$ factors through $\hat{Z}\{t\}$.

2.4.2. Analytification of a tangential point. For an algebraic variety V over k, we denote by $\bar{V}$ the base extension to $\bar{k} \subset \mathbb{C}$, and by V^{an} the complex analytic variety corresponding to V.

It is known that the category of unramified finite covers of V^{an} is canonically equivalent to the category of finite etale covers of $\bar{V}$ (SGA1 [11]). For a point Q of V^{an}, there is the associated fiber functor $F_Q : V_{fet}^{an} \to$ {Finite sets}. For another point R, the homotopical lift of a path in V^{an} from Q to R yields an isomorphic natural transformation from F_Q to F_R. This gives

$$\pi_1^{top}(V^{an}; Q, R) \to \pi_1(\bar{V}; F_Q, F_R),$$

which becomes an isomorphism if we complete the left set to $\lim_{\leftarrow N} \pi_1^{top}(V^{an}; Q, R)/N$, where N runs over the finite index normal subgroups of $\pi_1^{top}(V^{an}, Q)$.

In the case of a geometric tangential point, there is no corresponding topological point on V^{an}. So, we content ourselves with giving an isomorphism between the fiber functor coming from a tangential point and another fiber functor coming from a geometric point.

Let $\theta : \operatorname{Spec} \bar{k} \to \bar{V}$ be a geometric tangential point. Thus, we have an immersion $V \subset V^*$, open affine subset $U \subset V^*$, with $V \cap U = U - D$ where D is the union of the zeros of $t_1, \ldots, t_l \in \Gamma(U, \mathcal{O}_U)$, which are normally crossing. Let P be the point on U defined by $t_1 = \ldots = t_l = 0$. Since $t_i = 0$ are normal crossing, $(t_1, \ldots, t_l)$ gives a local coordinate at P, and induces an analytic isomorphism from an open neighborhood $\mathcal{N}_P$ of P in U^{an} to an open ball. Fix a point Q in this neighborhood, and identify $Q = (c_1, \ldots, c_l)$ by these coordinates. Let $\mathcal{N}_{P,1/N} \subset U_{1/N}^{an}$ be the inverse image of $\mathcal{N}_P \subset U^{an}$. This is again an open ball.

Fix a compatible system of $c_i^{1/N} \in \bar{k}$, $1 \le i \le l$, for $N \in \mathbb{N}$. Put $Q_N := (c_1^{1/N}, \ldots, c_l^{1/N}) \in \mathcal{N}_{P,1/N}$. Then, we can specify an isomorphic natural transformation from τ_{fet}^* to F_Q, as follows.

Let V, V^*, U be as in 2.1, W be a finite etale cover of V, and $U_{1/N}$, $N(W|_{U_{1/N}})$ be as in 2.4.1.

Then, the finite set $F_Q(W)$ is the fiber of $W \to V$ above $Q \in V$, which bijects to $F_{Q_N}(N(W|_{U_{1/N}}))$ by $U_{1/N} \to U$. Since $N(W|_{U_{1/N}})^{an} \to U_{1/N}^{an}$ is etale, the inverse image of $\mathcal{N}_{P,1/N}$ is isomorphic to the disjoint union of copies of $\mathcal{N}_{P,1/N}$. Now both $F_{Q_N}(N(W|_{U_{1/N}}))$ and $\theta_{fet}^* = F_P(N(W|_{U_{1/N}}))$ biject to this set of the copies. It is not difficult to see that this defines a natural transformation $\theta \to F_Q$.

Now we have fixed a "path" from θ to F_Q. There is a canonical way to choose Q and Q_N. We take c_i $(1 \le i \le l)$ to be small enough positive real numbers. We choose the $c_i^{1/N}$ to be all positive real. (This choice is then unique.) Let $Q' = (c_1', \ldots, c_l')$ be another such point. It is easy to see that the natural transformation $F_Q \to F_{Q'}$ defined by a path from Q to Q' in the positive real region makes the following diagram commute.

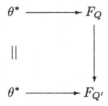

(This is because the positive real path between Q and Q' lifts to that between Q_N and Q_N' in $U_{1/N}$.)

Definition 2.3. We identify these fiber functors F_Q for any such Q, and denote it simply by θ^{an}.

2.4.3. Compatibility. We consider the case where $k = \bar{k}$ and Z' is a geometric point in 2.3. Let $\tau : Z \to V$ be a tangential morphism, and $\theta : Z' = \operatorname{Spec} k \to Z$ be a geometric tangential point, such that we have a composite tangential point $\tau\theta : \operatorname{Spec} k \to V$, as in 2.3. We denote by P the corresponding geometric point of V^{**} to $t_1 = \cdots = t_l = s_1 = \cdots = s_m = 0$. We assume another such $\theta' : \operatorname{Spec} k \to Z$ exists, so that we have a tangential morphism $\tau\theta'$. Then, we would like to understand the upper morphism in

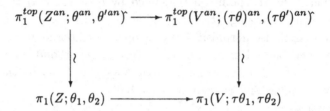

By definition of θ_1 and the composability of $\tau\theta_1$, we have a commutative diagram

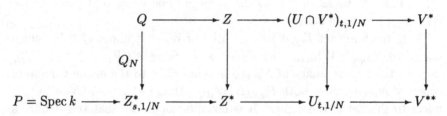

Here, $(U \cap V^*)_{1/N}$ etc. is defined by

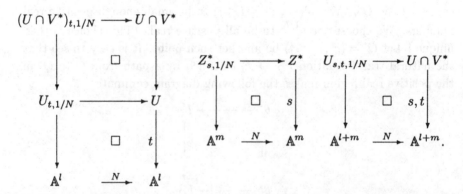

P is defined by $s_1 = \cdots = s_m = t_1 = \cdots t_l = 0$. Q is a point in Z chosen to analytify θ_1, whose coordinates are $(s_1, \ldots, s_m) = (c_1, \ldots, c_m)$. Since we fix the N-th roots of c_i's, we have Q_N on $Z^*_{s,1/N}$.

Let W be a finite etale cover of V. Take N as a common multiple of all ramification indices over V^{**}. Then, $\theta^* \circ \tau^*(W)$ is identified with the fiber of a suitable etale cover of $Z^*_{s,1/N}$ on the point P. This is identified with the fiber on $\tilde{P}$ in $N(W|_{U_{s,t,1/N}}) \to U_{s,t,1/N}$, where

$$\tilde{P} = \operatorname{Spec} k \to U_{s,t,1/N}$$

is defined by $s = t = 0$. We denote the functor taking the fiber on $\tilde{P}$ by $F_{\tilde{P}}$, which is identified with $(\tau\theta)^*$.

We fix $\tilde{Q} \in V^{**an}$, whose coordinate is small positive real numbers:

$$(s_1, \ldots, s_m, t_1, \ldots, t_l) = (c_1, \ldots, c_m, d_1, \ldots, d_l).$$

Then we have

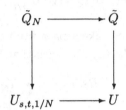

For θ', we have a similar construction to obtain $F_{P'}, F_{Q'}, F_{\tilde{P}'}, F_{\tilde{Q}'}$. The coordinates for θ' is given by $s'_1, \ldots, s'_m$, but we take the common $t_1, \ldots, t_l$ and $d_1, \ldots, d_l$.

What we want to see is the following. Let γ be a topological path from Q to Q' in Z^{an}. Then, which morphism can be substituted for the question mark in the diagram below, in order to make the diagram commute?

$$
\begin{array}{ccccccc}
F_P & \cong & F_Q & \xrightarrow{\gamma \circ \tau^*} & F_{Q'} & \cong & F_{P'} \\[2mm]
\| & & & & & & \| \\[4mm]
F_{\tilde{P}} & \cong & F_{\tilde{Q}} & \xrightarrow{\ ?\ } & F_{\tilde{Q}'} & \cong & F_{\tilde{P}'}.
\end{array}
$$

We can regard $F_P, F_Q, F_{\tilde{P}}, F_{\tilde{Q}}$, respectively, as a functor to take a fiber on $P, Q_N, \tilde{P}, \tilde{Q}_N \in U_{s,t,1/N}$, respectively, in the etale map $N(W|_{U_{s,t,1/N}}) \to U_{s,t,1/N}$. Let $\mathcal{N}_P \subset U^{an}$ be the open ball neighborhood of P, and $\mathcal{N}_{P,N}$ be its inverse image by $U_{s,t,1/N} \to U$, as in 2.4.2. Then, $P = \tilde{P}$, $Q_N = (c_1^{1/N}, \ldots, c_m^{1/N}, 0, \ldots, 0)$ and $\tilde{Q}_N = (c_1^{1/N}, \ldots, c_m^{1/N}, d_1^{1/N}, \ldots, d_l^{1/N})$ are all on $\mathcal{N}_{P,N}$. Since $\mathcal{N}_{P,N}$ is simply connected, all the functors F_P, F_Q and $F_{\tilde{Q}}$ are identified.

Similar things hold for θ'. Thus, the problem is reduced to obtaining

$$
\begin{array}{ccc}
F_Q & \xrightarrow{\gamma \circ \tau^*} & F_{Q'} \\[2mm]
\wr \downarrow & & \downarrow \wr \\[4mm]
F_{\tilde{Q}} & \xrightarrow{\ ?\ } & F_{\tilde{Q}'}
\end{array}
$$

We may consider these four functors as taking fibers in $N(W|_{U_{t,1/N}}) \to U_{t,1/N}$ on the points $R_N, R'_N, \tilde{R}_N, \tilde{R}'_N$, where

$$R_N = (c_1, \ldots, c_m, 0, \ldots, 0), \quad \tilde{R}_N = (c_1, \ldots, c_m, d_1^{1/N}, \ldots, d_l^{1/N})$$

in the (s, t) coordinate, and

$$R'_N = (c'_1, \ldots, c'_m, 0, \ldots, 0), \tilde{R}'_N = (c'_1, \ldots, c'_m, d_1^{1/N}, \ldots, d_l^{1/N})$$

in the (s', t) coordinate (note that $N(W|_{U_{t,1/N}}) \to U_{t,1/N}$ is etale on these points).

What we need to do is to make a path from $\tilde{Q}$ to $\tilde{Q}'$ in V^{an}, which lifts in $U_{t,1/N}$ to a path from $\tilde{R}_N$ to $\tilde{R}'_N$, such that the square of paths

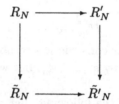

is homotopically trivial in $U_{t,1/N}$ for any N.

Let Z_ϵ be the inverse image of $\{(t_1, \ldots, t_l)|0 \leq t_1 \leq d_1, \ldots, 0 \leq t_l \leq d_l\}$ in $U \cap V^* \stackrel{(t_1, \ldots, t_l)}{\to} \mathbb{A}^l$. Let γ be the path from Q to Q' in $Z \subset Z_\epsilon$. Then, we take a path γ' from $\tilde{Q}$ to $\tilde{Q}'$ in V, which is homotopic to γ in Z_ϵ when composed with the paths $Q \to \tilde{Q}$, $Q' \to \tilde{Q}'$, given by $(t_1, \ldots, t_l) = (0, \ldots, 0) \to (d_1, \ldots, d_l)$. Now the following is clear.

Proposition 2.4. *The above γ' is the image of γ in*

$$\pi_1^{top}(Z^{an}; \theta^{an}, \theta'^{an})\gamma \to \pi_1^{top}(V^{an}; (\tau\theta)^{an}, (\tau\theta')^{an})\gamma.$$

In sum, the above homomorphism $\gamma \mapsto \gamma'$ is given by the translation of γ from Z to outside, by perturbing $t_1, \ldots, t_l$ (the parameters giving tangential morphism $\tau : Z \rightharpoonup V$) to small positive real, so that the perturbation is homotopically trivial in Z_ϵ.

Proof. Take the inverse image of $Z - Z_\epsilon$ by $(U \cap V^*)_{t,1/N} \to U \cap V^*$. The connected component with $t_1^{1/N}, \ldots, t_l^{1/N} > 0$ contains both $\tilde{R}_N$ and $\tilde{R}'_N$, and isomorphic to $Z - Z_\epsilon$. If we add Z to this component, then it becomes isomorphic to Z_ϵ, and γ' lifts to a path from $\tilde{R}_N$ to $\tilde{R}'_N$, which makes the square of paths homotopically trivial on $(U \cap V^*)_{t,1/N}$.   ◇

2.4.4. Identifiable tangential points. If two geometric tangential points have an infinitesimal homotopy, and the germs of the positive real regions coincide, then we can identify them as analytic tangential points, too. In this case, we say these two tangential points are *identifiable*, and identify through the canonical isomorphisms.

2.5. Example.

2.5.1. Projective line minus three points. For $V = \mathbb{P}^1_{01\infty} = \operatorname{Spec} \mathbb{Q}[t, 1/(1 - t)]$, we take $\mathbb{P}^1$ as V^*. Six $\mathbb{Q}$-rational tangential points are considered in Ihara [15] $t, -t, 1 - t, t - 1, 1/t, -1/t$, which were denoted by $\vec{01}, \vec{0\infty}, \vec{10}, \vec{\infty1}, \vec{\infty0}, \vec{1\infty}$ respectively. The symmetric group S_3 acts on these points. Precisely saying, we need canonical infinitesimal homotopies. For example the image of $1 - t$ by $0 \leftrightarrow \infty$ is $1 - 1/t = (t - 1)/t$, but $\mathbb{Q}\{(t - 1)/t\}$ and $\mathbb{Q}\{t - 1\}$ are canonically infinitesimally homotopic by $(t - 1)/t \mapsto (t - 1)/(1 + (t - 1)) = (t - 1) - (t - 1)^2 + (t - 1)^3 - \cdots$, so we may identify these two. We write $\tau_{01} : \mathbb{Q} \to \mathbb{P}^1_{01\infty}$ for the corresponding $\mathbb{Q}$-rational tangential base point, and $\bar{\tau}_{01}$ the geometric tangential point given by $\mathbb{Q} \hookrightarrow \overline{\mathbb{Q}}$. We have two generators $x, y \in \pi_1(\mathbb{P}^1_{01\infty}, \bar{\tau}_{01})$. We have a path $p \in \pi_1(\mathbb{P}^1_{01\infty}; \bar{\tau}_{01}, \bar{\tau}_{10})$. An element $\sigma \in G_{\mathbb{Q}}$ acts on p by $p \mapsto \tau_{10}(\sigma)p\tau_{01}(\sigma)^{-1} = pf_\sigma(x, y)$, where $f_\sigma(x, y) \in \pi_1(\mathbb{P}^1_{01\infty}, \bar{\tau}_{01})$ is an element of free profinite group of two generators x, y. This is the definition of $f_\sigma(x, y)$. Then, the action of σ is $\sigma : x \mapsto x^{\chi(\sigma)}, y \mapsto f_\sigma(x, y)^{-1} y^{\chi(\sigma)} f_\sigma(x, y)$.

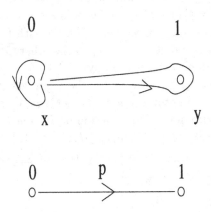

2.5.2. Another example. Let a be a rational number. We consider the (s, t)-plane minus a divisor: $V = \operatorname{Spec} k[s, t] - D$, where D is the union of the following five prime divisors: $s = 0, t = 0, t = 1, st = a, s(1 - t) = 1$.

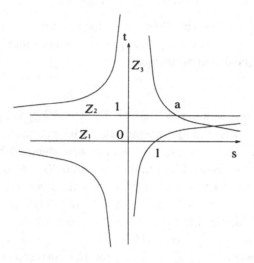

There are two horizontal $\mathbb{P}^1_{01\infty}$; namely, $t = 0, t = 1$, and one vertical $\mathbb{P}^1_{01\infty}$, $s = 0$. We name these Z_1, Z_2, Z_3, respectively. For Z_1 we choose t for the tangential morphism, for Z_2 we choose $1 - t$, and for Z_3 we choose s. Then, we have three tangential morphisms $\tau_i : Z_i \to V$. The composition of tangential morphisms $\mathbb{Q} \overset{\overline{01}}{\to} \mathbb{P}^1_{01\infty} \overset{\tau_i}{\to} V$ is, $\mathbb{Q}\{s,t\}$ for τ_1, $\mathbb{Q}\{as, 1-t\}$ for τ_2, and $\mathbb{Q}\{t,s\}$ for τ_3. The as for τ_2 comes from that Z_2 is isomorphic to $\mathbb{P}^1_{01\infty}$ if we take as as the coordinate. The image of $p : \vec{\tau}_{01} = \overline{\mathbb{Q}}\{t\} \to \vec{\tau}_{10} = \overline{\mathbb{Q}}\{1-t\}$ in Z_3 by τ_3 is a homotopy from $\overline{\mathbb{Q}}\{s,t\}$ to $\overline{\mathbb{Q}}\{s, 1-t\}$. Thus, unless $a \neq 1$, we have to fix a path (or homotopy) from the geometric tangential point $\overline{\mathbb{Q}}\{s, 1-t\}$ to the other $\overline{\mathbb{Q}}\{as, 1-t\}$ to patch Z_3 and Z_1. This is done by a homotopy $s^{1/N} \mapsto e^{-(\log a)/N}(as)^{1/N}$ from $\overline{\mathbb{Q}}\{s\}$ to $\overline{\mathbb{Q}}\{as\}$, named l_a. Here we choose the argument of $\log a$ to be πi if a is negative. $\tau_2(l_a)$ is a homotopy $\overline{\mathbb{Q}}\{s, 1-t\} \to \overline{\mathbb{Q}}\{as, 1-t\}$ compatible with Galois action. We write down Galois action on l_a. Let us define $\chi_a(\sigma) \in \widehat{\mathbb{Z}}$ by $\sigma(a^{1/N}) = e^{2\pi i/N \cdot \chi_a(\sigma)} a^{1/N}$. Then, $l_a^{-1} \sigma l_a \sigma^{-1} : \overline{\mathbb{Q}}\{s\} \to \overline{\mathbb{Q}}\{s\}$ is given by
$$s^{1/N} \overset{\sigma^{-1}}{\mapsto} s^{1/N} \overset{l_a}{\mapsto} a^{-1/N} s^{1/N} \overset{\sigma}{\mapsto} e^{-2\pi i/N \cdot \chi_a(\sigma)} a^{-1/N} s^{1/N} \overset{l_a^{-1}}{\mapsto} e^{-2\pi i/N \cdot \chi_a(\sigma)} s^{1/N}.$$
This coincides with $x^{\chi_a(\sigma)}$. Now, set

$$x_1 := \tau_1(x)$$
$$y_1 := \tau_1(y)$$
$$x_3 := \tau_3(x)$$
$$y_3 := \tau_3(y)$$
$$x_2 := \tau_3(p^{-1})\tau_2(l_a^{-1})\tau_2(x)\tau_2(l_a)\tau_3(p)$$
$$y_2 := \tau_3(p^{-1})\tau_2(l_a^{-1})\tau_2(y)\tau_2(l_a)\tau_3(p)$$

The actions of $\sigma \in G_{\mathbb{Q}}$ on x_1, y_1 and x_3, y_3 are same as that of $\mathbb{P}^1_{01\infty}$, since τ_1 and τ_3 are tangential morphism over $\mathbb{Q}$: $\sigma : x_i \mapsto x_i^{\chi(\sigma)}, y_i \mapsto$

$f_\sigma(x_i, y_i)^{-1} y_i^{\chi(\sigma)} f_\sigma(x_i, y_i)$ for $i = 1, 3$. For $i = 2$, we need the action on l_a. We have $\sigma : p \mapsto p f_\sigma(x, y)$ and $\sigma : l_a \mapsto l_a x^{\chi_a(\sigma)}$. By substitution, we have

$$x_2 \mapsto f_\sigma(x_3, y_3)^{-1} x_2^{\chi(\sigma)} f_\sigma(x_3, y_3)$$

and

$$y_2 \mapsto f_\sigma(x_3, y_3)^{-1} x_2^{\chi_a(\sigma)} f_\sigma(x_2, y_2)^{-1} y_2^{\chi(\sigma)} f_\sigma(x_2, y_2) x_2^{-\chi_a(\sigma)} f_\sigma(x_3, y_3).$$

This example shows that topologically identical objects may give rise to different Galois actions. Even a scaling in the embedding of $\mathbb{P}^1_{01\infty}$ influences on the Galois action.

§3. Galois action on Artin groups

3.1. Motivation

Let V be a geometrically connected variety over the rational number field $\mathbb{Q}$, and x be a $\mathbb{Q}$-rational point of V. Then, we have Galois representations on the fundamental group of V:

$$\rho_{V,x} : \mathrm{Gal}(\overline{\mathbb{Q}}/\mathbb{Q}) \to \mathrm{Aut}\,\hat{\pi}_1(V, x),$$

where $\hat{\pi}_1(V, x)$ denote the profinite completion of the topological fundamental group $\pi_1^{top}(V(\mathbb{C}), x)$. We may use $\mathbb{Q}$-rational tangential base point as x.

To obtain an approximation of the Galois group, it is a natural idea to consider all V, and consider systems of elements of $\mathrm{Aut}\,\hat{\pi}_1(V, x)$, one for each V, which commutes with any algebraic homomorphism $V \to W$ defined over $\mathbb{Q}$.

We may start with some restricted families of varieties and restricted families of morphisms. Grothendieck-Teichmüller group defined by Drinfel'd [9] is the one obtained from $M_{0,4}$ and $M_{0,5}$, and some natural morphisms (see Ihara [16]).

It would be also natural to consider the families of $M_{g,n}$ with all g, n. One open problem is whether a new relation occurs or not if we consider $g > 0$, not only $g = 0$. For $g = 2$, it is easy to prove that the Grothendieck-Teichmüller group acts on M_2, and hence no new relations will be obtained. So, the first nontrivial example is $g = 3$. The author knows nothing on this.

The author also suggests the possibility to consider the family of $V(\Phi)$ with root systems Φ, with some (still ambiguous) suitable morphisms.

One of the reasons is that these varieties have a moduli interpretation of the deformation spaces of singularities with finite group action [31].

This idea comes from a trial to find a geometric meaning of an interesting research by P. Lochak and L. Schneps [20]. There, they treated the families of groups Ar_n, Br_n, $n \geq 3$, with Br_n the braid group of n-strands and Ar_n the subgroup of Br_n consisting of braids that connects the first point with the first point, i.e., the inverse image of the stabilizer of the first element by the morphism $Br_n \to S_n$. Their main result is: if we consider Ar_n, Br_n $n \geq 3$ and suitable group homomorphisms, then we can characterize Grothendieck-Teichmüller group as the automorphism group of a tower of these groups. (In fact, $n = 3, 4$ suffices.)

The author just notes that this somewhat strange group Ar_n is the fundamental group of $V(B_n) \cong V(C_n)$. The author does not see what are the corresponding algebraic homomorphisms to those group homomorphisms considered in [20].

We shall write down the Galois action on $\hat{\pi}_1 V(\Phi)$, in terms of that on $\mathbb{P}^1_{01\infty}$ for $\Phi = A_n$, B_n, C_n, D_n or E_7. The author did not study E_6, E_8, F_4 and G_2. Since A_n case was treated in [17] and B, C, D basically follow from this, we first treat the most nontrivial case E_7.

3.2. Tangential base point on $V(E_7)$.

3.2.1. E_7 root system.
An E_7 root system is constructed as follows. Consider seven vectors $e_2 - e_3, e_3 - e_4, e_4 - e_5, e_5 - e_6, e_6 + e_7, -\frac{1}{2}\sum_{i=1}^{8} e_i$ in $\mathbf{A}^8$, where $e_1, \ldots, e_8$ denote the unit vectors. Let U be the seven dimensional subspace of $\mathbf{A}^8$ spanned by these vectors. Then, these seven vectors are fundamental roots of E_7 root system. Let $U(E_7)$ denote U minus the reflection hyperplanes. If we use the coordinate $\sum_{i=2}^{7} x_i e_i + x_8(e_1 + e_8)$ of U, then the corresponding Weyl Chamber C is $-x_6 < x_7 < x_6 < x_5 < \cdots < x_2$, $x_8 < -\frac{1}{2}(x_2 + \cdots + x_7)$ By a general result of Brieskorn [6], the following are the generators of E_7 Artin group.

In this figure, the moves of the configuration of the points give a path in $U(E_7)(\mathbb{C})$ from C to T_iC, where T_i is an element of the Weyl group corresponding to τ_i. After taking the quotient by the Weyl group, these give the desired elements of $\pi_1^{top}(V(E_7), C)$, with it Dynkin diagram as follows:

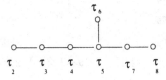

3.2.2. Blow up. Let $\mathbb{A}^7{}_t$ denote seven dimensional affine space with coordinates $t_2, \ldots, t_8$, and $\mathbb{A}^7{}_x$ that with coordinates $x_2, \ldots, x_8$. We shall define $\mathbb{A}^7{}_t \to \mathbb{A}^7{}_x$, which induces

$$
\begin{array}{ccc}
U(E_7) & \lhook\joinrel\longrightarrow & \mathbb{A}^7{}_t = V^* \\
\| & & \downarrow \\
U(E_7) & \lhook\joinrel\longrightarrow & \mathbb{A}^7{}_x,
\end{array}
$$

so that the upper morphism gives a $\mathbb{Q}$-rational tangential point $(U(E_7), V^*, Y, t_2, \ldots, t_8, \operatorname{Spec} \mathbb{Q})$, for a suitable open affine $Y \subset V^*$.

$\mathbb{A}^7{}_t \to \mathbb{A}^7{}_x$ is defined as follows:

$$
\begin{aligned}
x_2 &= t_2 \\
x_3 &= t_2 t_3 \\
x_4 &= t_2 t_3 t_4 \\
x_5 &= t_2 t_3 t_4 t_5 \\
x_6 &= t_2 t_3 t_4 t_5 t_6 \\
x_7 &= t_2 t_3 t_4 t_5 t_6 t_7 \\
x_8 &= -\frac{1}{2}(t_2 + t_2 t_3 + t_2 t_3 t_4 + t_2 t_3 t_4 t_5 + t_2 t_3 t_4 t_5 t_6 \\
&\quad + t_2 t_3 t_4 t_5 t_6 t_7 + 2 t_2 t_3 t_4 t_5 t_6 (t_7 + 1) t_8).
\end{aligned}
$$

These parameters are chosen so that on $U(E_7)$ we have the following inverse map:

$$t_2 = x_2$$
$$t_3 = x_3/x_2$$
$$t_4 = x_4/x_3$$
$$t_5 = x_5/x_4$$
$$t_6 = x_6/x_5$$
$$t_7 = x_7/x_6$$
$$t_8 = -(x_2 + x_3 + \cdots + x_7 + 2x_8)/2(x_6 + x_7).$$

The removed divisors are

$$x_i \pm x_j = 0 \quad (2 \leq i < j \leq 7)$$

$$\sum_{i=2}^{7} \pm x_i + 2x_8 = 0 \quad \text{(the number of minus is even.)}$$

in $\mathbf{A}^7{}_x$. The corresponding divisors in $\mathbf{A}^7{}_t$ are:

$$t_2 t_3 \cdots t_i (1 \pm t_{i+1} \cdots t_j) = 0 \quad (2 \leq i < j \leq 7)$$

$$\sum_{i=2}^{7} \pm t_2 \cdots t_i = \sum_{i=2}^{7} t_2 \cdots t_i + 2t_2 \cdots t_6 (1 + t_7) t_8,$$

where the number of minus in the lower identity is even. We denote by D the set of the prime divisors contained in the above-defined locus in $\mathbf{A}^7{}_t$. So, D contains the divisors $t_i = 0$, $(2 \leq i \leq 8, i \neq 7)$. After removing these, there are no prime divisors on $(t_2, \ldots, t_8) = (0, \ldots, 0)$. This is easy to see for the former type of divisors. For the latter, let k be the minimum such that the minus sign is chosen in the left side of the second equation. If $k \leq 5$, then $-t_2 \cdots t_k \pm t_2 \cdots t_{k+1} \pm \cdots = t_2 \cdots t_k + t_2 \cdots t_{k+1} + \cdots$ holds. Both sides can be divided by the first term (since we already removed $t_i = 0$). Thus we get $-2 = t_{k+1} \times f(t)$, where $f(t)$ is a polynomial in $t_{k+2}, \ldots, t_8$ with constant term 2 or 0. These do not lie at the origin. If $k = 6$, then automatically the minus sign is chosen for $t_2 \cdots t_7$. Then, the corresponding divisor after divided by $t_2 \cdots t_6$ is $-1 - t_7 = 1 + t_7 + 2(1 + t_7)t_8$, i.e., $(1 + t_7)(1 + t_8) = 0$. If k does not exist, then all signs are plus, and we have $(1 + t_7)t_8 = 0$.

Since $t_i = 0$ $(2 \leq i \leq 8)$ are normal crossing, we defined a tangential morphism $\operatorname{Spec} \mathbb{Q} \to U(E_7)$ by $(U(E_7), V^*, Y, t_2, \ldots, t_8, \operatorname{Spec} \mathbb{Q})$ with Y the complement in V^* of $D - \{t_i = 0 | 2 \leq i \leq 8, i \neq 7\}$.

Note that $t_7 = 0$ is not a removed divisor, and thus we need some modification for the definition of tangential morphism. However, this is easy: just do not take $t_7^{1/N}$, but simply t_7 in the definition of the associated functor between the etale site in 2.1. Then it is easy to see that $(t_1, \ldots, t_6, ct_7, t_8)$ for any $c \in \mathbb{Q}$ gives the same tangential morphism. We denote this tangential point by $\mathcal{O}$, as origin.

Now we have seven tangential morphisms μ_i $(2 \leq i \leq 8)$, which correspond to $(t_1, \ldots, t_{i-1}, t_{i+1}, \ldots, t_8)$. It is not difficult to show that this gives a tangential morphism from $\mathbb{A}^1 - \{0, \pm 1\}$ to $U(E_7)$ for μ_i, $1 \leq i \leq 6$, from $\mathbb{A}^1 - \{\pm 1\}$ to $U(E_7)$ for μ_7, and from $\mathbb{A}^1 - \{-1, 0\}$ to $U(E_7)$ for μ_8.

This can be seen as follows. For each μ_i, $t_j = 0$ holds for every $2 \leq j \leq 8$ except one j, namely, $j = i$. Prime divisors which can intersect with such one dimensional locus are: of the first type, $1 \pm t_j = 0$ $(3 \leq j \leq 7)$ and of the second type, $2 \leq k \leq 5$, $-2 = t_{k+1}f(t)$, where $f(t)$ is a polynomial in $t_{k+2}, \ldots, t_8$ with constant term 0 or 2, or, $(t_7 + 1)(t_8 + 1)$ we saw just above. For the former, surely $1 \pm t_j$ is removed in the codomain of μ_j. For the second, if the constant term of $f(t)$ is zero, then this does not intersect with the locus $t_2 = \cdots = t_{j-1} = t_{j+1} = \cdots = t_8 = 0$, and if the constant term is two, then this intersects at $t_j = -1$. Thus, in any way, μ_j defines a tangential morphism with codomain described above. Note also that, in the above investigation, we noticed that $t_j - 1 = 0$ and $t_i = 0$ $(i \neq j)$ are all the divisors on $(0, 0, \ldots, 0, 1, 0, \ldots 0)$. (The troublesome divisors are $-2 = t_{k+1} \times f(t)$, where $f(t)$ is a polynomial in $t_{k+2}, \ldots, t_8$ with constant term 2 or 0. These may pass through $(0, 0, \ldots, 0, -1, 0, \ldots, 0)$, but not through $(0, 0, \ldots, 0, 1, 0, \ldots, 0)$.) Thus, we can define a tangential point $(t_2, \ldots, t_{j-1}, t_j - 1, t_{j+1}, \ldots, t_8)$ on $U(E_7)$.

For t_7, all the divisors on $(0, 0, \ldots, 0, -1, 0)$ are $t_7 + 1 = 0$, $t_i = 0$ $(1 \leq i \leq 8, i \neq 7)$. Again a trouble may lie in $-2 = t_{k+1} \times f(t)$, but this divisor exists for $k \leq 5$. Thus we have a tangential point $(t_2, \ldots, t_6, t_7 + 1, t_8)$.

3.3. $\mathbb{P}^1_{01\infty}$ and $\mathbb{P}^1_{0\pm 1\infty}$.

We consider a double cover

$$\mathbb{A}^1 - \{0, \pm 1\} \to \mathbb{A}^1 - \{0, 1\}$$

given by $t \mapsto t^2$. Then, $\pi_1^{top}(\mathbb{A}^1 - \{0, \pm 1\}, \vec{01})$ has three generators as shown below.

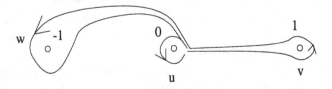

It is easy to see that $u = x^2$, $v = y$, $w = x^{-1}yx$ in the inclusion $\pi_1^{top}(\mathbb{A}^1 - \{0, \pm 1\}, \vec{01}) \to \pi_1^{top}(\mathbb{A}^1 - \{0, 1\}, \vec{01})$. Let p denote the path from $\vec{01}$ to $\vec{10}$

on the $(0,1)$ interval of $\mathbb{A}^1 - \{0, \pm 1\}$. An element $\sigma \in G_\mathbb{Q}$ acts on this path by $p \mapsto p g_\sigma(u, v, w)$. Here, $g_\sigma(u, v, w)$ is a proword. In $\pi_1(\mathbb{A}^1 - \{0,1\})$, we have $g_\sigma(u, v, w) = f_\sigma(x, y)$. Thus, $g_\sigma(\xi^2, \eta, \xi^{-1}\eta\xi) = f_\sigma(\xi, \eta)$ for any two elements ξ, η in any profinite group.

We also define elements of groupoids in the figure below.

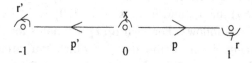

We use the following notation. For an odd number χ, we denote by r^χ the following path.

Here, the point goes around $(\chi - 1)/2$ times counterclockwise. By passing to the profinite completion, we define r^χ for $\chi \in \widehat{\mathbb{Z}}^\times$.

3.4. Galois action on each generator.

3.4.1. τ_i $(i = 2, 3, 4, 5)$. Let i be a number $2 \leq i \leq 5$. Then, roughly, τ_i is realized by moving t_{i+1} in the figure below, i.e., rp in the previous section. (See 3.2.1 for a global picture. If t_{i+1} moves as below, then $x_{i+1}, \ldots, x_7$ move proportionally, and the ordering of the points x_i and x_{i+1} is reversed and the other orderings are stable.)

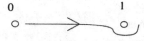

Let $\mu_{i+1,t}$ be the tangential morphism given by $(t_2, t_3, \ldots, t_i, t_{i+2}, \ldots, t_8)$ from $\mathbb{A}^1 - \{0, \pm 1\}$ to $U(E_7)$. Then, $\mu_{i+1,t}(rp)$ is a topological path on $U(E_7)$, with $t_j = \epsilon$ for $j \neq i + 1$, and t_{i+1} moves along rp. Since ϵ is positive small real number, $x_7, x_6, \ldots, x_{i+2}$ moves around the small positive real number and don't change their order. Only x_{i+1} moves to the right to x_i, and other x_j $(j \leq i)$ does not move. x_8 moves but does not go beyond $-(x_2 + \cdots + x_7)/2$. This gives τ_i.

The tangential morphism $\mu_{i+1,t}$ maps the starting (tangential) point (t_{i+1}) of rp on $\mathbb{A}^1 - \{0, \pm 1\}$ to the tangential point $(t_2, t_3, t_4, t_5, \ldots, t_8)$ on $U(E_7)$, i.e., the origin $\mathcal{O}$. It maps the terminal (tangential) point $(t_{i+1} - 1)$ to $(t_2, t_3, \ldots, t_i, t_{i+1} - 1, t_{i+2}, \ldots, t_8)$.

If we contract the fundamental Weyl chamber (since it is simply connected), then the image of this path is surely τ_{i+1} after $U(E_7)$ divided by the Weyl group. However, to obtain the correct Galois action, we need to connect the end (tangential) point of this path to the tangential base point $T_i\mathcal{O}$. (Remember that T_i is the element of the Weyl group corresponding to τ_i.)

T_i exchanges x_i and x_{i+1}. Let us assume $2 \le i \le 4$ for a while. Then the coordinates giving $T_i\mathcal{O}$ is $s_2, s_3, \ldots, s_8$, with $s_2 = t_2, s_3 = t_3, \ldots, s_i = t_i t_{i+1}, s_{i+1} = 1/t_{i+1}, s_{i+2} = t_{i+1}t_{i+2}, s_{i+3} = t_{i+3}, \ldots, s_8 = t_8$. (This can be calculated by exchanging x_{i+1} and x_i in the defining identities of t's. For $i = 5$, $s_8 \ne t_8$, and we shall consider this case soon later.) We denote by $\mu_{i+1,s}$ the tangential morphism from $\mathbb{A}^1 - \{0, \pm1\}$ to $U(E_7)$ given by $(s_2, s_3, \ldots, s_i, s_{i+2}, \ldots, s_8)$. The end tangential point of $\mu_{i+1,t}(rp)$ is given by $(t_2, t_3, \ldots, t_i, t_{i+1} - 1, t_{i+2}, \ldots, t_8)$.

Since $s_i = t_i(1 + (t_{i+1} - 1)), 1 - s_{i+1} = (t_{i+1} - 1)/(1 + (t_{i+1} - 1)), s_{i+2} = t_{i+2}(1 + (t_{i+1} - 1)), s_{i+3} = t_{i+3}, \ldots, s_8 = t_8$, this tangential point is identifiable with $(s_2, s_3, \ldots, s_i, 1 - s_{i+1}, s_{i+2}, \ldots, s_8)$. Now we move s_{i+1} from $1 - \epsilon$ to ϵ, then we return to the tangential point given by $(s_2, s_3, \ldots, s_8)$. Thus, the path τ_i from the tangential point (t_i) to (s_i) is obtained by the composition of the following three paths: t_{i+1}: ϵ to $1 - \epsilon$ (on the interval $(0,1)$), $1 - \epsilon$ to $1 + \epsilon$ (rotate counterclockwise), s_{i+1}:$1 - \epsilon$ to ϵ (on the interval $(0,1)$). Here, the end point of the first path is identified with the starting point of the last path via infinitesimal homotopy.

Then,

$$\tau_i = \mu_{i+1,s}(p^{-1})\mu_{i+1,t}(r)\mu_{i+1,t}(p).$$

Thus, the image of this element by $\sigma \in G_{\mathbb{Q}}$ is

$$\mu_{i+1,s}(g_\sigma(u,v,w)^{-1}p^{-1})\mu_{i+1,t}(r^{\chi(\sigma)})\mu_{i+1,t}(pg_\sigma(u,v,w)).$$

We define ξ_i by $\xi_i := \mu_{i+1,t}(x)$. This path connects the fundamental Weyl chamber to another chamber, and can be considered as an element of $\pi_1(V(E_7))$, where $V(E_7) = U(E_7)/W(E_7)$.

One can show that

$$\xi_i^2 = \mu_{i+1,t}(u), \tau_i^2 = \mu_{i+1,t}(v), \xi_i^{-1}\tau_i^2\xi_i = \mu_{i+1,t}(w).$$

A difficulty in the proof is in the jump along Weyl chambers. The element ξ_i in the Artin group yields an element of the Weyl group, denoted by X_i, which is given explicitly by

$$(x_1, \ldots, x_8) \mapsto (x_2, \ldots, x_i, -x_{i+1}, -x_{i+2}, \ldots, -x_6, \pm x_7, x_8)$$

Here, $\pm$ of x_7 is chosen so that the number of minus is even. The tangential point $X_i\mathcal{O}$ is given by $(t_1, t_2, \ldots, t_i, -t_{i+1}, t_{i+2}, \ldots, \pm t_7, t_8)$. Here $t_7 = 0$ does not ramify, so $\pm$ does not affect. To show the last identity $\xi_i^{-1}\tau_i^2\xi_i = \mu_{i+1,t}(w)$, for example, we have to show that the move of t_{i+1} in the next figure

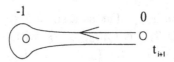

coincide with τ_i^2, after identifying by X_i the Weyl chamber containing $X_i\mathcal{O}$ with that containing $\mathcal{O}$. But, explicitly X_i exchanges v and the above path, and hence both give τ_i^2.

Since $\tau_i^{\chi(\sigma)} = \mu_{i+1,s}(p^{-1})\mu_{i+1,t}(r^{\chi(\sigma)})\mu_{i+1,t}(p)$, we have

$$\sigma : \tau_i \mapsto f_\sigma(\xi_i, \tau_i^2)^{-1}\tau_i^{\chi(\sigma)}f_\sigma(\xi_i, \tau_i^2).$$

Now, ξ_{i+1} commutes with both ξ_i and τ_i^2. Since $f_\sigma(x, y)$ is in the commutator, we have $f_\sigma(\xi_i, \tau_i^2) = f_\sigma(\eta_i, \tau_i^2)$, where $\eta_i = \xi_i\xi_{i+1}^{-1}$. This path η_i is given by moving t_{i+1} along x at the same time moving t_{i+2} along x'. Thus only x_{i+1} moves and other x_j are fixed. We have a formula $\eta_i = \tau_{i+1}\tau_{i+2}\cdots\tau_4\tau_5(\tau_6\tau_7)\tau_5\tau_4\cdots\tau_{i+2}\tau_{i+1}$. This can be proved by chasing the Weyl chambers.

For $i = 5$, we need a small modification, since the s-coordinate is then $s_2 = t_2, \ldots, s_4 = t_4, s_5 = t_5t_6, s_6 = t_6^{-1}, s_7 = t_6t_7$, but

$$\begin{aligned}
s_8 &= -(x_2 + \cdots x_7 + 2x_8)/2(x_5 + x_7) = (x_6 + x_7)t_8/(x_5 + x_7) \\
&= t_6(1 + t_7)t_8/(1 + t_6t_7).
\end{aligned}$$

Here we need to check the existence of an infinitesimal homotopy between $(t_2, \ldots, t_5, t_6 - 1, t_7, t_8)$ and $(s_2, \ldots, s_5, 1 - s_6, s_7, s_8)$. The only difference is s_8, and

$$s_8 = t_8(1 + t_7)(1 + (t_6 - 1))/(1 + (1 + (t_6 - 1))t_7)$$

shows the infinitesimal homotopy. The remaining arguments follow in the same way.

Thus, we have

$$\sigma : \tau_i \mapsto f(\eta_i, \tau_i^2)^{-1}\tau_i^{\chi(\sigma)}f(\eta_i, \tau_i^2)$$

for $2 \leq i \leq 5$, where $\eta_i = \tau_{i+1}\tau_{i+2}\cdots\tau_4\tau_5\tau_6\tau_7\tau_5\tau_4\cdots\tau_{i+2}\tau_{i+1}$.

3.4.2. τ_6. τ_6 is the image of rp by $\mu_{7,t}$, similarly to the above. T_6 exchanges x_6 and x_7 and maps t-coordinates to s-coordinates by $s_2 = t_2, \ldots, s_5 = t_5, s_6 = t_6 t_7, s_7 = 1/t_7, s_8 = t_8$. Now the situation is completely analogous to the previous section, and we have

$$\tau_6 = \mu_{7,s}(p^{-1})\mu_{7,t}(rp)$$

and

$$\sigma : \tau_6 \mapsto g_\sigma(\mu_{7,s}(u), \mu_{7,s}(v), \mu_{7,s}(w))^{-1}\tau_6^{\chi(\sigma)}g_\sigma(\mu_{7,t}(u), \mu_{7,t}(v), \mu_{7,t}(w)).$$

A difference from the previous section is that we don't have a good ξ_{i+1} for $i = 6$. Thus we need a direct computation. We have $u = 1$ in $\pi_1(\mathbb{A}^1 - \{\pm 1\})$. $\mu_{7,t}(v) = \tau_6^2$. However, $\mu_{7,t}(w)$ requires some calculation. This is the path in $U(E_7)$ shown in the figure below.

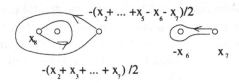

This is actually $(\tau_7 \tau_8 \tau_7)^2$. This can be checked by a straightforward chasing of the Weyl Chambers where this path passes over. It is left to the reader. Note that $(\tau_7 \tau_8 \tau_7)^2$ commute with both τ_7 and τ_8, by the braid relations.

Thus, we have

$$\tau_6 \mapsto g_\sigma(1, \tau_6^2, (\tau_7 \tau_8 \tau_7)^2)^{-1}\tau_6^{\chi(\sigma)}g_\sigma(1, \tau_6^2, (\tau_7 \tau_8 \tau_7)^2).$$

Now τ_6 commutes with both τ_σ^2 and $(\tau_7 \tau_8 \tau_7)^2$, hence g_σ cancels out. Thus we have

$$\tau_6 \mapsto \tau_6^{\chi(\sigma)}.$$

3.4.3. τ_7 and τ_8. τ_7 is the most complicated element. If we move t_7 along $r'p'$ (see 3.3), then x_i's move as follows.

To make this τ_7, we have to move x_8 to the left. This can be done as follows. Let s be the image of t by T_7. T_7 maps $x_6 \mapsto -x_7$ and $x_7 \mapsto -x_6$.

Thus we have $s_2 = t_2, \ldots, s_5 = t_5, s_6 = -x_7/x_5 = -t_6 t_7$, $s_7 = x_6/x_7 = 1/t_7$, $s_8 = -1/2 \cdot (x_2 + \cdots + x_5 - x_6 - x_7 + 2x_8)/(-x_6 - x_7) = -(t_8 + 1)$. Now, τ_7 is realized by

$$\tau_7 = \mu_{8,s}(\alpha)\mu_{7,s'}(p'^{-1})\mu_{7,t}(r'p').$$

Here, $\mu_{7,s'}$ is the tangential morphism coming from $(s_2, s_3, \ldots, s_6, -(s_8+1))$, and α is the following path:

That is, $x^{-1}p^{-1}r'^{-1}$ in 3.3. The right most $\mu_{7,t}(r'p')$ moves $x_2, \ldots, x_8$ as in the figure in 3.4.2, $\mu_{7,s'}(p'^{-1})$ changes the ratio, then $\mu_{8,s}(\alpha)$ unwind x_8 to get τ_7. For these three paths, the end tangential point is identifiable, via an infinitesimal homotopy, with the start tangential point of the next path. The end of the first path is $(t_2, \ldots, t_6, -(t_7 + 1), t_8)$, and the start of the second path is $(s_2, \ldots, s_6, s_7 + 1, -(s_8 + 1))$. These are identifiable. The next point is $(s_2, \ldots, s_7, -(s_8 + 1))$, which is compatible with both $\mu_{7,s'}$ and $\mu_{8,s}$.

Now the Galois action is described as follows:

$$\sigma(\tau_7) = \mu_{8,s}(\sigma(\alpha))\mu_{7,s'}(\sigma(p'^{-1}))\mu_{7,t}(\sigma(r'p')).$$

We shall calculate these three terms from the right. Now

$$\sigma(r'p') = r'^{\chi(\sigma)}p'g_\sigma(1, w, v),$$

where the exchange of w and v comes from the relation between p and p'. We have

$$\mu_{7,t}(v) = \tau_6^2$$
$$\mu_{7,t}(w) = \mu_{7,t}(p'^{-1}r'^2 p') = (\tau_7\tau_8\tau_7)^2.$$

Then

$$\mu_{7,t}(\sigma(r'p')) = \mu_{7,t}(r'p'p'^{-1}r'^{\chi(\sigma)-1}p'g_\sigma(1, w, v))$$
$$= \mu_{7,t}(r'p')(\tau_7\tau_8\tau_7)^{\chi(\sigma)-1}g_\sigma(1, (\tau_7\tau_8\tau_7)^2, \tau_6^2).$$

The middle term is

$$\mu_{7,s'}(\sigma(p'^{-1})) = \mu_{7,s'}(g_\sigma(1, w, u)^{-1}p'^{-1})$$
$$= \mu_{8,s}(\alpha)^{-1} \cdot \mu_{8,s}(\alpha)\mu_{7,s'}(g_\sigma(1, w, u))^{-1}\mu_{8,s}(\alpha)^{-1}\mu_{8,s}(\alpha)\mu_{7,s'}(p'^{-1}).$$

Now, we have

$$\mu_{8,s}(\alpha)\mu_{7,s'}(v)\mu_{8,s}(\alpha)^{-1} = \tau_6^2$$
$$\mu_{8,s}(\alpha)\mu_{7,s'}(w)\mu_{8,s}(\alpha)^{-1} = (\tau_7\tau_8\tau_7)^2.$$

The former is proved using the commutativity of τ_6 with τ_7, τ_8, and the latter is proved by drawing a picture. Then we have

$$\mu_{7,s'}(\sigma(p'^{-1})) = \mu_{8,s}(\alpha)^{-1} \cdot g_\sigma(1, (\tau_7\tau_8\tau_7)^2, \tau_6^2))^{-1} \mu_{8,s}(\alpha)\mu_{7,s'}(p'^{-1}).$$

The first term is obtained as follows.

$$\begin{aligned}
\sigma(\alpha) &= x^{-\chi(\sigma)}p'^{-1}f_\sigma(r'^2, p'x^2p'^{-1})r'^{-\chi(\sigma)} \\
&= x^{-\chi(\sigma)+1}x^{-1}f_\sigma(p'^{-1}r'^2p', x^2)x \cdot x^{-1}p'^{-1}r'^{-\chi(\sigma)+1}p'x \cdot x^{-1}p'^{-1}r'^{-1}.
\end{aligned}$$

Then we have

$$\mu_{8,s}(x^{-1}p'^{-1}r'^2p'x) = (\tau_8^{-1}\tau_7^2\tau_8).$$

and

$$\mu_{8,s}(x^2) = \tau_8^2.$$

Thus,

$$\mu_{8,s}(\sigma(\alpha)) = \tau_8^{-\chi(\sigma)+1}f_\sigma(\tau_8^{-1}\tau_7^2\tau_8, \tau_8^2) \cdot (\tau_8^{-1}\tau_7^2\tau_8)^{\frac{-\chi(\sigma)+1}{2}} \cdot \mu_{8,s}(\alpha).$$

Gathering all together, we have

$$\begin{aligned}
\tau_7 \mapsto &\tau_8^{-\chi(\sigma)+1}f_\sigma(\tau_8^{-1}\tau_7^2\tau_8, \tau_8^2)(\tau_8^{-1}\tau_7^2\tau_8)^{\frac{-\chi(\sigma)+1}{2}}\mu_{8,s}(\alpha)\times \\
&\mu_{8,s}(\alpha)^{-1}g_\sigma(1, (\tau_7\tau_8\tau_7)^2, \tau_6^2))^{-1}\mu_{8,s}(\alpha)\mu_{7,s'}(p'^{-1})\times \\
&\mu_7(r'p')(\tau_7\tau_8\tau_7)^{\chi(\sigma)-1}g_\sigma(1, (\tau_7\tau_8\tau_7)^2, \tau_6^2).
\end{aligned}$$

Thus we get

$$\begin{aligned}
\tau_7 \mapsto &\tau_8^{-\chi(\sigma)+1}f_\sigma(\tau_8^{-1}\tau_7^2\tau_8, \tau_8^2)(\tau_8^{-1}\tau_7^2\tau_8)^{\frac{-\chi(\sigma)+1}{2}} \\
&g_\sigma(1, (\tau_7\tau_8\tau_7)^2, \tau_6^2))^{-1}\tau_7(\tau_7\tau_8\tau_7)^{\chi(\sigma)-1}g_\sigma(1, (\tau_7\tau_8\tau_7)^2, \tau_6^2).
\end{aligned}$$

Since $(\tau_7\tau_8\tau_7)^2$, τ_6, τ_7 are commutative, g_σ cancels out, and we have the following result:

$$\tau_7 \mapsto \tau_8^{-\chi(\sigma)+1}f_\sigma(\tau_8^{-1}\tau_7^2\tau_8, \tau_8^2)(\tau_8^{-1}\tau_7^2\tau_8)^{\frac{-\chi(\sigma)+1}{2}}\tau_7(\tau_7\tau_8\tau_7)^{\chi(\sigma)-1}.$$

To simplify[3] this complicated expression, we need 2-cycle and 3-cycle defining relations of Grothendieck-Teichmüller group [9] [15] [16].

Let x, y, z be elements of a profinite group with $xyz = 1$. Then,

(I): $f_\sigma(x, y)f_\sigma(y, x) = 1$

---

[3] In the previous version of this manuscript, the author stopped the calculation here, and the above expression was the final form. The following simplification was given by L. Schneps, who got her insight from the comparison with Nakamura's result [27].

(II):$f_\sigma(z,x)z^m f_\sigma(y,z)y^m f_\sigma(x,y)x^m = 1$

hold, where $m = \frac{\chi(\sigma)-1}{2} \in \widehat{\mathbb{Z}}$. Set temporarily only in this paragraph $x = \tau_8^2$, $y = \tau_8^{-1}\tau_7^2\tau_8$. Then $z = (xy)^{-1} = \tau_8^{-1}\tau_7^{-2}\tau_8^{-1}$ and $x = \tau_7^{-1}y\tau_7$ hold. Put $c = (\tau_7\tau_8\tau_7)^2$. This is in the center of the group generated by τ_7 and τ_8, and $zc = \tau_7^2$ holds.

Now, the image of τ_7 becomes

$$x^{-m}f_\sigma(y,x)y^{-m}\tau_7 c^m$$

$$\overset{(I)}{=} x^{-m}f_\sigma(x,y)^{-1}y^{-m}\tau_7 c^m$$

$$\overset{(II)}{=} f_\sigma(z,x)z^m f_\sigma(y,z)\tau_7 c^m$$

$$= f_\sigma(z,x)z^m\tau_7 f_\sigma(\tau_7^{-1}y\tau_7, \tau_7^{-1}z\tau_7)c^m$$

$$= f_\sigma(z,x)z^m\tau_7 f_\sigma(x,z)c^m$$

$$= f_\sigma(z,x)(zc)^m\tau_7 f_\sigma(x,z)$$

$$\overset{(I)}{=} f_\sigma(x,z)^{-1}\tau_7^{\chi(\sigma)}f_\sigma(x,z)$$

$$= f_\sigma(\tau_8^2,\tau_7^2)^{-1}\tau_7^{\chi(\sigma)}f_\sigma(\tau_8^2,\tau_7^2).$$

The last step used $f_\sigma(x,z) = f_\sigma(x,\tau_7^2 c^{-1}) = f_\sigma(x,\tau_7^2)$. This follows from that $f_\sigma(x,y)$ lies in the closure of the commutator subgroup of the discrete free group (see for example Lemma 2.5 in [22]). The image of τ_8 by $\sigma \in G_\mathbb{Q}$ is easy. τ_8 is realized as $\mu_{8,t}(x)$, hence

$$\sigma : \tau_8 \mapsto \tau_8^{\chi(\sigma)}.$$

3.5. The result on E_7 and M_3.

In summary, we have proved the following:

Theorem 3.1. $\sigma \in G_\mathbb{Q}$ *acts on the profinite Artin group* $\pi_1^{alg}(\overline{V(E_7)}, \mathcal{O})$ *as follows:*

$$\sigma : \tau_i \mapsto f(\eta_i,\tau_i^2)^{-1}\tau_i^{\chi(\sigma)}f(\eta_i,\tau_i^2)$$

for $2 \leq i \leq 5$,

$$\tau_6 \mapsto \tau_6^{\chi(\sigma)},$$

$$\tau_7 \mapsto f_\sigma(\tau_8^2,\tau_7^2)^{-1}\tau_7^{\chi(\sigma)}f_\sigma(\tau_8^2,\tau_7^2),$$

and

$$\tau_8 \mapsto \tau_8^{\chi(\sigma)},$$

where $\tau_2,\ldots,\tau_8$ *are the following generators of* E_7 *Artin group,*

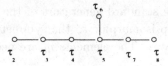

and $\eta_i = \tau_{i+1}\tau_{i+2}\cdots\tau_4\tau_5\tau_6\tau_7\tau_5\tau_4\cdots\tau_{i+2}\tau_{i+1}$.

If we take the image of $\mathcal{O}$ in $V(E_7) \to \mathcal{M}_3$, then we get the same action for $\tau_2, \ldots, \tau_8$, where each generator is a Dehn Twist as follows.

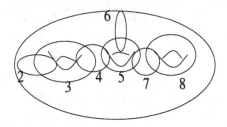

3.6. A_n, B_n, C_n, D_n cases.

3.6.1. A_n. This case is exactly same as the action on profinite braid groups given in Ihara-Matsumoto [17]. Let $e_1, \ldots, e_{n+1}$ be the unit vectors of the $n + 1$-dimensional vector space. We consider the n-dimensional subspace $U := \{(x_1, x_2, \ldots, x_{n+1}) | x_1 + x_2 + \cdots + x_{n+1} = 0\}$.

The root system is $e_i - e_j$ ($1 \leq i \neq j \leq n + 1$). The fundamental roots are $e_1 - e_2, e_2 - e_3, \ldots, e_n - e_{n+1}$. The Weyl chamber C is the set of vectors that has positive inner product with any of the fundamental roots, i.e., $C = \{(x_1, \ldots, x_{n+1}) | x_1 > x_2 > x_3 > \cdots > x_{n+1}\}$. There are n walls of this Weyl chamber, which are given by $x_1 > \cdots > x_i = x_{i+1} > x_{i+2} > \cdots > x_{n+1}$ ($i = 1, 2, \ldots, n$), corresponding to the i-th fundamental root. We denote by $U(A_n)$ the complement of the union of the reflection hyperplanes.

The corresponding Artin group is generated by τ_i, $1 \leq i \leq n$.

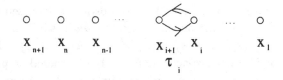

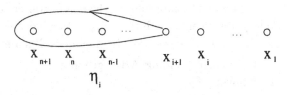

These are actually the standard generators of the braid group, after divided by the Weyl group isomorphic to the symmetric group S_{n+1} as the permutation of coordinates. The elements η_i are necessary to write down the Galois action.

The "suitable" parameters are $t_i := (x_i - x_{n+1})/(x_{i-1} - x_{n+1})$ $(2 \le i \le n)$, $t_1 := (x_1 - x_{n+1})$. We see easily that $x_i - x_{n+1} = t_1 t_2 \cdots t_i$. Then, we get the same construction as in 3.2.2:

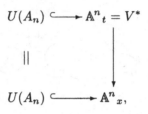

and get a $\mathbb{Q}$-rational tangential point $(t_1, \ldots, t_n)$ on $U(A_n)$. The tangential morphisms given by $(t_1, \ldots, t_i, t_{i+2}, \ldots, t_n)$ are easily checked to be from $\mathbb{P}^1_{01\infty}$ to $U(A_n)$. Similar arguments to those in 3.4.1 show that the Galois action is given by

$$\sigma : \tau_i \mapsto f_\sigma(\eta_i, \tau_i^2)^{-1} \tau_i^{\chi(\sigma)} f_\sigma(\eta_i, \tau_i^2) \quad (1 \le i \le n),$$

with $\eta_n = 1$.

3.6.2. B_n, C_n. These cases can be reduced to the A_n case. The B_n root system is $\{\pm e_i \pm e_j | 1 \le i \ne j \le n\} \cup \{\pm e_i | 1 \le i \le n\}$. The Weyl group is same as that of C_n, and hence $V(B_n) = V(C_n)$. The vector space spanned by the roots minus the reflection planes is: $U = \{(x_1, \ldots, x_n) | x_i \pm x_j \ne 0, x_i \ne 0\}$. The Weyl group is the group of permutation matrices with components $0, \pm 1$, and it is isomorphic to the semidirect product $(\mathbb{Z}/2)^n \rtimes S_n$. By putting $s_i := x_i^2$, we see easily that the quotient $U^{(\mathbb{Z}/2)^n}$ is $\{(s_1, \ldots, s_n) | s_i \ne s_j (1 \le i \ne j \le n), s_i \ne 0\}$. This is nothing but the configuration space $F_n(\mathbb{A}^1 - 0)$ of ordered n points on the affine line minus origin. Thus, by dividing by S_n, we have $V(B_n) = V(C_n) =$ configuration space of unordered n points on $\mathbb{A}^1 - 0$. The fundamental group of this space is the subgroup of the $n+1$-strands braid group, which stabilizes one point after mapping to S_{n+1}. It is notable that these are the groups used in [20] to characterize Grothendieck-Teichmüller group.

The generators of the corresponding Artin group are τ_i $(1 \le i \le n-1)$ and η_{n-1} as in the following figure.

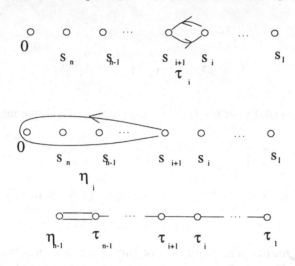

If we take the parameters $t_1 = s_1, t_2 = s_2/s_1, \ldots, t_n = s_n/s_{n-1}$, then we have a $\mathbb{Q}$-rational tangential point on $U^{(\mathbb{Z}/2)^n} - \cup H_\alpha$, and hence on $V(B_n)$. A similar argument with A_n case shows that the generators are mapped by:

$$\eta_{n-1} \mapsto \eta_{n-1}^{\chi(\sigma)}$$
$$\tau_i \mapsto f_\sigma(\eta_i, \tau_i^2)^{-1} \tau_i^{\chi(\sigma)} f_\sigma(\eta_i, \tau_i^2) \quad (1 \le i \le n-1).$$

3.6.3. D_n. The root system is $\{\pm e_i \pm e_j | 1 \le i \ne j \le n\}$. The vector space spanned by roots minus reflection planes is: $U = \{(x_1, \ldots, x_n) | x_i \pm x_j \ne 0\}$. The Weyl group is $(\mathbb{Z}/2)^{n-1} \rtimes S_n$, which is the set of permutation matrices, with coefficients $0, \pm 1$, where the number of -1 is even. It is easy to see that the quotient $U^{(\mathbb{Z}/2)^{n-1}}$ is $\{(s_1, \ldots, s_n, t) | t^2 = s_1 \cdots s_n, s_i \ne s_j\}$, by putting $s_i := x_i$ and $t := x_1 \cdots x_n$. Then we divide this space by S_n. The result is $\{(u_1, \ldots, u_n, t) | t^2 = u_n, \ x^n - u_1 x^{n-1} + u_2 x^{n-2} - \cdots + (-1)^n u_n$ has no multiple root$\}$. This is a double cover of the configuration space of ordered n points, which ramifies only along the locus $u_n = 0$.

Thus, we obtain $\pi_1(V(D_n) - \{u_n = 0\})$ as the subgroup of index two of $\pi_1(V(B_n))$, which is the kernel of the composition $\pi_1(V(B_n)) \to \mathbb{Z} \to \mathbb{Z}/2$ (the left arrow maps $\eta_{n-1} \mapsto 1$, $\tau_i \mapsto 0$ $(1 \le i \le n-1)$). By van Kampen's theorem, $\pi_1(V(D_n))$ is its quotient by normal subgroup generated by the square of the path which goes around the divisor $u_n = 0$. Thus, $\eta_{n-1}^2 = 1$ is the newly added relation, and $\pi_1(V(D_n))$ is generated by $\tau_1, \ldots, \tau_{n-1}, \tau'_{n-1} := \eta_{n-1} \tau_{n-1} \eta_{n-1}$. It can be checked that τ_{n-1} and τ'_{n-1} commute, and the group is isomorphic to the Artin group of D_n type.

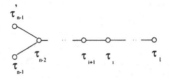

By the compatibility of the Galois action with an algebraic morphism, we have

$$\tau_{n-1} \mapsto \tau_{n-1}^{\chi(\sigma)}$$

$$\tau_{n-1}' \mapsto \tau_{n-1}'^{\chi(\sigma)}$$

$$\tau_i \mapsto f_\sigma(\eta_i, \tau_i^2)^{-1} \tau_i^{\chi(\sigma)} f_\sigma(\eta_i, \tau_i^2) \quad (1 \le i \le n-2).$$

Note that $\eta_i = \tau_{i+1}\tau_{i+2}\cdots\tau_{n-2}\tau_{n-1}\tau_{n-1}'\tau_{n-2}\cdots\tau_{i+2}\tau_{i+1}$.

Acknowledgment. The main idea of this research comes from the joint work with Y. Ihara, who constantly encourages me by giving invaluable comments. T. Oda gave me a motivation to study Galois actions on mapping class groups. As for the result related with the deformation of singularities, I am indebted to P. Slodowy, K. Saito, and J. Matsuzawa who taught me much when I was in RIMS. S. Morita, T. Kitano, A. Ishida and other topologists I met at the Riemann Surface workshop in Hokkaido University informed me of the topology of Dehn twists. A. Tamagawa helped me on the notion of tangential morphisms. H. Nakamura kindly informed me of his beautiful result on the Galois action on the mapping class groups. His stimulation enables me to get rid of $\mathbb{P}^1_{0\pm1\infty}$ in my result. L. Schneps kindly showed me the simplification of the expression in Theorem 3.1, using the two and three cycle relations of the Grothendieck-Teichmüller group in a nice way. This is celebrating, because the original complicated expression is dramatically shortened to a familiar expression. The anonymous referee showed me a number of important improvements. I would also like to thank all my friends working on algebraic fundamental groups, in particular P. Lochak and L. Schneps, who inspire me constantly.

Lastly, I would like to dedicate this research to the memory of categorist N. Yoneda, who motivated me to start mathematics when I was a student in a computer department.

References

[1]   G.V. Belyĭ, On Galois extensions of a maximal cyclotomic field, Math USSR Izv. **14** (1980), 247-256.

[2]   J. Birman, *Braids, links, and mapping class groups*, Ann. of Math. Studies **82**, Princeton Univ. Press 1975.

[3] J. Birman, Mapping class groups of surfaces, in AMS Contemporary Math. **78** *Braids* (1988), 13–43.

[4] N. Bourbaki, *Groupes et algèbres de Lie*, Ch. 2 et 3, Éléments de Mathématique, 1972.

[5] E. Brieskorn, Singular elements of semisimple algebraic groups, in *Actes Congrès Intern. Math.* (1970), t. 2, 279–284.

[6] E. Brieskorn, Die Fundamentalgruppe des Raumes der regulären Orbits einer endlichen komplexen Spiegelungsgruppe, *Inventiones Math.* **12** (1971), 57–61.

[7] P. Deligne, Le groupe fondamental de la droite projective moins trois points, in *Galois groups over* $\mathbb{Q}$, Publ. MSRI **16** (1989), 79–298.

[8] P. Deligne and D. Mumford, The irreducibility of the space of curves of given genus, *Publ. IHES* **36** (1969), 75–109.

[9] V.G. Drinfel'd, On quasitriangular quasi-Hopf algebras and a group closely connected with $\mathrm{Gal}(\overline{\mathbb{Q}}/\mathbb{Q})$, *Algebra i Analiz* **2** (1990), 114–148; English transl. *Leningrad Math. J.* **2** (1991), 829–860.

[10] K. Fujiwara, Theory of tubular neighborhood in etale topology, *Duke Math.* **80** (1995), 15–57.

[11] A. Grothendieck, *Revêtement Etales et Groupe Fondamental (SGA 1)*, Lecture Notes in Math. **224**, Springer-Verlag, 1971.

[12] A. Grothendieck, *Esquisse d'un programme,* typescript (1984), *Geometric Galois Actions*, London Math. Soc. Lecture Note Series, Cambridge University Press, 1997.

[13] S. Humphries, Generators for the mapping class group, in *Topology of Low-Dimensional Manifolds*, Lecture Notes in Math. **722**, Springer-Verlag 1979, 44–47.

[14] Y. Ihara, Profinite braid groups, Galois representations and Complex multiplications, *Ann. Math.* **123** (1986), 43–106.

[15] Y. Ihara, Braids, Galois groups, and some arithmetic functions, Proceedings of the ICM 90 (I), 1991, 99-120.

[16] Y. Ihara, On the embedding of $\mathrm{Gal}(\overline{\mathbb{Q}}/\mathbb{Q})$ into $\widehat{GT}$, in *The Grothendieck Theory of Dessins d'Enfants*, London Math. Soc. Lecture Note Series **200**, Cambridge Univ. Press, 1994, pp. 289–305.

[17] Y. Ihara and M. Matsumoto, On Galois Actions on Profinite Completions of Braid Groups, in AMS Contemporary Math. **186** *Recent Developments in the Inverse Galois Problem* (1994), 173–200.

[18] Y. Ihara and H. Nakamura, On Deformation of Maximally Degenerate Stable Marked Curves and Oda's Problem, to appear in *J. reine angew. Math.*

[19] A. Ishida, The structure of subgroup of mapping class groups generated by two Dehn twists, preprint 1995 (in Japanese), preprint 1996

218 Makoto Matsumoto

(in English).

[20] P. Lochak and L. Schneps, The Grothendieck-Teichmüller group and automorphisms of braid groups, in *The Grothendieck Theory of Dessins d'Enfants*, London Math. Soc. Lecture Note Series **200**, Cambridge Univ. Press, 1994, pp. 323–358.

[21] E. Looijenga, Cohomology of $\mathcal{M}_3$ and $\mathcal{M}_3^1$, *Contemporary Mathematics* **150** (1993), 205–228.

[22] M. Matsumoto, Galois representations on profinite braid groups on curves, *J. reine angew. Math.* **474** (1996), 169–219.

[23] S. Mochizuki, The local pro-p Grothendieck conjecture for hyperbolic curves, RIMS preprint 1045 (1995).

[24] S. Mochizuki, The local pro-p anabelian geometry of curves, RIMS Preprint 1097 (1996),

[25] H. Nakamura, Galois rigidity of pure sphere braid groups and profinite calculus, *J. Math. Sci. Univ. Tokyo* **1** (1994), 71–136.

[26] H. Nakamura, Coupling of universal monodromy representations of Galois-Teichmüller modular groups, *Math. Ann.* **304** (1996), 99–119.

[27] H. Nakamura, Galois representations in the profinite Teichmüller modular groups, *Geometric Galois Actions*, London Math. Soc. Lecture Note Series, Cambridge University Press, 1997.

[28] T. Oda, Etale homotopy type of the moduli spaces of algebraic curves, preprint (1989), *Geometric Galois Actions*, London Math. Soc. Lecture Note Series, Cambridge University Press, 1997.

[29] T. Oda, The universal monodromy representations on the pro-nilpotent fundamental groups of algebraic curves, Mathematische Arbeitstagung (Neue Serie) 9–15 Juni 1993, Max-Planck-Institute preprint MPI/93-57.

[30] A. Shiho, Tangential map and Log scheme, in Hokkaido Univ. Tech. Rep. Ser. in Math. **47**, Oct. 1996, 28–36 (in Japanese).

[31] P. Slodowy, *Simple Singularities and Simple Algebraic Groups*, Lecture Notes in Math. **815**, Springer-Verlag 1980.

[32] T. A. Springer, *Invariant Theory*, Lecture Notes in Math. **585**, Springer-Verlag 1977.

[33] A. Tamagawa, The Grothendieck conjecture for affine curves, RIMS preprint 1064 (1996), to appear in *Comp. Math.*

[34] B. Wajnryb, A simple presentation for the mapping class group of an orientable surface, *Israel J. Math.* **45** (1983), 157–174.

Department of Mathematics, Keio University
3-14-1 Hiyoshi Kohoku-ku 223 Japan
e-mail: matumoto@math.keio.ac.jp

Monodromy of Iterated Integrals

and

Non-abelian Unipotent Periods

Zdzisław Wojtkowiak

Contents

§0. Introduction.

0.1. Let X be a smooth projective algebraic variety defined over a number field k. Let $\sigma : k \hookrightarrow \mathbb{C}$ be an embedding. We set $X_\mathbb{C} = X \underset{k}{\times} \mathbb{C}$. Let $X(\mathbb{C})$ be the set of $\mathbb{C}$-points of $X_\mathbb{C}$ with its complex topology. There is a canonical isomorphism

$$p_{\text{comp}} : H_B^n(X(\mathbb{C})) \otimes \mathbb{C} \xrightarrow{\approx} H_{DR}^n(X) \underset{k}{\otimes} \mathbb{C}$$

between Betti (singular) cohomology and algebraic De Rham cohomology. The period matrix (p_{ij}) is defined by equations $\omega_i = \sum p_{ji}\sigma_j$, where $\{\omega_i\}$ and $\{\sigma_i\}$ are bases of $H^n_{DR}(X)$ and $H^n_B(X(\mathbb{C}))$.

In this paper we study periods for fundamental groups. The main motivation of the paper is the construction considered in [D2]. We describe it briefly. First we recall that there is a canonical way of associating a class $cl_B(Z)$ in $H^{2p}_B(X(\mathbb{C}))(p) := H^{2p}_B(X(\mathbb{C})) \otimes \mathbb{Q}((2\pi i)^p)$ and a class $cl_{DR}(Z)$ in $H^{2p}_{DR}(X)$ with an algebraic cycle Z on X of codimension p. The class $cl_B(Z)$ is a $(2\pi i)^{dim Z}$ multiple of the Poincaré dual of the homology class determined by Z.

We also recall that T is a G-torsor if T is equipped with a free transitive action of G. Let $S \subset T$. We say that S is a subtorsor of T if there is a subgroup H of G such that S is an H-torsor under the natural action of H.

Let us assume that X is an abelian variety. Let $\nu \in \mathbf{G_m}(\mathbb{Q})$ act on $\mathbb{Q}(2\pi i)$ as multiplication by ν^{-1}. Then the group $GL(H^1_B(X(\mathbb{C}))) \times \mathbf{G_m}$ acts on $H^r_B(X(\mathbb{C})^m)(n)$ for any r, n and m because $H^r_B(X(\mathbb{C})^m)$ can be expressed via $H^1_B(X(\mathbb{C}))$. Let G be the subgroup of $GL(H^1_B(X(\mathbb{C}))) \times \mathbf{G_m}$ fixing all tensors of the form $cl_B(Z)$, Z an algebraic cycle on some X^m.

Let P be a functor from k-algebras to sets such that an element of $P(A)$ is an isomorphism $p : H^1_B(X(\mathbb{C})) \otimes A \to H^1_{DR}(X) \underset{k}{\otimes} A$ mapping $cl_B(Z) \otimes 1$ to $cl_{DR}(Z) \otimes 1$ for all algebraic cycles Z on all X^n. The isomorphism p_{comp} belongs to $P(\mathbb{C})$. The functor P is represented by an algebraic variety over k, denoted also by P. Let $G_k := G \times k$ be obtained from G by an extension of scalars. The group $G_k(A)$ - the group of A-points of G_k - acts on $P(A)$ by composition and P becomes a G_k-torsor. It is a subtorsor of the $GL(H^1_B(X(\mathbb{C})) \otimes k)$-torsor $\mathrm{Iso}(H^1_B(X(\mathbb{C})) \otimes k, H^1_{DR}(X))$.

Let T be the smallest subtorsor defined over k of the torsor $\mathrm{Iso}(H^1_B(X(\mathbb{C})) \otimes k, H^1_{DR}(X))$, which contains p_{comp} as a $\mathbb{C}$-point and let $\mathcal{G}$ be the corresponding subgroup (defined over k) of $GL(H^1_B(X(\mathbb{C})) \otimes k)$. Observe that $T \subset P$ because both are subtorsors defined over k, both contain p_{comp} as a $\mathbb{C}$-point and T is the smallest one with these properties.

Let $Z(p_{\text{comp}})$ denote the k-Zariski closure of p_{comp} in $\mathrm{Iso}(H^1_B(X(\mathbb{C}) \otimes k, H^1_{DR}(X))$ i.e. the smallest Zariski closed subset defined over k, which contains p_{comp} as a $\mathbb{C}$-point. Then we have

$$Z(p_{\text{comp}}) \subset T \subset P$$

and

$$\mathcal{G} \subset G.$$

In order to calculate the dimension of $Z(p_{\text{comp}})$ one needs to know the transcendence degree of the field generated by the periods p_{ij} over k. In

[D2] (Remark 1.8.) Deligne asks whether $Z(p_{\text{comp}}) = P$. He mentioned that Chudnowsky has shown it when X is an elliptic curve with complex multiplication.

On the other hand to calculate T and $\mathcal{G}$ seems to be an easier task. The requirement that T be a subtorsor of $\text{Iso}(H_B^1(X(\mathbb{C})) \otimes k, H_{DR}^1(X))$ is very strong and usually only relatively weak information about the periods p_{ij} is necessary to calculate T and $\mathcal{G}$. We give an obvious example. If $X = \mathbb{P}_{\mathbb{Q}}^1$ is the projective line over $\mathbb{Q}$ and $p_{\text{comp}} : H_B^2(\mathbb{P}_{\mathbb{Q}}^1(\mathbb{C})) \otimes \mathbb{C} \xrightarrow{\approx} H_{DR}^2(\mathbb{P}_{\mathbb{Q}}^1) \otimes \mathbb{C}$ then in order to show that $Z(p_{\text{comp}})$ is isomorphic to $\mathbb{P}_{\mathbb{Q}}^1 \setminus \{0, \infty\}$ we must know that $2\pi i$ is transcendental.

Observe that the group $\mathbf{G_m}$ has no non-trivial algebraic subgroups other than the finite groups μ_k of roots of unity. Notice that for any rational number $a \neq 0$ the set of all roots of an equation $x^k - a = 0$ is a μ_k-torsor and these are the only μ_k-torsors. Hence already the fact that $2\pi i$ is not a kth-root of a rational number for any $k \in \mathbb{N}$ implies that T is isomorphic to $\mathbb{P}_{\mathbb{Q}}^1 \setminus \{0, \infty\}$ and $\mathcal{G} = \mathbf{G_m}$.

In this paper we construct analogues of T and $\mathcal{G}$ for fundamental groups. On the other hand we have no analogue of P and G.

0.2. Let us briefly discuss the content of the paper. Let us assume that X is a smooth quasi-projective geometrically connected scheme of finite type over a number field k. Let x be a k-point of X. In [W1] (see also [D4]) we defined affine connected pro-unipotent group schemes over k and $\mathbb{Q}$ respectively: $\pi_1^{\text{DR}}(X, x)$ - the algebraic De Rham fundamental group and $\pi_1^B(X(\mathbb{C}), x)$ - the Betti fundamental group. We have also the inclusion (of $\mathbb{Q}$-points into $\mathbb{C}$-points)

$$\Phi_x : \pi_1^B(X(\mathbb{C}), x)(\mathbb{Q}) \to \pi_1^{\text{DR}}(X, x)(\mathbb{C})$$

such that the induced homomorphism on $\mathbb{C}$-points

$$\varphi_x : \pi_1^B(X(\mathbb{C}), x)(\mathbb{C}) \to \pi_1^{\text{DR}}(X, x)(\mathbb{C})$$

is an isomorphism (see [W1] Theorem 7.4.).

The affine pro-algebraic scheme over k

$$\text{Iso} := \text{Iso}(\pi_1^B(X(\mathbb{C}), x) \times k, \pi_1^{\text{DR}}(X, x))$$

is an $\text{Aut}(\pi_1^{\text{DR}}(X, x))$-torsor. Let $Z(\varphi_x)$ be the k-Zariski closure of φ_x in Iso. Let $T(\varphi_x)$ be the smallest subtorsor of Iso, defined over k, which contains φ_x as a $\mathbb{C}$-point. Let $G_{\text{DR}}(\varphi_x) \subset \text{Aut}(\pi_1^{\text{DR}}(X, x))$ be the corresponding subgroup. One can ask whether

$$(0.1.) \qquad\qquad Z(\varphi_x) = T(\varphi_x).$$

The reader can compare this question with the question of Deligne in [D2]
(Remark 1.8) and Conjecture 1 in [An] which says that any polynomial
relation with cofficients in $\bar{\mathbb{Q}}$ between periods for the nth cohomology of
a smooth projective algebraic variety defined over $\bar{\mathbb{Q}}$ comes from algebraic
cycles (see also [G]).

We shall denote by $\mathcal{G}_{\text{DR}}(X)$ the image of $G_{\text{DR}}(\varphi_x)$ in $\text{Out}(\pi_1^{\text{DR}}(X,x))$.
Our aim is to calculate the group $\mathcal{G}_{\text{DR}}(X)$ or at least to get some information
about this group.

The calculation of the homomorphism φ_x is equivalent to the calculation
of the monodromy of all iterated integrals on X. Hence in the first eight
sections we are studying the monodromy of iterated integrals. General
properties of the canonical unipotent connection are established in section
1. In section 3 we present a "naive" approach to the Deligne tangential
base point. In sections 5 and 6 we describe the monodromy of iterated
integrals on $\mathbb{P}^1(\mathbb{C}) \setminus \{a_1, \ldots, a_{n+1}\}$ and in more details on $\mathbb{P}^1(\mathbb{C}) \setminus \{0,1,\infty\}$.
We describe below one of the main results of these sections.

Let us set $V = P_{\mathbb{C}}^1 \setminus \{0,1,\infty\}$. Let $Lie(V)$ be a free Lie algebra over $\mathbb{C}$ on
two generators X and Y. Let us set $L(V) := \varprojlim_n \left(\text{Lie}\,(V)/\Gamma^n \text{Lie}\,(V) \right)$. We
equip $L(V)$ with the multiplication given by the Baker-Campbell-Hausdorff
formula and the obtained group we denote by $\pi(V)$. The one form $\frac{dz}{z} \otimes X + \frac{dz}{z-1} \otimes Y$ defines a connection form on the principal $\pi(V)$-bundle $V(\mathbb{C}) \times \pi(V) \to V(\mathbb{C})$. We denote by $\theta_x : \pi_1(V(\mathbb{C}),x) \to \pi(V)$ the monodromy
homomorphism at the base point x. (This is the monodromy homomorphism
of all iterated integrals on V.) Let us take the vector $\overrightarrow{01}$ as a tangential base
point. Then we have:

Theorem A. *Let S_0, S_1 and S_∞ be the elements of $\pi_1(\mathbb{P}^1(\mathbb{C})\setminus\{0,1,\infty\}, \overrightarrow{01})$
shown in the following figure (loops around 0, 1 and ∞ respectively, with
$S_0 \cdot S_1 \cdot S_\infty = 1$). Then the monodromy homomorphism*

$$\theta_{\overrightarrow{01}} : \pi_1(\mathbb{P}^1(\mathbb{C}) \setminus \{0,1,\infty\}, \overrightarrow{01}) \to \pi(V)$$

is given by

$$\theta_{\overrightarrow{01}}(S_0) = (-2\pi i)X$$
$$\theta_{\overrightarrow{01}}(S_1) = (\alpha_{01}^{10}(X,Y))^{-1} \cdot (-2\pi i)Y \cdot \alpha_{01}^{10}(X,Y)$$
$$\theta_{\overrightarrow{01}}(S_\infty) = (-\pi i X) \cdot (\alpha_{01}^{10}(X,Z))^{-1} \cdot (-2\pi i)Z \cdot \alpha_{01}^{10}(X,Z) \cdot \pi i X$$

where $Z = -X - Y$ and $\alpha_{01}^{10}(X,Y) \in (\pi(V),\pi(V))$.

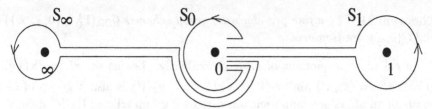

The element $\alpha_{01}^{10}(X,Y) \in \pi(V)$ is the same one which appears in [D5] and in [Dr]. In section 8 we give a proof that the element $\alpha_{01}^{10}(X,Y)$ satisfies the Drinfeld-Ihara $\mathbb{Z}/5$-cycle relation. Our proof is analogous to the proof in [I2] and it is similar to the proof of the $\mathbb{Z}/2$ and $\mathbb{Z}/3$ relations in [D5]. It is different from the proof in [Dr]. In section 7 we discuss the monodromy of iterated integrals on configuration spaces. The main result (stated somewhat imprecisely) is: the monodromy representation of iterated integrals on a configuration space of n points in $\mathbb{C}$ is determined up to conjugacy by the monodromy homomorphism

$$\theta_{\overrightarrow{01}} : \pi_1(\mathbb{P}^1(\mathbb{C}) \setminus \{0,1,\infty\}, \overrightarrow{01}) \to \pi(V).$$

Let us now describe the main results of the second part of the paper. One shows that there is an affine pro-algebraic pro-unipotent group scheme $\Pi(V)$ over $\mathbb{Q}$ such that the group of $\mathbb{C}$-points of $\Pi(V)$ is $\pi(V)$. Let us set

$$\mathcal{T}(\mathbb{C}) := \{(x,y,z) \in (\pi(V))^3 |$$
$$\exists \alpha \in \mathbb{C}^*, \quad x = \alpha X, \quad y \approx \alpha Y, \quad z \sim \alpha Z, \quad x \cdot y \cdot z = 0\}$$

and

$$\mathrm{Aut}^*(\Pi(V))(\mathbb{C}) :=$$
$$\{f \in \mathrm{Aut}(\pi(V)) | \exists \alpha \in \mathbb{C}^*, \quad f(X) = \alpha X, \quad f(Y) \approx \alpha Y, \quad f(Z) \sim \alpha Z\}.$$

(Here $\approx$ means conjugation by an element of $(\pi(V), \pi(V))$, whereas $\sim$ means conjugation by an element of $\pi(V)$.) One shows that $\mathcal{T}(\mathbb{C})$ (resp. $\mathrm{Aut}^*(\Pi(V))(\mathbb{C})$) is a set (resp. a group) of $\mathbb{C}$-points of an affine pro-algebraic scheme (resp. group scheme) over $\mathbb{Q}$, which we denote by $\mathcal{T}$ (resp. by $\mathrm{Aut}^*(\Pi(V))$. The group $\mathrm{Aut}^*(\Pi(V))$ is a subgroup of the group of automorphisms of $\Pi(V)$. Moreover $\mathcal{T}$ is an $\mathrm{Aut}^*(\Pi(V))$-torsor. Let $\theta_{\overrightarrow{01}} : \pi_1(V(\mathbb{C}), \overrightarrow{01}) \to \pi(V)$ be the monodromy homomorphism from Theorem A. Observe that the triple

$$\left(\theta_{\overrightarrow{01}}(S_0), \ \theta_{\overrightarrow{01}}(S_1), \ \theta_{\overrightarrow{01}}(S_\infty)\right) \in \mathcal{T}(\mathbb{C}).$$

We define $T(\theta_{\overrightarrow{01}})$ to be the smallest subtorsor defined over $\mathbb{Q}$ of $\mathcal{T}$ such that the triple $\left(\theta_{\overrightarrow{01}}(S_0), \ \theta_{\overrightarrow{01}}(S_1), \ \theta_{\overrightarrow{01}}(S_\infty)\right)$ is a $\mathbb{C}$-point of $T(\theta_{\overrightarrow{01}})$. Let $G(\theta_{\overrightarrow{01}})$ be the corresponding group. We have:

Theorem B. *The affine pro-algebraic group schemes* $\mathcal{G}_{DR}(\mathbb{P}^1_{\mathbb{Q}} \setminus \{0,1,\infty\})$ *and* $G(\theta_{\overrightarrow{01}})$ *are isomorphic.*

We calculate a quotient of $G(\theta_{\overrightarrow{01}})$ explicitly. Let us set $\pi' := (\pi(V), \pi(V))$, $\pi'' := (\pi', \pi')$ and $\pi_2(V) := \pi(V)/\pi''$. $\pi_2(V)$ is also a group of $\mathbb{C}$-points of an affine pro-unipotent pro-algebraic group scheme $\Pi_2(V)$ defined over $\mathbb{Q}$. We can also regard $\pi_2(V)$ as a Lie algebra. The elements X, Y and $(YX)X^{i-1}Y^{j-1} := (\dots(Y,X)X)\dots X)Y)\dots Y)$ $i = 1,2,3\dots, j = 1,2,3,\dots$ form a linear topological basis of $\pi_2(V)$ considered as a Lie algebra. Let ω be an element of this Lie algebra. If the polynomials $\exp(\sum_{i=1,j=1} f_{i,j}X^iY^j)$ and $1 + \sum_{n=1,m=1} F_{n,m}X^nY^m$ are equal then we set

$$(\omega, \exp(\dots)) := \omega + \sum_{n=1,m=1} F_{n,m}(\dots((\dots(\omega,X)\dots X)Y)\dots Y).$$

Let us set

$$\mathcal{T}_2(C) := \big\{(x,y,z) \in \big(\pi_2(V)\big)^3 \,\big|$$
$$\exists \alpha \in \mathbb{C}^*, \quad x = \alpha X, \quad y \approx \alpha Y, \quad z \sim \alpha Z, \quad x \cdot y \cdot z = 0\big\}$$

and

$$\mathrm{Aut}^*\big(\Pi_2(V)\big)(\mathbb{C}) :=$$
$$\big\{f \in \mathrm{Aut}\big(\pi_2(V)\big)\big| \, \exists \alpha \in \mathbb{C}^*, \quad f(X) = \alpha X, \quad f(Y) \approx \alpha Y, \quad f(Z) \sim \alpha Z\big\}.$$

As before there are affine pro-algebraic schemes $\mathcal{T}_2$ and $\mathrm{Aut}^*(\Pi_2(V))$, whose sets of $\mathbb{C}$-points are describe above. $\mathrm{Aut}^*(\Pi_2(V))$ is a group scheme and $\mathcal{T}_2$ is an $\mathrm{Aut}^*(\Pi_2(V))$-torsor. Let $\theta : \pi_1(V, \overrightarrow{01}) \to \pi_2(V)$ be the composition of $\theta_{\overrightarrow{01}}$ with the quotient map $\pi(V) \to \pi_2(V)$. The triple $\big(\theta(S_0), \theta(S_1), \theta(S_\infty)\big) \in \mathcal{T}_2(\mathbb{C})$. We define $T(\theta)$ to be the smallest subtorsor defined over $\mathbb{Q}$ of $\mathcal{T}_2$ such that the triple $\big(\theta(S_0), \theta(S_1), \theta(S_\infty)\big)$ is a $\mathbb{C}$-point of $T(\theta)$. Let $G(\theta)$ be the corresponding subgroup of $\mathrm{Aut}^*(\Pi_2(V))$. Then we have:

Theorem C. *The affine pro-algebraic group scheme* $G(\theta)$ *is isomorphic to a quotient of* $\mathcal{G}_{DR}(\mathbb{P}^1_{\mathbb{Q}} \setminus \{0,1,\infty\})$.

Let $\mathcal{C} : \mathbf{G_m} \to \mathrm{Aut}^*(\Pi_2(V))$ be given by $\mathcal{C}(t)(X) = tX$, $\mathcal{C}(t)(Y) = tY$. Let H be the image of $\mathcal{C}$. The next theorem gives a description of the torsor $T(\theta)$ and the group $G(\theta)$.

Theorem D. *i) The torsor* $T(\theta)(\mathbb{C})$ *is given by*

$$\Big\{\alpha X, \Big(\alpha Y, \exp\Big(\sum_{k=1}^{\infty} \sum_{\substack{i+j=2k \\ i \geq 1, j \geq 1}} \frac{(2k-1)!}{i!\,j!} r_{2k}\alpha^{2k} X^iY^j + $$

$$\sum_{\substack{k=3 \\ k \text{ odd}}}^{\infty} \sum_{\substack{i+j=k \\ i\geq 1, j\geq 1}} \frac{(k-1)!}{i!\,j!}\Big((1-\varepsilon(k))\cdot c_k + \varepsilon(k)\cdot b_{1,k-1}\Big)X^i Y^j\Big)\Big)\Big|$$

$$\alpha \in \mathbb{C}^*, \quad -\zeta(2k) = r_{2k}\cdot(2\pi i)^{2k}, \quad \underset{k}{\forall}\; b_{1,k-1}\in\mathbb{C}, \quad c_k = -\zeta(k)$$

$$\text{and} \quad \varepsilon(k)=0 \quad \text{if} \quad \zeta(k)\in\mathbb{Q}, \quad \varepsilon(k)=1 \quad \text{if} \quad \zeta(k)\notin\mathbb{Q}\Big\}.$$

ii) The corresponding group $G(\theta)(\mathbb{C})$ is given by

$$\Big\{f\in \operatorname{Aut}^*(\pi_2(V))\;\Big|\;f(X)=tX, f(Y)=$$

$$(tY, \exp\Big(\sum_{\substack{k=3 \\ k\text{-odd}}}\sum_{\substack{i+j=k \\ i\geq 1,j\geq 1}} \frac{(k-1)!}{i!\,j!}(1-\varepsilon(k))\cdot c_k\cdot(1-t^k)+\varepsilon(k)\beta_{1,k-1}\Big)X^i Y^j\Big)$$

$$\text{with } t\in\mathbb{C}^*,\; \beta_{1,k-1}\in\mathbb{C}\Big\}.$$

iii) Let G be the smallest closed affine pro-algebraic subgroup of $\operatorname{Aut}^*(\Pi_2(V))$ defined over $\mathbb{Q}$, which contains $G(\theta)$ and H. Then

$$G(\mathbb{C}) = \{f\in\operatorname{Aut}^*(\pi_2(V))\;|\;f(X)=tX,\; f(Y)=$$

$$(tY,\; \exp\Big(\sum_{\substack{k=3 \\ k\text{ odd}}}^{\infty}\sum_{\substack{i+j=k \\ i\geq 1,j\geq 1}} \frac{(k-1)!}{i!\,j!}\beta_{1,k-1}X^i Y^j\Big)) \text{ with } t\in\mathbb{C}^*,\; \beta_{1,k-1}\in\mathbb{C}\}.$$

Corollary E. *The group $G(\theta)$ contains the group H if and only if all numbers $\zeta(2k+1)$ for $k=1,2,\dots$ are irrational.*

Let $\mathcal{C}: \mathbf{G_m}\to\operatorname{Aut}^*(\Pi(V))$ be given by $\mathcal{C}(t)(X)=tX$, $\mathcal{C}(t)(Y)=tY$. Let $\mathcal{H}$ be the image of $\mathcal{C}$. Let G be the smallest subgroup of $\operatorname{Aut}^*\Pi(V)$ defined over $\mathbb{Q}$ which contains $G(\theta_{\overrightarrow{01}})$ and $\mathcal{H}$. Let T be a subtorsor of $\mathcal{T}$ defined over $\mathbb{Q}$, whose corresponding group is G and such that the triple

$$(\theta_{\overrightarrow{01}}(S_0),\; \theta_{\overrightarrow{01}}(S_1),\; \theta_{\overrightarrow{01}}(S_\infty))$$

is a $\mathbb{C}$-point of T.

Let Z be the $\mathbb{Q}$-Zariski closure of the triple

$$(\theta_{\overrightarrow{01}}(S_0),\; \theta_{\overrightarrow{01}}(S_1),\; \theta_{\overrightarrow{01}}(S_\infty))$$

in $\mathcal{T}$. Then we have

$$Z\subset T(\theta_{\overrightarrow{01}})\subset T.$$

Question. *Is it true that*

$$Z = T(\theta_{\overrightarrow{01}}) = T?$$

In section 15 we state some conjectures relating the group $\mathcal{G}_{DR}(X)$ to the Galois representations on fundamental groups.

Deligne in [D4] considers the motivic Galois group of the tannakian category of mixed motives generated by the Tate motive $\mathbb{Q}(1)$. The group $\mathcal{G}_{DR}(\mathbb{P}^1(\mathbb{C}) \setminus \{0, 1, \infty\})$ from this paper should be a quotient of the group considered by Deligne. Perhaps they are even equal. In this paper there is missing a motivic interpretation of $\mathcal{G}_{DR}(X)$. We are also not discussing the mixed Hodge structures on the fundamental groups, though it would be very natural to do this here. We hope to study these topics in a future paper.

Acknowledgments. I would like to thank very much Professor Deligne, who once showed me the one-form considered in Section 1 in the case of $\mathbb{C} \setminus \{0, 1\}$. Thanks are due to Professor Y. Ihara for showing me his proof of the 5-cycle relation, which helped us to find analogous one for unipotent periods. I would like to express thanks to Professor Hubbuck for his invitation to Aberdeen, where Section 8 was written and where in May 1993 I gave seminar talks on 5-cycle relation for unipotent periods.

I would like to thank very much Professor Y. Ihara for his invitation to Kyoto, and Professors Oda and Matsumoto and Tamagawa for useful discussions and comments during my seminar talks.

Thanks are due to Professor L. Lewin, who once invited me to write a chapter in the book on polylogarithms and suggested also including results on monodromy of iterated integrals. This encouraged me very much to continue to work on this subject (see preprints [W2] and [W3] of which some parts are included in the present paper).

Finally, I would like to thank Leila Schneps and Pierre Lochak for their help in the editing of the paper and their encouragements, and the referees for their comments, remarks and interest.

This paper is a revised version of [W5].

§1. Canonical connection with logarithmic singularities.

In this section we introduce the canonical connection with logarithmic singularities. The results of this section are well-known and are due to Aomoto, Chen, Deligne, Hain and others. Our approach is slightly different from the one in [HZ]. We consider the one-form of the canonical unipotent connection as a connection form of a principal G-bundle, where G is isomorphic to (a quotient of) $\pi_1^{DR}(X, x)$. We found this approach useful studying functional equations of iterated integrals (see [W4]). It is also well adapted to study the monodromy. We also use it in work in progress concerning a generalization of the Zagier conjecture on polylogarithms to arbitrary iterated integrals.

1.0. Connection on a trivial principal G-bundle. Let G be a Lie group. Let us consider the principal G-bundle

1.0.1. $$p : V \times G \to V.$$

Let θ be the canonical one form on G i.e. the left invariant one-form with values in the Lie algebra $\mathfrak{g}$ of G such that $\theta_e(X) = X$ for $X \in \mathfrak{g}$. Let $pr : V \times G \to G$ be a projection. Then the one-form $pr^*\theta$ is a connection form on the principal G-bundle 1.0.1. Let $\omega \in \Omega^1(V) \otimes \mathfrak{g}$ be a one form on V with values in $\mathfrak{g}$. Let $\tilde{\omega}$ be a one-form on $V \times G$ vanishing on vertical vectors and such that

i) $\tilde{\omega}_{(v,e)}(Y) = \omega_v(pr_*Y)$, where $(v, e) \in V \times G$, e is the neutral element of G and $Y \in T_{(v,e)}(V \times G)$.

ii) $\tilde{\omega}_{(v,g)}((R_g)_*Y) = ad(g^{-1})\tilde{\omega}_{(v,e)}(Y)$, where R_g is the right action of g on $V \times G$.

It follows from [L] (Théorème, p. 68) that $pr^*\theta + \tilde{\omega}$ is the connection form of a certain connection on the principal G-bundle 1.0.1. If ω is integrable then $pr^*\theta + \tilde{\omega}$ is also integrable. Abusing the notation we shall also denote the one-form $pr^*\theta + \tilde{\omega}$ by ω and we shall say that ω is a connection form on the principal G- bundle 1.0.1.

1.1. Let X be a smooth projective scheme of finite type over a field k of characteristic zero. Let D be a divisor with normal crossings in X and let $V = X \setminus D$. Let

$$A^*(V) := \Gamma(X, \Omega_X^*\langle \log D \rangle)$$

be the differential algebra of global sections of the algebraic De Rham complex on X with logarithmic singularities along D. It follows from [D1] Corollaire 3.2.14 that each element of $A^*(V)$ is closed and the natural map $A^*(V) \to H_{DR}^*(V)$ is injective.

We shall denote by $\wedge^2(A^1(V))$ the exterior product of the vector space $A^1(V)$ with itself and by $A^1(V) \wedge A^1(V)$ the image of $\wedge^2(A^1(V))$ in $A^2(V)$. Let $H(V) := (A^1(V))^*$ and $R(V) := (A^1(V) \wedge A^1(V))^*$ be dual vector spaces. The map $\wedge^2(A^1(V)) \to A^1(V) \wedge A^1(V)$ induces a map $R(V) \to \wedge^2(H(V))$. Let $\mathrm{Lie}\,(H(V))$ be the free Lie algebra over k on $H(V)$. Observe that $R(V)$ is contained in degree 2 terms of $\mathrm{Lie}\,(H(V))$. Let $(R(V))$ be the Lie ideal generated by $R(V)$. We set

$$\mathrm{Lie}\,(V) := {}^{\mathrm{Lie}\,(H(V))}\!/\!{(R(V))}$$

and

$$L(V) := \varprojlim_n \left(\mathrm{Lie}\,(V) / \Gamma^n \mathrm{Lie}\,(V) \right).$$

Definition 1.1.1. The Lie algebra $L(V)$ equipped with the multiplication given by the Baker-Campbell-Hausdorff formula is a group. We shall denote this group by $\pi(V)$.

Observe that $\pi(V)$ is a set of k-points of an affine pro-algebraic pro-unipotent group defined over k. The Lie algebra of $\pi(V)$ can be identified with $L(V)$. Observe that $H(V)$ can be regarded as included in $L(V)$. Let $T[H(V)]$ be the tensor algebra over k on $H(V)$ and let $(R(V))$ be the ideal of $T[H(V)]$ generated by $R(V)$. Let $\mathbb{Q}(V) := {}^{T[H(V)]}\!/\!{(R(V))}$ be the quotient algebra and let $\widehat{\mathbb{Q}}(V)$ be its completion with respect to the augmentation ideal $I := \ker(\mathbb{Q}(V) \to k)$, i.e. $\widehat{\mathbb{Q}}(V) := \varprojlim_n \left(\mathbb{Q}(V)/I^n \right)$.

Definition 1.1.2. We denote by $P(V)$ the group of invertible elements in $\widehat{\mathbb{Q}}(V)$, whose constant term equals 1.

Observe that the vector space $H(V)$ can be regarded as included in the Lie algebra of $P(V)$.

We identify the elements of $L(V)$ with Lie elements (which can be of infinite length) in $P(V)$. The exponential series defines an injective homomorphism

$$\exp : \pi(V) \to P(V).$$

The inverse of $\exp$ is defined on the subgroup $\exp(\pi(V))$ of $P(V)$ and it is given by the formula

$$\log z = (z-1) - \tfrac{1}{2}(z-1)^2 + \tfrac{1}{3}(z-1)^3 - \tfrac{1}{4}(z-1)^4 + \ldots.$$

We define a one form ω_V on V with values in the Lie algebra $L(V)$ (and in the Lie algebra of $P(V)$) in the following way:

Definition 1.1.3. The form ω_V corresponds to the identity homomorphism id $_{A^1(V)}$ under the natural isomorphism

$$A^1(V) \otimes H(V) = A^1(V) \otimes (A^1(V))^* \approx \mathrm{Hom}\,(A^1(V), A^1(V))).$$

(see [D4] and [H] for the definition of the form ω_V).

Lemma 1.2. (see [D4], [H]) *The one-form ω_V is integrable.*

Proof. The facts that $\mathrm{Lie}(V)$ is a quotient of a free Lie algebra by the ideal $(R(V))$ and that each element of $A^1(V)$ is closed imply integrability.   ◇

Let us assume that k is the field of complex numbers $\mathbb{C}$. Then $V(\mathbb{C})$ is a complex variety with the standard complex topology. Abusing the notation we shall denote it by V. Let us consider the principal bundles

$$(1)\ \ V \times P(V) \to V \text{ and }\ \ (2)\ \ V \times \pi(V) \to V.$$

We equip them with the connection given by the one-form ω_V. Let us fix a path γ on V from x to z_0. We define a germ of flat sections of the trivial bundle (1) (resp. (2)) near z_0: $\Lambda_V(z;x,\gamma) \in P(V)$ (resp. $L_V(z;x,\gamma) \in \pi(V)$) to be a function of z moving in a simply connected neighbourhood of z_0 with values in $P(V)$ (resp. $\pi(V)$). We require it to take the value 1 (resp. 0) at x along the flat section over the path γ.

Definition 1.3. Let $x \in V$ and let $\alpha \in \pi_1(V,x)$ be a loop. We define a homomorphism

$$\theta_{x,V} : \pi_1(V,x) \to P(V)\ \ \ (\text{resp. } \theta_{x,V} : \pi_1(V,x) \to \pi(V))$$

by the formula

$$\theta_{x,V}(\alpha) := \Lambda_V(\alpha(1);x,\alpha)\ \ \ (\text{resp. } \theta_{x,V}(\alpha) := L_V(\alpha(1);x,\alpha))$$

and we call it the monodromy homomorphism of the form ω_V at the base point x. (The homomorphism $\theta_{x,V}$ we shall also denote by θ_x.)

Proposition 1.4. *Let $x_1, x_2 \in V$ and let γ be a smooth path from x_1 to x_2. The path γ gives an identification of $\pi_1(V,x_1)$ with $\pi_1(V,x_2)$. Via this identification the monodromy homomorphisms $\theta_{x_1,V}$ and $\theta_{x_2,V}$ of the form ω_V are conjugate.*

Proof. This is a property of a connection on a principal fibre bundle.   ◇

The next result is a special case of a result of Chen (see [Ch2], Theorem 2.1.1).

Proposition 1.5. *If $A^1(V) \to H^1_{DR}(V)$ is an isomorphism then the monodromy homomorphism $\theta_{x,V} : \pi_1(V,x) \to \pi(V)$ induces an isomorphism of the Malcev $\mathbb{C}$-completion of $\pi_1(V,x)$ into $\pi(V)$.*

Remark. If the map $A^1(V) \to H^1_{DR}(V)$ is an isomorphism then $H^1(V,\mathbb{C})$ has a Hodge type $(1,1)$. Examples of such varieties are Zariski open subsets of simply connected smooth projective varieties.

Let X_i (for $i = 1,2$) be smooth, projective schemes of finite type over k. Let D_i be divisors with normal crossings in X_i (for $i = 1,2$). Let $V_i = X_i \backslash D_i$ (for $i = 1,2$) and let $f : X_1 \to X_2$ be a morphism such that $f^{-1}(D_2) = D_1$. Then f induces $f^* : A^1(V_2) \to A^1(V_1)$. Let $f_* : H(V_1) \to H(V_2)$ be the dual map. This map induces group homomorphisms

$$f_* : P(V_1) \to P(V_2) \quad \text{and} \quad f_* : \pi(V_1) \to \pi(V_2).$$

Lemma 1.6. *We have*

$$(id \otimes f_*)(\omega_{V_1}) = (f^* \otimes id)(\omega_{V_2}).$$

The lemma follows from the definition of ω_{V_i} as $id_{A^1(V_i)}$.

Corollary 1.7. *We have*

$$f_*(\Lambda_{V_1}(z; x, \gamma)) = \Lambda_{V_2}(f(z); f(x), f(\gamma))$$

and

$$f_*(L_{V_1}(z; x, \gamma)) = L_{V_2}(f(z); f(x), f(\gamma)).$$

Let $\omega_1, \ldots, \omega_n$ be a basis of $A^1(V)$. Let $X_1, \ldots, X_n$ be the dual basis of $H(V)$. Let $\mathbb{C}\{\{X_1, \ldots, X_n\}\}$ be the algebra of formal power series in non-commuting variables $X_1, \ldots, X_n$. Let $\mathbb{C}\{\{X_1, \ldots, X_n\}\}^*$ be the multiplicative group of these power series whose constant term is 1. Then the algebra $\widehat{\mathbb{Q}}(V)$ is a quotient of $\mathbb{C}\{\{X_1, \ldots, X_n\}\}$ by the ideal generated by $R(V)$. The group $P(V)$ is a quotient group of $\mathbb{C}\{\{X_1, \ldots, X_n\}\}^*$.

If $\alpha(X_1, \ldots, X_n)$ is a formal power series in non-commuting variables $X_1, \ldots, X_n$, we shall also denote by $\alpha(X_1, \ldots, X_n)$ its image in $P(V)$.

Proposition 1.8. *Let γ be a path on X from x to z. We have*

i) $\Lambda_V(z; x, \gamma) = 1 + \sum ((-1)^k \int_{x,\gamma}^z \omega_{i_1}, \ldots, \omega_{i_k} X_{i_k} \cdot \ldots \cdot X_{i_1}) \in P(V).$

(The summation is over all non-commutative monomials in variables $X_1, \ldots,$
X_n, the interated integrals are calculated along the path γ.)

ii) $L_V(z; x, \gamma) = \log(\Lambda_V(z; x, \gamma))$.

Proof. We equip the principal bundle $V \times \mathbb{C}\{\{X_1, \ldots, X_n\}\}^* \to V$ with
the connection given by a one-form $\sum\limits_{i=1}^{n} \omega_i \otimes X_i$. Using the definitions in 1.0
we calculate the horizontal subspaces and then the system of differential
equations for horizontal sections. We get that horizontal sections are given
by

$$z \to (z, 1 + \sum((-1)^k \int_{x,\gamma}^{z} \omega_{i_1}, \ldots, \omega_{i_k} X_{i_k} \cdot \ldots \cdot X_{i_1}).$$

Hence the point i) follows. Observe that $\exp : \pi(V) \to P(V)$ identifies
$\omega_V \in A^1(V) \otimes \mathrm{Lie}\,(\pi(V))$ with $\omega_V \in A^1(V) \otimes \mathrm{Lie}\,(P(V))$. Hence the point
ii) follows.   ◇

§2. The Gauss-Manin connection associated with the morphism $X^{\Delta[1]} \to X^{\partial\Delta[1]}$ of cosimplicial schemes.

2.1.   We review here some constructions from [W1]. Let I be a unit interval
and let X be an arc-connected and locally arc-connected topological space.
Let $p : X^I \to X \times X$ be given by $p(\omega) = (\omega(0), \omega(1))$. This is a fibration.
Applying the connected component functor to each fibre we get a local
system of sets $p : P \to X \times X$, such that the fibre over $(x, x) \in X \times X$ is
$\pi_1(X, x)$ and the restriction to $X \times \{x\}$ is a universal covering space of X.
   We want to do an analogous construction in algebraic geometry. Let us
notice that the inclusion of simplicial sets

$$\partial\Delta[1] \hookrightarrow \Delta[1]$$

induces a morphism of cosimplicial spaces

$$X^{\Delta[1]} \to X^{\partial\Delta[1]},$$

whose geometric realization is $p : X^I \to X \times X$. We use this observation
to construct an analogue of the local system $p : P \to X \times X$ in algebraic
geometry.
   Let V be a smooth quasi-projective scheme over a field k of characteristic
zero. The inclusion of simplicial sets $\partial\Delta[1] \hookrightarrow \Delta[1]$ induces a morphism of
cosimplicial schemes

$$\rho^{\bullet} : V^{\Delta[1]} \to V^{\partial\Delta[1]}.$$

Assume that k is the field of complex numbers $\mathbb{C}$. As in the case of a morphism of algebraic varieties $X \to S$, the cohomology of the fibres $H^i(\rho^{\bullet\,-1}(x,y); \mathbb{C})$ form a local system of complex vector spaces. Imitating the construction of Katz and Oda (see [KO]) we equipped the sheaves $H^i(tR\rho_*^{\bullet}\Omega^*_{(V^{\Delta[1]})/(V^{\partial\Delta[1]})})$ of the relative De Rham cohomology with the integrable connection $d_{\mathbb{C}}$ such that the horizontal sections form a local system described above. ($R\rho_*^{\bullet}$ is the component-wise derived functor of ρ_* and t is the functor which associates a total complex to a bicomplex.) $H^0(\rho^{\bullet\,-1}(x,x), \mathbb{Q})$ is the ring of polynomial functions on the Malcev rational completion of the fundamental group $\pi_1(V,x)$. The representation of $\pi_1(V,x) \times \pi_1(V,x)$ on the vector space $H^0(\rho^{\bullet\,-1}(x,x); \mathbb{Q})$ is induced by

$$\varphi : \pi_1(V,x) \times \pi_1(V,x) \to (\text{bijections of } \pi_1(V,x))$$

given by $\varphi(\alpha,\beta)(g) = \alpha \cdot g \cdot \beta^{-1}$. Therefore the sheaf $H^0(tR\rho_*^{\bullet}\Omega^*_{(V^{\Delta[1]})/(V^{\partial\Delta[1]})})$ equipped with the connection $d_{\mathbb{C}}$ is an analogue of the local system $P \to X \times X$ in algebraic geometry. This suggests the following definitions of the Betti and De Rham fundamental groups. We set

$$\pi_1^B(V,x) := \text{Spec } H^0(\rho^{\bullet\,-1}(x,x); \mathbb{Q}).$$

If V is a smooth quasi-projective scheme over a field k of characteristic zero and x is a k-point of V then we set:

$$\pi_1^{DR}(V,x) := \text{Spec } H^0_{DR}(\rho^{\bullet\,-1}(x,x)).$$

2.2. Let V be such as in 1.1. We assume additionally that $A^1(V) \approx H^1_{DR}(V)$.

2.2.1. The inclusion of complexes

$$(\mathbb{C} \xrightarrow{0} A^1(V) \xrightarrow{0} A^1(V) \wedge A^1(V) \to 0) \hookrightarrow \Omega^*(V).$$

induces an isomorphism

$$H^0 := H^0\big(\text{Bar}(\mathbb{C} \xrightarrow{0} A^1(V) \xrightarrow{0} A^1(V) \wedge A^1(V) \to 0)\big) \approx H^0_{DR}(\rho^{\bullet\,-1}(x,x)).$$

(The bar construction is chosen to be compatible with the inclusion into the total complex of the De Rham complex of $\rho^{\bullet\,-1}(x,x)$. The isomorphism means that to calculate $H^0_{DR}(\rho^{\bullet\,-1}(x,x))$ we need only $A^1(V)$ and $A^1(V) \wedge A^1(V)$.)

We recall that $H(V) = (A^1(V))^*$, $T[H(V)]$ is the tensor algebra on $H(V)$, $R(V) := (A^1(V) \wedge A^1(V))^*$ is contained in $\wedge^2 H(V)$ and $Q(V) := T[H(V)]/(R(V))$. Let us set

$$Q(V)^* := \varinjlim_n \left(Q(V)/I^n \right)^*,$$

where I is the augmentation ideal of $Q(V)$. Recall that

$$\hat{Q}(V) := \varprojlim_n \left(Q(V)/I^n \right)$$

and $P(V) \subset \hat{Q}(V)$.

2.2.2. Notice that the symmetric algebra on $Q(V)^*$ is the algebra of polynomial functions on $\hat{Q}(V)$.

Let us fix a basis $\omega_1, \dots, \omega_n$ of $A^1(V)$. Let $X_1, \dots, X_n$ be the dual basis of $H(V)$. Then $T[H(V)] = \mathbb{C}\{X_1, \dots, X_n\}$ is the polynomial algebra on non-commuting variables $X_1, \dots, X_n$. Let $(X_{i_1} \cdot \dots \cdot X_{i_k})^*$ denotes the linear form on $T[H(V)]$ which to f in $T[H(V)]$ associates its coefficient at the monomial $X_{i_1} \cdot \dots \cdot X_{i_k}$. Let T^* be the vector space generated by all $(X_{i_1} \cdot \dots \cdot X_{i_k})^*$ including 1^*. Observe that $Q(V)^*$ is a subspace of T^*. We equipped T^* with a shuffle multiplication and a Hopf algebra structure given by $(X_{i_1} \cdot \dots \cdot X_{i_p})^* \cdot (X_{i_{p+1}} \cdot \dots \cdot X_{i_{p+q}})^* = \sum_{\sigma \in Sh(p,q)} (X_{i_{\sigma(1)}} \cdot \dots \cdot X_{i_{\sigma(p+q)}})^*$

and $(X_{i_1} \cdot \dots \cdot X_{i_k})^* \to \sum_{l=0}^{k} (X_{i_1} \cdot \dots \cdot X_{i_l})^* \otimes (X_{i_{l+1}} \cdot \dots \cdot X_{i_k})^*$. The shuffle multiplication and the Hopf algebra structure on T^* induce the analogous structures on $Q(V)^*$. H^0 is also equipped with a shuffle multiplication and a Hopf algebra structure which on the form level are given by the same formulas.

Lemma 2.2.3. *There is an isomorphism of Hopf algebras*

$$H^0 \xrightarrow{\approx} Q(V)^*$$

given by $\omega_{i_1} \otimes \dots \otimes \omega_{i_k} \to (X_{i_1} \cdot \dots \cdot X_{i_k})^*$.

Proof. We identify the linear form $(X_{i_1} \cdot \dots \cdot X_{i_k})^*$ with the vector $(X_{i_1} \otimes \dots \otimes X_{i_k})^* = \omega_{i_1} \otimes \dots \otimes \omega_{i_k}$. Observe that H^0 is a subspace of $\mathbb{C} \oplus A^1(V) \oplus (A^1(V) \otimes A^1(V)) \oplus (A^1(V) \otimes A^1(V) \otimes A^1(V)) \oplus \dots = T^*$. Both algebras H^0 and $Q(V)^*$ have a natural grading. The part of H^0 in degree 2 is equal to $\ker(A^1(V) \otimes A^1(V) \to A^1(V) \wedge A^1(V))$. This is also the part of $Q(V)^*$ in degree 2. One checks that H^0 and $Q(V)^*$ are isomorphic in each degree.

Let $Sym(T^*)$ be the symmetric algebra on T^*. Let $Sh(V)$ be the ideal in $Sym(T^*)$ generated by the shuffle product relations

$$(X_{i_1} \cdot \ldots \cdot X_{i_p})^* \cdot (X_{i_{p+1}} \cdot \ldots \cdot X_{i_{p+q}})^* = \sum_{\sigma \in Sh(p,q)} (X_{i_{\sigma(1)}} \cdot \ldots \cdot X_{i_{\sigma(p+q)}})^*.$$

Then $Sym(T^*)$ divided by $Sh(V)$ is isomorphic to T^* equipped with shuffle multiplication.

2.2.4. This implies that the symmetric algebra on $\mathbb{Q}(V)^*$ divided by the ideal $Sh(V) \cap \mathbb{Q}(V)^*$ is isomorphic to $\mathbb{Q}(V)^*$equipped with the shuffle multiplication.

Let $\mathrm{Alg}\,(\pi(V))$ be the algebra of polynomial functions on $\pi(V)$. It follows from a theorem of R. Ree (see [R] Theorem 2.5) that a formal power series $f \in P(V)$ is in the image of $\exp : \pi(V) \to P(V) \subset \widehat{\mathbb{Q}}(V)$ if and only if its coefficients (i.e. the coefficients of some lifting to $\mathbb{C}\{\{X_1, \ldots, X_n\}\}$) satisfy shuffle product relations. Therefore 2.2.2 and 2.2.4 imply that

$$\exp : \pi(V) \to P(V)$$

induces an isomorphism of Hopf algebras

2.2.5 $R : \mathbb{Q}(V)^* \xrightarrow{\approx} \mathrm{Alg}\,(\pi(V)).$

Combining 2.2.1, Lemma 2.2.3 and 2.2.5 we get:

Lemma 2.3.   *The Hopf algebras $H^0_{DR}(\rho^{\bullet\,-1}(x,x))$ and $\mathrm{Alg}\,(\pi(V))$ are isomorphic.*

Corollary 2.4.   *The group schemes $\pi_1^{DR}(V,x)$ and $\mathrm{Spec}\big(\mathrm{Alg}\,(\pi(V))\big)$ are isomorphic. We denote by $u : \pi_1^{DR}(V,x) \to \mathrm{Spec}\big(\mathrm{Alg}\,(\pi(V))\big)$ the isomorphism induced by isomorphisms 2.2.1, 2.2.3 and 2.2.5.*

2.5. Let k be the field of complex numbers $\mathbb{C}$. We shall assume that $A^1(V) \approx H^1_{DR}(V)$.

In section 1 we considered the principal $\pi(V)$-bundle equipped with the integrable connection given by ω_V. We shall relate it to the objects described in 2.1. The group $\pi(V)$ acts on $\mathrm{Alg}\,(\pi(V))$ on the left by the formula

a) $g(f)(x) := f(x \cdot g)$ b) $g(f)(x) := f(g^{-1}x)$

We form the associated vector bundle

$$V \times \mathrm{Alg}\,(\pi(V)) \approx V \times \pi(V) \underset{\pi(V)}{\times} \mathrm{Alg}\,(\pi(V)) \to V$$

and we equip it with the connection induced by ω_V. We shall denote this connection by ω'_V in the case a) and by $'\omega_V$ in the case b). Let $\Omega^*_{(V^{\Delta[1]})/(V^{\partial\Delta[1]})}$ be the algebraic De Rham complex of smooth relative differentials on $V^{\Delta[1]}$ (i.e. on each $V^{\Delta[1]_n}$).

Let $\mathcal{H}^0_x$ (respectively $_x\mathcal{H}^0$) be the restriction of the vector bundle $H^0(tR\rho^\bullet_*\Omega^*_{(V^{\Delta[1]})/(V^{\partial\Delta[1]})})$ to $X \times \{x\}$ (resp. $\{x\} \times X$.) Both vector bundles are equipped with the induced connection, which we also denote by $d_{\mathbb{C}}$.

Our main result in this section is the following theorem.

Theorem 2.6. *Let V be as in 2.1. Then the algebraic vector bundles equipped with connections*

$$(\mathcal{H}^0_x, d_{\mathbb{C}}) \quad \text{and} \quad (V \times \text{Alg}(\pi(V)) \to V, \omega'_V)$$

are isomorphic in the category of algebraic vector bundles equipped with algebraic integrable connections.

Remark. $(_x\mathcal{H}^0, d_{\mathbb{C}})$ and $(V \times \text{Alg}(\pi(V)) \to V, '\omega_V)$ are also isomorphic.

Proof. We shall calculate horizontal sections of the connection $d_{\mathbb{C}}$ on $\mathcal{H} :=$ $H^0(tR\rho^\bullet_*\Omega^*_{(V^{\Delta[1]})/(V^{\partial\Delta[1]})})$. We recall the construction of $d_{\mathbb{C}}$. Following Katz and Oda (see [KO]) the filtration $\{F^i_n\}_i$ on $\Omega^*_{V^{n+2}}$ is defined by

$$F^i_n\Omega^*_{V^{n+2}} := \text{image}\,(\Omega^{*-i}_{V^{n+2}} \otimes (\rho^n)^*\Omega^i_{V\times V} \to \Omega^*_{V^{n+2}}).$$

The filtrations $\{F^i_n\}_i$ for $n = 0, 1, 2, \ldots$ give a filtration F^i of $\Omega^*_{V^{\Delta[1]}}$. The connection $d_{\mathbb{C}}$ is the connecting homomorphism of the functor $H^0(tR\rho^\bullet_*(\ ))$ for the short exact sequence of complexes on $V^{\Delta[1]}$

$$0 \to F^1/F^2 \to F^0/F^2 \to F^0/F^1 \to 0.$$

On V^{n+2} we have

i) $\quad F^0/F^1 \approx \Omega^*_{V^{n+2}}/_{V \times V}$ ii) $\quad F^0/F^2 \approx \Omega^*_{V^{n+2}}/_{\Omega^{*-2}_{V^{n+2}}} \otimes (\rho^n)^*\Omega^2_{V\times V}.$

Observe that $\mathcal{O}_V \otimes \Omega^*_{V^n} \otimes \mathcal{O}_V \subset F^0/F^2$.

Let us assume that $\sum \omega_{i_1} \otimes \ldots \otimes \omega_{i_m} \in A(V)^{\otimes m}$ is closed in the total complex of $\Omega^*(\rho^{\bullet-1}(x,x))$. Its cohomology class c belongs to $H^0_{DR}(\rho^{\bullet-1}(x,x))$, the fibre of $\mathcal{H}$ over $(x,x) \in V \times V$.

We face two problems. We must extend c to a continuous section of $\mathcal{H}$ and we must show that this section is horizontal. Let $\mathcal{T} := \text{Tot}(n \to \mathcal{O}_V \otimes \Omega^*(V^n) \otimes \mathcal{O}_V)$ be the total complex of the bicomplex $(p,q) \to \mathcal{O}_V \otimes$

$\Omega^p(V^q) \otimes \mathcal{O}_V$. The element $\omega_{i_1} \otimes \ldots \otimes \omega_{i_m}$ can be interpreted as a global continuous section of $\mathcal{T}$. One verifies that the class of the element

$$\sum_{\substack{0 \leq k, l \leq m \\ k+l \leq m}} \left(\int_x \omega_{i_1}, \ldots, \omega_{i_k} \right) \otimes \omega_{i_{k+1}} \otimes \ldots \otimes \omega_{i_{m-l}} \otimes (-1)^l \left(\int_x \omega_{i_m}, \ldots, \omega_{i_{m-l+1}} \right)$$

$$\in \sum \mathcal{O}_V \otimes \Omega^{m-1-k}(V^{m-1-k}) \otimes \mathcal{O}_V$$

is a horizontal section of $\mathcal{H}^{an}$. ($\mathcal{H}^{an} := \mathcal{H} \otimes_{\mathcal{O}_V} \mathcal{O}_V^{an}$). Hence we get:

2.7. Let $(\alpha, \beta) \in \pi_1(V \times V, (x, x))$. Then the representation of $\pi_1(V \times V, (x, x))$ on the vector space $H^0_{DR}(\rho^{\bullet -1}(x, x); \mathbb{Q})$ is given by

$$(\alpha, \beta) : \omega_{i_1} \otimes \ldots \otimes \omega_{i_m} \to$$

$$\sum_{\substack{0 \leq k, l \leq m \\ k+l \leq m}} \left(\int_\alpha \omega_{i_1}, \ldots, \omega_{i_k} \right) \omega_{i_{k+1}} \otimes \ldots \otimes \omega_{i_{m-l}} \left(\int_{\beta^{-1}} \omega_{i_{m-l+1}}, \ldots, \omega_{i_m} \right).$$

By 2.2.5 the Hopf algebra $\mathbb{Q}(V)^*$ is isomorphic to $\mathrm{Alg}\,(\pi(V))$. Hence we shall consider ω'_V on the associated bundle $V \times \mathbb{Q}(V)^* \to V$. It follows from Proposition 1.9.i) that the monodromy representation is given by

$$\pi_1(V, x) \ni \alpha : (X_{i_1} \cdot \ldots \cdot X_{i_k})^* \to \sum_l (X_{i_1} \cdot \ldots \cdot X_{i_l})^* \left(\int_{\alpha^{-1}} \omega_{i_{l+1}}, \ldots, \omega_{i_k} \right)$$

for the connection ω'_V, and by

$$\pi_1(V, x) \ni \alpha : (X_{i_1} \cdot \ldots \cdot X_{i_k})^* \to \sum_l \left(\int_\alpha \omega_{i_1} \cdot \ldots \cdot \omega_{i_l} \right) (X_{i_{l+1}}, \ldots, X_{i_k})^*$$

for the connection $'\omega_V$.

The isomorphism $H^0 \xrightarrow{\sim} \mathbb{Q}(V)^*$ defined by $\omega_{i_1} \otimes \ldots \otimes \omega_{i_k} \to (X_{i_1} \cdot \ldots \cdot X_{i_k})^*$ gives an isomorphism of monodromy representations of holomorphic vector bundles $(\mathcal{H}^0_x, d_{\mathbb{C}})^{an}$ and $(V \times \mathbb{Q}(V)^* \to V, \omega'_V)^{an}$ (resp. $(_x\mathcal{H}^0, d_{\mathbb{C}})^{an}$ and $(V \times \mathbb{Q}(V)^* \to V, '\omega_V)^{an})$ at the point $x \in V$. (( )^{an} is an analytic object corresponding to an algebraic one).

Observe that both connections are regular. One can check this directly for the connection ω'_V. The connection $d_{\mathbb{C}}$ is a successive extension of trivial connections. The monodromy representations at the point x are isomorphic. Hence the theorem follows from [D3] Théorème 5.9.

2.8. The cohomology class c in $H^0_{DR}(\rho^{\bullet -1}(x, x))$ is a polynomial function on $\pi_1^{DR}(V, x)(\mathbb{C})$. Let $\alpha \in \pi_1(V, x)$. There is a canonical map $\pi_1(V, x) \to$

$\pi_1^{DR}(V,x)(C)$ (the monodromy homomorphism) and it follows from 2.7 that the value of c on the image of α is equal to $\sum \int_{\alpha^{-1}} w_{i_1}, \ldots, w_{i_m}$.

§3. Homotopy relative tangential base points on $\mathbb{P}^1(\mathbb{C})\backslash\{a_1, \cdots, a_{n+1}\}$.

3.1. In this section we define a tangential base point on the complex projective line minus a finite number of points.

Definition 3.1.1. Let $X = \mathbb{P}^1(\mathbb{C}) \setminus \{a_1, ..., a_{n+1}\}$. A tangential base point on X is a non-zero tangent vector $\vec{v}$ at one of the missing points a_i. We shall denote by $\vec{v}$ (or by v) the tangential base point defined by the tangent vector $\vec{v}$.

If $X = \mathbb{C}\setminus\{a_1, ..., a_n\}$ we identify the tangent space at a_i with the vector space $\mathbb{C}$ and the tangent vector $\vec{v}$ at a_i with the vector $\overrightarrow{a_i x}$, where $x = a_i + \vec{v}$. The complex projective line $\mathbb{P}^1(\mathbb{C})$ has two canonical charts $\mathbb{C} \to \mathbb{P}^1(\mathbb{C})$, $z \to [z : 1]$ and $\mathbb{C} \to \mathbb{P}^1(\mathbb{C})$, $z \to [1 : z]$ which allow to identify tangent vectors to $\mathbb{P}^1(\mathbb{C})$ with tangent vectors to $\mathbb{C}$.

Let k be a subfield of $\mathbb{C}$. We say that a tangential base point $\vec{v}$ at a_i is a tangential base k-point if $a_i \in k$ and $a_i + \vec{v} \in k$.

Definition 3.1.2. Let $X = \mathbb{P}^1(\mathbb{C}) \setminus \{a_1, ..., a_{n+1}\}$ and let $\vec{v}$ be a tangential base point (at the missing point a_i). A path on X starting from the tangential base point $\vec{v}$ is a smooth path in $\mathbb{P}^1(\mathbb{C})$, satisfying the following conditions:

i) the image of the open interval $(0, 1)$ is contained in X;

ii) the tangent vector to the path at a_i is equal to $\vec{v}$.

We say that a path γ ends at a tangential base point $\vec{v}$ if the path γ^{-1} ($\gamma^{-1}(t) := \gamma(1-t)$) starts from the tangential base point $\vec{v}$.

If a path γ on $\mathbb{P}^1(\mathbb{C})$ starts from $\vec{v}$ or ends at $\vec{w}$ or starts from $\vec{v}$ and ends at $\vec{w}$ and if $\gamma((0, 1)) \subset X$ then we say that γ is a path on X.

Definition 3.1.3. A homotopy between two paths γ and γ' on X starting from a tangential base point $\vec{v}$ is a smooth homotopy H_s between the paths γ and γ' on $\mathbb{P}^1(\mathbb{C})$ satisfying the following conditions:

i) $H_s((0,1)) \subset X$ for $s \in [0, 1]$;

ii) the tangent vector of the path H_s at the starting point is $\vec{v}$ for $s \in [0, 1]$.

A homotopy between two paths γ and γ' which end at a tangential base point $\vec{v}$ is a homotopy between the paths γ^{-1} and γ'^{-1} which start at $\vec{v}$.

According to this definition, the paths α, β on $\mathbb{C} \setminus \{0\}$ shown in the following figure are not homotopic.

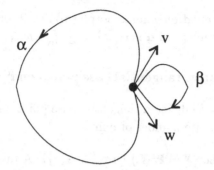

Definition 3.1.4. Let γ be a path starting from a tangential base point $\vec{v}$ (a tangent vector at a_i) and ending at the same tangential base point $\vec{v}$. We say that γ is null homotopic if there is a smooth homotopy H_s between γ and the constant map into a_i such that:

i) H_s is a path on X starting from $\vec{v}$ and ending at $\vec{v}$ for $s < 1$;

ii) H_1 is the constant map into a_i.

If γ is a path on X we denote by $[\gamma]$ the homotopy class of γ. Suppose that a path φ on X ends at a tangential base point $\vec{v}$ and a path ψ on X starts from $\vec{v}$. We define a composition of homotopy classes $[\psi] \cdot [\varphi]$ in the following way. Changing φ and ψ in their homotopy classes we can assume that φ and ψ coincide near a_i. Let $\varepsilon > 0$ be small. Let $\varphi_{1-\varepsilon} := \varphi_{|[0,1-\varepsilon]}$ and $\psi_\varepsilon := \psi_{|[\varepsilon,1]}$. We can assume that $\varphi(1 - \varepsilon) = \psi(\varepsilon)$. We define the composition of paths $\psi \cdot \varphi$ to be the composition $\psi_\varepsilon \cdot \varphi_{1-\varepsilon}$ (with a necessary reparametrization and smoothing). While $\psi \cdot \varphi$ depends on ε, its homotopy class is well defined and we set

$$[\psi] \cdot [\varphi] := [\psi \cdot \varphi].$$

Let $\hat{X} = X \cup \bigcup_{i=1}^{n+1} (T_{a_i}(\mathbb{P}^1(\mathbb{C}))\backslash\{0\})$ be a sum of X and the tangent spaces minus zero at the missing points. We define a groupoid P over $\hat{X} \times \hat{X}$ in the following way. If $(x,y) \in \hat{X} \times \hat{X}$ then $P_{x,y}$ is the set of homotopy classes of paths on X starting from x and ending at y. We set $P = \cup_{(x,y)\in\hat{X}\times\hat{X}}P_{x,y}$. We define a projection $p : P \to \hat{X} \times \hat{X}$ by $p(P_{x,y}) = (x,y)$. The partial composition of homotopy classes makes $p : P \to \hat{X} \times \hat{X}$ into a groupoid over $\hat{X} \times \hat{X}$. The fibre over (x,x), $\pi_1(X,x) := p^{-1}(x,x)$ is the fundamental group of X at a base point $x \in \hat{X}$. Observe that x can be a tangential base point $\vec{v}$.

3.2. We shall construct functions $\Lambda_X(z;\vec{v},\gamma)$ if $\vec{v}$ is a tangential base point. Let us set $X = \mathbb{C}\backslash\{a_1,\ldots,a_n\}$. Let $x_0 \in \mathbb{C}$ and let $\delta : [0,1] \ni t \to a_i + t \cdot (x_0 - a_i)$ be an interval joining a_i and x_0. Let γ be a path from a_i to $z \in X$ (not passing through any a_k, $k = 1,\ldots n$) tangent to δ in a_i. We assume that in a small neighbourhood of a_i the path γ coincides with δ. We identify $\vec{v} = x_0 - a_i$ with a tangent vector to $\mathbb{C}$ in a_i. Let $\omega_1 = \frac{dz}{z-a_1}, \ldots, \omega_n = \frac{dz}{z-a_n}$. We set

$$\Lambda_{a_i,\vec{v}}(\alpha_1, \ldots, \alpha_k)(z) := \int_{a_i,\gamma}^{z} \omega_{\alpha_1}, \ldots, \omega_{\alpha_k} \quad \text{if} \quad \alpha_1 \neq i.$$

Let $\varepsilon \in im(\delta)$ be near a_i. Let γ_ε be a part of γ from ε to z, and let δ_ε be a part of δ from ε to x_0. We set

$$\Lambda_{a_i,\vec{v}}(i, \ldots, i)(z) = \int_{x_0, \gamma_\varepsilon + \delta_\varepsilon^{-1}}^{z} \frac{dt}{t - a_i}, \ldots, \frac{dt}{t - a_i},$$

$$\Lambda_{a_i,\vec{v}}(i, \ldots, i, \alpha_{k+1})(z) := \lim_{\varepsilon \to a_i} \int_{\varepsilon, \gamma_\varepsilon}^{z} \Lambda_{a_i,\vec{v}}(i, \ldots, i)(t)\omega_{\alpha_{k+1}} \quad \text{if} \quad \alpha_{k+1} \neq i$$

and

$$\Lambda_{a_i,\vec{v}}(i, \ldots, i, \alpha_{k+1} \ldots, \alpha_{l+1})(z) := \int_{a_1,\gamma}^{z} \Lambda_{a_i,\vec{v}}(i, \ldots, i, \alpha_{k+1} \ldots, \alpha_l)(t)\omega_{\alpha_{l+1}}$$

if $\alpha_{k+1} \neq i$, and $l \geq k + 1$.

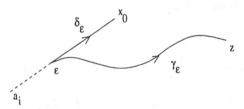

Lemma 3.2.1. *The integrals* $\Lambda_{a_i,\vec{v}}(\alpha_1, \ldots, \alpha_k)(z)$ *exist and they are analytic, multivalued functions on* X.

Proof. Assume that $\alpha_t \neq i$ for $t \leq l$ and $\alpha_{l+1} = i$. The function $g(z) := \int_{a_i}^{z} \omega_{\alpha_1}, \ldots, \omega_{\alpha_l}$ is analytic multivalued on $X \cup \{a_i\}$ and vanishes in a_i. Hence the integral $g_1(z) := \int_{a_i}^{z} g(z)\frac{dz}{z-a_i}$ exists, the function $g_1(z)$ is analytic multivalued on $X \cup \{a_i\}$ and vanishes in a_i. Hence by induction we get that $\Lambda_{a_i,v}(\alpha_1, \ldots, \alpha_n)(z)$ exists and it is analytic, multivalued on $X \cup \{a_i\}$.

Assume now that $\alpha_t = i$ for $t \leq l$ and $\alpha_{l+1} \neq i$. Without loss of generality we can assume that $a_i = 0$ and $x_0 = 1$. Observe that

$$\lim_{\varepsilon \to 0} \int_{x_0, \gamma_\varepsilon \circ (\delta_\varepsilon)^{-1}}^{z} z^n (\log z)^m dz = z^{n+1}\left(\sum_{i=0}^{m} \beta_i (\log z)^{m-i}\right)$$

where β_i are rational numbers. The function $z^q(\log z)^p$ for q and p positive
integers, is analytic multivalued on X, continuous on any small cone with
a vertex in a_i (0 in this case) and it vanishes in a_i. The function $\frac{1}{z-a_j}$ for
$j \neq i$ is bounded on any sufficiently small neighbourhood of a_i. Hence the
integral

$$\lim_{\epsilon \to a_i} \int_{\epsilon, \gamma_\epsilon}^{z} (\Lambda_{a_i, \vec{v}}(i, \ldots, i)(t)) \, \frac{dt}{t - a_j}$$

is an analytic multivalued function on V, continuous and univalued on any
small cone with a vertex in a_i and it vanishes in a_i. This also implies that
the last integral exists.   ◇

We recall that $X_1, \ldots, X_n$ are duals of $\frac{dz}{z-a_1}, \ldots, \frac{dz}{z-a_n}$. Let us set

$$\Lambda_X(z; \vec{v}, \gamma) = 1 + \sum (-1)^k \Lambda_{a_i, \vec{v}}(\alpha_1, \ldots, \alpha_k)(z) X_{\alpha_k} \cdot \ldots \cdot X_{\alpha_1}.$$

Observe that $\Lambda_X(z; \vec{v}, \gamma)$ depends on $\vec{v}$, i.e. on x_0.

Lemma 3.2.2. *The map*

$$X \ni z \to (z, \Lambda_X(z; \vec{v}, \gamma)) \in X \times P(X)$$

is horizontal with respect to ω_X.

Proof. We have

$$d\Lambda_{a_i, \vec{v}}(\alpha_1, \ldots, \alpha_k)(z) = \Lambda_{a_i, \vec{v}}(\alpha_1, \ldots, \alpha_{k-1})(z) \omega_{\alpha_k}.$$

Hence the functions $(-1)^k \Lambda_{a_i, \vec{v}}(\alpha_1, \ldots, \alpha_k)(z)$ satisfy the system of differential equations defining horizontal sections of the principal bundle $X \times$
$P(X) \to X$ equipped with the connection ω_X.   ◇

It remains to define functions $\Lambda_X(z; \vec{v}, \gamma)$ if $a_i = \infty$ or all a_i are different
from ∞. Let $f : Y = \mathbb{C} \backslash \{b_1, \ldots, b_n\} \to X = \mathbb{P}^1(\mathbb{C}) \backslash \{a_1, \ldots, a_{n+1}\}$ be a
regular map of the form $\frac{az+b}{cz+d}$ with $\det \begin{pmatrix} a & b \\ c & d \end{pmatrix} \neq 0$. Let $y \in Y$ and let
γ be a path on Y from y to z. Then it follows from corollary 1.7 that
$f_*(\Lambda_Y(z; y, \gamma)) = \Lambda_X(f(z); f(y), f(\gamma))$ if $y \in Y$. We shall use this fact to
define $\Lambda_X(z; \vec{v}, \gamma)$ where $\vec{v}$ is a tangent vector to $\mathbb{P}^1(\mathbb{C})$ in a_i and γ is a path
from a_i to z, which is tangent to $\vec{v}$.

We set

(3.2.2.1.) $\Lambda_X(z; \vec{v}, \gamma) := f_*(\Lambda_Y(f^{-1}(z); f_*^{-1}(\vec{v}), f^{-1}(\gamma)))$.

It is clear that $\Lambda_X(z; \vec{v}, \gamma)$ does not depend on the choice of f. Thus we have the following lemma.

Lemma 3.2.3. *Let $X = \mathbb{P}^1(\mathbb{C}) \setminus \{a_1, \ldots, a_{n+1}\}$ and let $\vec{v}$ be a tangential base point. Then the map*

$$X \ni z \to (z, \Lambda_X(z; \vec{v}, \gamma)) \in X \times P(X)$$

is horizontal with respect to ω_X.

We set $L_X(z; \vec{v}, \gamma) := \log\Lambda_X(z; \vec{v}, \gamma)$. If we are dealing with only one space X we shall usually omit the subscript X and we shall write $\Lambda(z; \vec{v}, \gamma)$ and $L(z; \vec{v}, \gamma)$, or $\Lambda_{\vec{v}}(z; \gamma)$ and $L_{\vec{v}}(z; \gamma)$, or even $\Lambda_{\vec{v}}(z)$ and $L_{\vec{v}}(z)$.

The constructions from 3.1 and 3.2 are summarized in the following proposition.

Proposition 3.2.4. *Let $v \in \hat{X}$. The functions $\Lambda_X(z; v, \gamma)$ and $L_X(z; v, \gamma)$ depend only on the homotopy class of γ in $P_{v,z}$.*

§4. Generators of $\pi_1(\mathbb{P}^1(\mathbb{C}) \backslash \{a_1, \ldots, a_{n+1}\}, x)$.

Let $X = \mathbb{P}^1(\mathbb{C}) \backslash \{a_1, \ldots, a_{n+1}\}$ and let $x \in \hat{X}$. Let $\vec{v_k}$ be a tangent vector in a_k. Then the loop around a_k at the base point $\vec{v_k}$ is the following element $S_{\vec{v_k}}$ of $\pi_1(X, \vec{v_k})$ (see picture).

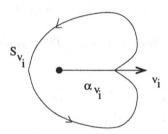

Now we shall describe how to choose generators of $\pi_1(X, x)$. Let us choose a tangent vector $\vec{v_i} \in T_{a_i}(\mathbb{P}^1(\mathbb{C})) \setminus \{0\}$ for each $i = 1, 2, \ldots, n+1$.

i) If $x \in X$ then we choose a family of paths $\Gamma = \{\gamma_i\}_{i=1}^{n+1}$ on X from x to each $\vec{v_i}$ such that any two paths do not intersect and no path self intersects. The indices are chosen in such a way that when we make a small circle around x in the opposite clockwise direction, starting from γ_1, we meet $\gamma_2, \gamma_3, \ldots, \gamma_{n+1}$.

ii) If $x = \vec{v_1}$ then we choose a family of paths $\Gamma = \{\gamma_i\}_{i=2}^{n+1}$ from x to each $\vec{v_i}$ such that making a small circle around a_1 in the opposite clockwise direction, starting from γ_2 we meet $\gamma_3, \ldots, \gamma_{n+1}$.

iii) We associate the following element S_i in $\pi_1(X, x)$ to the path γ_i. We move along γ_i, we make a small loop in the opposite clockwise direction around

a_i and we return to x along γ_i^{-1}. In the case ii) the element S_1 is the loop $S_{\bar{v_1}}$.

iv) We can choose any order of missing points to construct the elements $S_1, ..., S_{n+1}$.

The following lemma is obvious.

Lemma 4.1. *The elements* $S_1, S_2, \ldots, S_{n+1}$ *are generators of* $\pi_1(X, x)$. *We have* $S_{n+1} \cdot \ldots \cdot S_2 \cdot S_1 = 1$.

Definition 4.2. The ordered sequence $(S_1, \ldots, S_{n+1})$ of elements of $\pi_1(\mathbb{X}, x)$ obtained from the family of paths Γ as in i),...,iv) will be called a sequence of geometric generators of $\pi_1(X, x)$ associated to Γ.

§5. Monodromy of iterated integrals on $\mathbb{P}^1(\mathbb{C})\backslash\{a_1, \ldots, a_{n+1}\}$

Let $X = \mathbb{P}^1(\mathbb{C})\backslash\{a_1, \ldots, a_{n+1}\}$. We want to compare functions $\Lambda_v(z)$ for different base points. Let $x_1, x_2, x_3 \in \hat{X}$ and let $z_0 \in X$. Let δ_i for $i = 1, 2, 3$ be a path on X from x_i to z_0. Let us set $\gamma_{ij} := \delta_i^{-1} \circ \delta_j$.

Proposition 5.1. *Let us continue each function* $\Lambda_{x_i}(z)$ *along* δ_i *to the point* z_0. *There exist elements* $a_{x_j}^{x_i}(\gamma_{ij}) \in P(X)$ *such that*

$$\Lambda_{x_i}(z) \cdot a_{x_j}^{x_i}(\gamma_{ij}) = \Lambda_{x_j}(z)$$

for all z *in a small neighbourhood of* z_0. *The elements* $a_{x_j}^{x_i}(\gamma_{ij})$ *satisfy the following relations*

$$a_{x_i}^{x_i}(\gamma_{ii}) = 1,$$
$$a_{x_j}^{x_i}(\gamma_{ij}) \cdot a_{x_i}^{x_j}(\gamma_{ji}) = 1,$$
$$a_{x_j}^{x_i}(\gamma_{ij}) \cdot a_{x_k}^{x_j}(\gamma_{jk}) = a_{x_k}^{x_i}(\gamma_{ik}).$$

(Observe that the elements $a_{x_j}^{x_i}(\gamma_{ij})$ depend on the choice of paths γ_{ij}.)

Proof. The existence of $a_{x_j}^{x_i}(\gamma_{ij})$ follows from the fact that the $\Lambda_{x_i}(z)$'s are horizontal sections. The first two relations are obvious. The last relation follows from the equalities $\Lambda_{x_i}(z) \cdot a_{x_j}^{x_i}(\gamma_{ij}) = \Lambda_{x_j}(z)$, $\Lambda_{x_j}(z) \cdot a_{x_k}^{x_j}(\gamma_{jk}) = \Lambda_{x_k}(z)$ and $\Lambda_{x_i}(z) \cdot a_{x_k}^{x_i}(\gamma_{ik}) = \Lambda_{x_k}(z)$. $\diamond$

If γ is a path on X from v to w then we define $a_w^v(\gamma)$ by the formula

$$\Lambda_v(z) \cdot a_w^v(\gamma) = \Lambda_w(z),$$

where $\Lambda_v(z)$ and $\Lambda_w(z)$ are continued along γ.

Proposition 5.2. *Let $\vec{v_k} \in T_{a_k}\mathbb{P}^1(\mathbb{C})\backslash\{0\}$. Let $S_{\vec{v_k}}$ be a loop around a_k based at $\vec{v_k} \in T_{a_k}(\mathbb{P}^1(\mathbb{C}))\backslash\{0\}$ (see figure at the beginning of section 4). The monodromy of $\Lambda_{\vec{v_k}}(z)$ along $S_{\vec{v_k}}$ is given by*

$$S_{\vec{v_k}} : \Lambda_{\vec{v_k}}(z) \to \Lambda_{\vec{v_k}}(z) \cdot e^{-2\pi i X_k}$$

Proof. The monodromy of $\Lambda_{\vec{v_k}}(k^m)(z) := \Lambda_{\vec{v_k}}(k, k, \ldots, k)(z)$ along $S_{\vec{v_k}}$ is given by $S_{\vec{v_k}} : \Lambda_{\vec{v_k}}(k^m)(z) \to \Lambda_{\vec{v_k}}(k^m)(z) + \sum_{l=1}^{m} \Lambda_{\vec{v_k}}(k^{m-l})(z)\frac{(-2\pi i)^l}{l!}$. Let γ be a path on X from $\vec{v_k}$ to z and let $\alpha_1 \neq k$. We must analytically continue the function $\Lambda_{\vec{v_k}}(k^n, \alpha_1, \ldots, \alpha_p)(z)$ along the composition $\gamma \cdot S_{\vec{v_k}}$. The contribution along $S_{\vec{v_k}}$ can be made arbitrarily small. However the function $\Lambda_{\vec{v_k}}(k^m)(z)$ changes after the tour along $S_{\vec{v_k}}$. This implies that the monodromy of $\Lambda_{\vec{v_k}}(k^m, \alpha_1, \ldots, \alpha_p)(z)$ along $S_{\vec{v_k}}$ is given by

$$S_k : \Lambda_{\vec{v_k}}(k^m, \alpha_1, \ldots, \alpha_p)(z) \to \Lambda_{\vec{v_k}}(k^m, \alpha_1, \ldots, \alpha_p)(z)$$
$$+ \sum_{l=1}^{m} \Lambda_{\vec{v_k}}(k^{m-l}, \alpha_1, \ldots, \alpha_p)(z)\frac{(-2\pi i)^l}{l!}.$$

for $\alpha_1 \neq k$. The formula for the monodromy of $\Lambda_{\vec{v_k}}(z)$ along $S_{\vec{v_k}}$ follows from this. $\diamond$

Let $x \in \hat{X}$. Let us choose $\vec{v_i} \in T_{a_i}\mathbb{P}^1(\mathbb{C})\backslash\{0\}$ for $i = 1, 2, \ldots, n+1$. Let $(S_1, \ldots, S_{n+1})$ be a sequence of geometric generators of $\pi_1(X, x)$ associated to $\Gamma = \{\gamma_i\}_{i=1}^{n+1}$ (if $x \in X$) or $\Gamma = \{\gamma_i\}_{i=2}^{n+1}$ (if x is a tangential base point, i.e. $x = \vec{v_1}$), where Γ is a family of paths on X from x to $\vec{v_i}$ for $i = 1, 2, \ldots, n+1$ or $i = 2, \ldots, n+1$ as in section 4.

Theorem 5.3. *The monodromy of the function $\Lambda_x(z)$ along the loop S_k is given by*

$$S_k : \Lambda_x(z) \to \Lambda_x(z) \cdot (a_x^{\vec{v_k}}(\gamma_k))^{-1} \cdot e^{-2\pi i X_k} \cdot a_x^{\vec{v_k}}(\gamma_k).$$

Proof. It follows from Proposition 5.1 that

$$(*_1) \qquad\qquad \Lambda_x(z) = \Lambda_{\vec{v_k}}(z) \cdot a_x^{\vec{v_k}}(\gamma_k)$$

for z in the small neighbourhood of some $\gamma_k(t)$. This equality is preserved after the monodromy transformation along S_k, hence we have

$$(*_2) \qquad\qquad (\Lambda_x(z))^{S_k} = (\Lambda_{\vec{v_k}}(z))^{S_k} \cdot a_x^{\vec{v_k}}(\gamma_k),$$

244   Zdzisław Wojtkowiak

where ( )^{S_k} denotes the function ( ) after the monodromy transformation along $S_{v_{\bar{k}}}$. Observe that $(\Lambda_{v_{\bar{k}}}(z))^{S_k} = (\Lambda_{v_{\bar{k}}}(z))^{S_{v_{\bar{k}}}}$. It follows from Proposition 5.2 that

$(*_3)$ $$(\Lambda_{v_{\bar{k}}}(z))^{S_k} = \Lambda_{v_{\bar{k}}}(z) \cdot e^{-2\pi i X_k}.$$

Substituting $(*_3)$ in $(*_2)$ and then substituting $(*_1)$ for $\Lambda_{v_{\bar{k}}}(z)$ yields the formula for $(\Lambda_x(z))^{S_k}$. $\Diamond$

Corollary 5.4. *The monodromy of the function $L_x(z)$ along the loop S_k is given by*

$$S_k : L_x(z) \to L_x(z) \cdot (\alpha_x^{v_{\bar{k}}}(\gamma_k))^{-1} \cdot (-2\pi i X_k) \cdot \alpha_x^{v_{\bar{k}}}(\gamma_k)$$

where $\alpha_x^{v_{\bar{k}}}(\gamma_k) = \log(a_x^{v_{\bar{k}}}(\gamma_k))$.

Proof. The corollary follows immediately from Proposition 1.8. ii). $\Diamond$

Definition 5.5. Let $x \in \hat{X}$. Define a homomorphism

$$\theta_{x,X} : \pi_1(X,x) \to P(X) \ (\text{resp.}\, \theta_{x,X} : \pi_1(X,x) \to \pi(X))$$

by the formula

$$\theta_{x,X}(S_k) = (a_x^{v_{\bar{k}}}(\gamma_k))^{-1} \cdot e^{-2\pi i X_k} \cdot a_x^{v_{\bar{k}}}(\gamma_k) \ (\text{resp.}\ \theta_{x,X}(S_k)$$
$$= (\alpha_x^{v_{\bar{k}}}(\gamma_k))^{-1} \cdot (-2\pi i X_k) \cdot \alpha_x^{v_{\bar{k}}}(\gamma_k));$$

it is called the monodromy homomorphism of the form ω_X at the base point x.

Observe that x can be a tangential base point. If $x \in X$ then the homomorphisms from Definitions 1.3 and 5.5 coincide.

Let $v, v' \in \hat{X}$ and let γ be a path on X from v to v'. If $\pi_1(X,v)$ and $\pi_1(X,v')$ are identified via the path γ, then the homomorphisms $\theta_{v,X}$ and $\theta_{v',X}$ are conjugate.

Proposition 5.6. *Let*

$$f : X = \mathbb{P}^1(\mathbb{C})\backslash\{a_1,\ldots,a_{n+1}\} \to Y = \mathbb{P}^1(\mathbb{C})\backslash\{b_1,\ldots,b_{n+1}\}$$

be a regular bijective map. Then for any $v, w \in \hat{X}$, any path γ from v to z and any path δ from v to w we have

$$f_*(\Lambda_X(z;v,\gamma)) = \Lambda_Y(f(z); f(v), f(\gamma))$$

and

$$f_*(a_w^v(\delta)) = a_{f(w)}^{f(v)}(f(\delta)).$$

(notation: $f(v) := f_*(v)$ if v is a tangent vector).

Proof. The proposition follows from the definitions of $\Lambda_X(z; v, \gamma)$ and a_w^v for tangent vectors, and from Corollary 1.7. $\diamond$

§6. Calculations.

Let $V = \mathbb{P}^1(\mathbb{C}) \backslash \{0, 1, \infty\}$. The forms $\frac{dz}{z}$ and $\frac{dz}{z-1}$ form a basis of $A^1(V)$. Let $X := (\frac{dz}{z})^*$ and $Y := (\frac{dz}{z-1})^*$ be the dual basis of $(A^1(V))^*$. Set $Z := -X - Y$. The group $P(V)$ is the group of invertible power series in non-commuting variables X and Y with constant term equal to 1.

Fix a path $\gamma_1 = $ interval $[0, 1]$ from $\vec{01}$ to $\vec{10}$. It follows from Proposition 5.1 that along the path γ_1 we have

(0) $$\Lambda_{\vec{10}}(z) \cdot a_{\vec{01}}^{\vec{10}}(X, Y) = \Lambda_{\vec{01}}(z).$$

Let $f(z) = 1 - z$. It follows from Proposition 5.6 that

$$f_*(a_{01}^{10}(X, Y)) = a_{10}^{01}(X, Y).$$

(We omit the arrow over vectors, when it does not cause confusion.)

Proposition 5.1 implies that

(1) $$a_{01}^{10}(X, Y) \cdot a_{10}^{01}(X, Y) = 1.$$

Observe that $f_*(X) = Y$ and $f_*(Y) = X$. Hence we get Deligne's formula

(2) $$a_{01}^{10}(X, Y) \cdot a_{01}^{10}(Y, X) = 1.$$

(The proof of (2) given here essentially repeats that of Deligne (see[D5]).)

Let us fix a path $\gamma_\infty = $ interval $[0, \varepsilon] + $ arc from ε to $-\varepsilon$ passing through $(-i) \cdot \varepsilon + $ interval $[-\varepsilon, \infty]$ from $\vec{01}$ to $\vec{\infty 0}$ where $\varepsilon > 0$ (see figure).

Let S_0 (around 0), S_1 (around 1) and S_∞ (around ∞) be geometric generators of $\pi_1(V, \vec{01})$ associated to the family $\{\gamma_1, \gamma_\infty\}$ (see picture on page 5). Then we have $S_0 \cdot S_1 \cdot S_\infty = 1$.

Theorem 6.1. *The monodromy of $\Lambda_{\overline{01}}(z)$ is given by the following formulas:*

$$S_0 : \Lambda_{\overline{01}}(z) \to \Lambda_{\overline{01}}(z) \cdot e^{(-2\pi i)X},$$
$$(3)\ S_1 : \Lambda_{\overline{01}}(z) \to \Lambda_{\overline{01}}(z) \cdot (a_{01}^{10}(X,Y))^{-1} \cdot e^{(-2\pi i)Y} \cdot a_{01}^{10}(X,Y),$$
$$S_\infty : \Lambda_{\overline{01}}(z) \to \Lambda_{\overline{01}}(z) \cdot e^{-\pi i X} \cdot (a_{01}^{10}(X,Z))^{-1} \cdot e^{(-2\pi i)Z} \cdot a_{01}^{10}(X,Z) \cdot e^{\pi i X}.$$

Proof. The formulas for the monodromy along S_0 and S_1 follows from Theorem 5.3. The monodromy along S_∞ needs some explanation. By Theorem 5.3 it is given by the formula $S_\infty : \Lambda_{\overline{01}}(z) \to \Lambda_{\overline{01}}(z) \cdot (a_{01}^{\infty 0}(X,Y))^{-1} \cdot e^{(-2\pi i)Z} \cdot a_{01}^{\infty 0}(X,Y)$. By Proposition 5.1 $a_{01}^{\infty 0}(X,Y) = a_{0\infty}^{\infty 0}(X,Y) \cdot a_{01}^{0\infty}(X,Y)$. One calculates using 3.2.3 that $a_{01}^{0\infty}(X,Y) = e^{\pi i \cdot X}$. Let $f(z) = \frac{z}{z-1}$. We have $f_*(X) = X$ and $f_*(Y) = Z$. It follows from Proposition 5.6 that

$$a_{01}^{10}(X,Z) = f_*(a_{01}^{10}(X,Y)) = a_{0\infty}^{\infty 0}(X,Y).$$

Hence we get the formula describing the monodromy along S_∞. $\Diamond$

We recall that the Lie algebra $L(V)$ is the completion of the free Lie algebra over $\mathbb{C}$ on two generators X and Y. The group $\pi(V)$ is $L(V)$ equipped with the multiplication given by the Baker-Campbell-Hausdorff formula. Let us set $\alpha(X,Y) := \alpha_{01}^{10}(X,Y) := \log(a_{01}^{10}(X,Y))$. It follows from Corollary 5.4 that the monodromy of $L_{\overline{01}}(z)$ is given by the following formulas:

$$S_0 : L_{\overline{01}}(z) \to L_{\overline{01}}(z) \cdot (-2\pi i)X,$$
$$(4)\ S_1 : L_{\overline{01}}(z) \to L_{\overline{01}}(z) \cdot \alpha(X,Y)^{-1} \cdot (-2\pi i)Y \cdot \alpha(X,Y),$$
$$S_\infty : L_{\overline{01}}(z) \to L_{\overline{01}}(z) \cdot (-\pi i)X \cdot \alpha(X,Z)^{-1} \cdot (-2\pi i)Z \cdot \alpha(X,Z) \cdot (\pi i)X.$$

Observe that **Theorem A** (in the introduction) follows from the formulas (4). Let us compute the coefficients of $a_{01}^{10}(X,Y)$ and $\alpha(X,Y)$. If w is a monomial in X and Y we denote by $a(w)$ the coefficient of $a_{01}^{10}(X,Y)$ at w. Let X be the first Lie basis element and let Y be the second element. We choose a basis of a free Lie algebra on X and Y as in [MKS] pages 324-325. If w is an element of this basis, let $\alpha(w)$ be the coefficient of w of $\alpha(X,Y)$. Observe that the coefficients $a(X^n)$ and $a(Y^n)$ vanish. We assume that z approaches 1 in the formula (0). Then it follows from the formula (0), the formula (1) and the formula (2) that

$$(5)\ a(X^n Y) = (-1)^n \zeta(n+1), \quad a(Y^n X) = (-1)^{n+1} \zeta(n+1)$$
$$(6)\ a(X^i Y^j) = \int_0^1 (-\frac{dz}{z-1})^j, (-\frac{dz}{z})^i, \quad a(Y^j X^i) = \int_0^1 (-\frac{dz}{z})^i, (-\frac{dz}{z-1})^j.$$

and

$$a(X^i Y^j) + a(Y^i X^j) = 0.$$

(If ω is a one-form then $\omega^i := \omega, \omega, \ldots, \omega$ i-times.) Let us set $\alpha_{i,j} := \alpha((Y, X) X^{i-1} Y^{j-1})$, where $w X^i Y^j = ((\ldots ((w, X) X) \ldots) X) Y) \ldots) Y)$. Comparing the coefficients at $Y^j X^i$ and $X^i Y^j$ in $\alpha(X, Y)$ and in $\log(a_{0,1}^{1,0}(X, Y))$ we get

$$\alpha_{i,j} = (-1)^{j-1} a(Y^j X^i) = (-1)^i a(X^i Y^j).$$

It follows from the equality $a(X^i Y^j) + a(Y^i X^j) = 0$ that $\alpha_{i,j} = \alpha_{j,i}$. Observe that we also have $a(X^i Y^j) + (-1)^{i+j} a(Y^j X^i) = 0$. This last equality can also be obtained from formula (1.6.2) in [Ch1].

Let us set $\pi' := (\pi(V), \pi(V))$, $\pi'' := (\pi', \pi')$ and $\pi_2(V) := \pi(V)/\pi''$. It follows from (3) that the monodromy homomorphism $\theta_{\vec{01}} : \pi_1(V, \vec{01}) \to \pi_2(V)$ is given by

$$
\begin{aligned}
S_0 &\to (-2\pi i) X, \\
(7) \qquad S_1 &\to (-2\pi i) Y + [-2\pi i Y, \alpha(X, Y)] \\
&= (-2\pi i) Y + \sum_{i=0, j=0}^{\infty} (2\pi i) \alpha_{i+1, j+1}((YX) X^i Y^{j+1}).
\end{aligned}
$$

Let us set $F(z) = \frac{(\log(1-z))^m}{m!}$. The formula

$$
\int_0^z F(z) \frac{dz}{z}, \left(\frac{dz}{z}\right)^n = \sum_{i=0}^{n} \frac{(-1)^{n-i}}{(n-i)! i!} \left(\int_0^z F(z) (\log z)^{n-i} \frac{dz}{z}\right) (\log z)^i
$$

(which can be proved by induction) implies that

$$
(8) \qquad \alpha_{n+1, m} = \frac{(-1)^n (-1)^m}{n! m!} \int_0^1 (\log(1-z))^m (\log z)^n \frac{dz}{z}.
$$

6.2. Suppose that $n_i \geq 1$ for $i = 1, \ldots, k-1$ and $n_k \geq 2$. Then we have

$$
a(X^{n_k - 1} Y \ldots X^{n_2 - 1} Y X^{n_1 - 1} Y) = \sum_{i_1, \ldots, i_k = 1}^{\infty} \frac{1}{i_1^{n_1} (i_1 + i_2)^{n_2} \ldots (i_1 + \ldots + i_k)^{n_k}}.
$$

These numbers are values of the functions

$$
\sum_{i_1, \ldots, i_k = 1}^{\infty} \frac{z^{i_1 + \ldots + i_k}}{i_1^{s_1} (i_1 + i_2)^{s_2} \ldots (i_1 + \ldots + i_k)^{s_k}}
$$

at $z = 1$ and $(s_1, \ldots, s_k) = (n_1, \ldots, n_k)$.

§7. Configuration spaces.

Let $V = \mathbb{P}^1(\mathbb{C}) \backslash \{a_1, \ldots, a_{n+1}\}$ and $V' = \mathbb{P}^1(\mathbb{C}) \backslash \{a'_1, \ldots, a'_{n+1}\}$. If the sequences $(a, x) := (a_1, \ldots, a_{n+1}, x)$ and $(a', x') := (a'_1, \ldots, a'_{n+1}, x')$ are close then the groups $\pi_1(V, x)$ and $\pi_1(V', x')$ are canonically isomorphic. We shall study how the monodromy homomorphisms $\theta_{x,a} := \theta_{x,V}$ and $\theta_{x',a'} := \theta_{x',V'}$ from sections 1 and 5

$$
\begin{array}{ccc}
\theta_{x,a}: & \pi_1(V, x) & \to & \pi(V) \\
& \wr & & \| \\
\theta_{x',a'}: & \pi_1(V', x') & \to & \pi(V')
\end{array}
$$

depend on a and a'.

If T is a topological space we set $T_*^n = \{(t_1, \ldots, t_n) \in T^n \mid t_i \neq t_j$ if $i \neq j\}$. Let $W_n := C_*^n$. The space $A^1(W_n)$ of global holomorphic one-forms on W_n with logarithmic singularities is spanned by $\omega_{ij} = \frac{dz_i - dz_j}{z_i - z_j}$ for $i, j \in \{1, 2, \ldots, n\}$ and $i < j$. Let $X_{ij} = (\frac{dz_i - dz_j}{z_i - z_j})^*$ be their formal duals. We set $X_{ji} = X_{ij}$ and $X_{ii} = 0$. In $A^2(W_n)$ we have

$$
\omega_{ki} \wedge \omega_{ij} + \omega_{jk} \wedge \omega_{ki} + \omega_{ij} \wedge \omega_{jk} = 0
$$

for i, j, k different. These are the only relations between the 2-forms $\omega_{ij} \wedge \omega_{lk}$ (this result is due to V.I.Arnold).

Dualizing the map

$$
\wedge^2(A^1(W_n)) \to A^1(W_n) \wedge A^1(W_n)
$$

we find that $R(W_n)$ is generated by

$$
[X_{ij}, X_{ik} + X_{jk}] \text{ with } i, j, k \text{ different}
$$

and

$$
[X_{ij}, X_{kl}] \text{ with } i, j, k, l \text{ different (see [C])}.
$$

Let $x = (x_1, \ldots, x_n, x_{n+1}) \in W_{n+1}$ be a base point. Let $p_i : W_{n+1} \to W_n$ $(i = 1, \ldots, n+1)$ be a projection $p_i(z_1, \ldots, z_{n+1}) = (z_1, \ldots, \hat{z}_i, \ldots, z_{n+1})$, let

$$
x(i) := (x_1, \ldots, x_{i-1}, x_{i+1}, \ldots, x_{n+1})
$$

and let

$$
V(i, x) := p_i^{-1}(x(i)) = \mathbb{C} \backslash \{x_1, \ldots, \hat{x}_i, \ldots, x_{n+1}\}.
$$

($\hat{z}$ means z is omitted). Let $k_i : V(i,x) \to W_{n+1}$ be given by $k_i(z) = (x_1,\ldots,x_{i-1},z,x_{i+1},\ldots,x_{n+1})$. The inclusion k_i induces

$$(k_i)_* : P(V(i,x)) \to P(W_{n+1})$$

and

$$(k_i)_* : \pi(V(i,x)) \to \pi(W_{n+1}).$$

One calculates that

$$(k_i)^*\left(\frac{dz_i - dz_j}{z_i - z_j}\right) = \frac{dz}{z - x_j} \text{ and } (k_i)^*\left(\frac{dz_l - dz_j}{z_l - z_j}\right) = 0 \text{ if } l \neq i \text{ and } j \neq i.$$

This implies that $(k_i)_*(X_j) = X_{ij}$ for $j = 1,2,\ldots,i-1,i+1,\ldots,n+1$ where X_j is the formal dual of $\frac{dz}{z-x_j}$. It follows from the form of the generators of $R(W_n)$ that the Lie subalgebra of $\mathrm{Lie}(W_n)$ generated by $X_{i,1},\ldots X_{i,i-1}, X_{i,i+1},\ldots,X_{i,n+1}$ is free and it is a Lie ideal of $\mathrm{Lie}(W_n)$. Hence the map $(k_i)_*$ is injective and its image, $(k_i)_*(\pi(V(i,x)))$ is a normal subgroup of $\pi(W_{n+1})$.

Let $x = (x_1,\ldots,x_n,x_{n+1}) \in W_{n+1}$ and $x' = (x'_1,\ldots,x'_n,x'_{n+1}) \in W_{n+1}$. Let us set $V := V(n+1,x)$ and $V' =: V(n+1,x')$. We choose a family of non-intersecting paths $\gamma_1,\ldots,\gamma_n,\gamma_{n+1}$ in $\mathbb{C}$ from x_1 to $x'_1,\ldots,x_n$ to x'_n and x_{n+1} to x'_{n+1}. We shall identify $\pi_1(V,x_{n+1})$ and $\pi_1(V',x'_{n+1})$ in the following way. Observe that $\gamma = (\gamma_1,\ldots,\gamma_n,\gamma_{n+1})$ is a path in W_{n+1} from x to x'. The identification isomorphism $\gamma. : \pi_1(V,x_{n+1}) \to \pi_1(V',x'_{n+1})$ is the unique isomorphism making the following diagram commute

$$\begin{array}{ccc} \pi_1(V,x_{n+1}) & \xrightarrow{(k_{n+1})_*} & \pi_1(W_{n+1},x) \\ \downarrow \gamma. & & \downarrow \gamma_\# \\ \pi_1(V',x'_{n+1}) & \xrightarrow{(k_{n+1})_*} & \pi_1(W_{n+1},x'). \end{array}$$

($\gamma_\#$ is induced by the path γ in a standard way).

Proposition 7.1. *After the identification of the fundamental groups of $V = \mathbb{C}\backslash\{x_1,\ldots,x_n\}$ and $V' = \mathbb{C}\backslash\{x'_1,\ldots,x'_n\}$ via γ, the monodromy homomorphisms*

$$\theta_{x_{n+1},V} : \pi_1(V,x_{n+1}) \to \pi(V) \text{ and } \theta_{x'_{n+1},V'} : \pi_1(V',x'_{n+1}) \to \pi(V') = \pi(V)$$

are conjugated by an element of the group $\pi(W_{n+1})$. (The group $(k_{n+1})_*\pi(V)$ is a normal subgroup of $\pi(W_{n+1})$ so $\pi(W_{n+1})$ acts on $\pi(V)$ by conjugation.)

250   Zdzisław Wojtkowiak

Proof. The result is a consequence of the following commutative diagram:

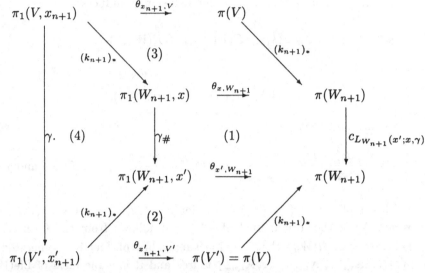

where $c_{L_{W_{n+1}}(x';x,\gamma)}$ is a conjugation by the element $L_{W_{n+1}}(x';x,\gamma)$. It follows from Proposition 1.4 that the square (1) commutes. Corollary 1.7 implies that (2) and (3) commutes. The square (4) commutes by the construction. ◇

Corollary 7.2. Let $x = (x_1,\ldots,x_{n+1}) \in W_{n+1}$. Set $V(i) := V(i,x)$. Let a_{ij} be a geometric generator of $\pi_1(V(i), x_i)$, which is a loop around the point x_j. Let A_{ij} be its image in $\pi_1(W_{n+1}, x)$. Then $\theta_{x,W_{n+1}}(A_{ij})$ is conjugate to $(-2\pi i)X_{ij}$ in the group $\pi(W_{n+1})$.

Proof. It follows from Theorem 5.3 that $\theta_{x_i,V(i)}(a_{ij})$ is conjugated to $(-2\pi i)X_{ij}$ in the group $\pi(V(i))$. Hence the corollary follows from Corollary 1.7. ◇

Now we shall study the relation between the monodromy representations for the configuration spaces $(\mathbb{P}^1(\mathbb{C})\backslash\{0,1,\infty\})^n_*$ and $(\mathbb{P}^1(\mathbb{C})\backslash\{0,1,\infty\})^m_*$. We shall use the Ihara result (see [I1] The Injectivity Theorem (i)).

Let us set $Y_n = (\mathbb{P}^1(\mathbb{C}))^n_*$ and $\mathcal{Y}_n = (\mathbb{P}^1(\mathbb{C})\backslash\{0,1,\infty\})^{n-3}_*$. The group $PGL_2(\mathbb{C})$ acts diagonally on Y_n and $\mathcal{Y}_n = Y_n/PGL_2(\mathbb{C})$. The identification is given by the map $(0,1,\infty,x_1,x_2,\ldots,x_{n-3}) \to (x_1,x_2,\ldots,x_{n-3})$. Let $\psi_k : W_{n-1} \to \mathcal{Y}_n$ be the composition of the map $(x_1,\ldots,x_{k-1},x_{k+1},\ldots,x_n) \to (x_1,\ldots,x_{k-1},\infty,x_{k+1},\ldots,x_n)$ and the projection $Y_n \to \mathcal{Y}_n$. The map ψ_k induces $(\psi_k)_* : H(W_{n-1}) \to H(\mathcal{Y}_n)$. Let us set $X_{ij} = (\psi_k)_*(X_{ij})$ where $X_{ij} = (\frac{dx_i - dx_j}{x_i - x_j})^* \in H(W_{n-1})$. (We use the same notation for $X_{ij} \in H(W_{n-1})$ and its image in $H(\mathcal{Y}_n)$. Notice also that it can be shown by computation that X_{ij} in $H(\mathcal{Y}_n)$ does not depend on the choice of ψ_k.)

Let $A_{ij} \in \pi_1(W_{n-1}, x)$ be as in Corollary 7.2. We shall also denote the image of A_{ij} in $\mathcal{Y}_n$ by A_{ij}.

Corollary 7.3. *The element* $\theta_{y, \mathcal{Y}_n}(A_{ij})$ *is conjugate to* $(-2\pi i)X_{ij}$ *in* $\pi(\mathcal{Y}_n)$.

Proof. This follows from Corollary 7.2 and the commutative diagram

$$
\begin{array}{ccc}
\theta_{x, W_{n-1}}: & \pi_1(W_{n-1}, x) & \to & \pi(W_{n-1}) \\
& \downarrow (\psi_k)_* & & \downarrow (\psi_k)_* \\
\theta_{y, \mathcal{Y}_n}: & \pi_1(\mathcal{Y}_n, y = \psi_k(x)) & \to & \pi(\mathcal{Y}_n).
\end{array}
$$

$\diamond$

Let $\mathrm{Aut}(\pi(\mathcal{Y}_n))$ be the group of algebraic automorphisms of a complex pro-algebraic group $\pi(\mathcal{Y}_n)$. Let $\mathrm{Aut}^*(\pi(\mathcal{Y}_n))$ be a subgroup of $\mathrm{Aut}(\pi(\mathcal{Y}_n))$ defined in the following way:

$$
\mathrm{Aut}^*(\pi(\mathcal{Y}_n)) = \{ f \in \mathrm{Aut}(\pi(\mathcal{Y}_n)) | \exists \alpha_f \in \mathbb{C}^*, f(X_{ij}) \sim \alpha_f X_{ij} \}.
$$

($\sim$ means "is conjugate to".)

Let us set

$$
T^n(\mathbb{C}) = \{ \varphi \in \mathrm{Hom}(\pi_1(\mathcal{Y}_n, y); \pi(\mathcal{Y}_n)) | \exists \alpha_\varphi \in \mathbb{C}^*, \forall A_{ij}, \varphi(A_{ij}) \sim \alpha_\varphi X_{ij} \}
$$

($A_{ij} \in \pi_1(\mathcal{Y}_n, y)$ are as in Corollary 7.3.). The group $\mathrm{Aut}(\pi(\mathcal{Y}_n))$ acts freely on $T^n(\mathbb{C})$. We indicate arguments that the action is transitive. The homomorphism $\varphi \in T^n(\mathbb{C})$ factors through the Malcev $\mathbb{C}$-completion of $\pi_1(\mathcal{Y}_n, y)$ and it becomes an isomorphism $\varphi_\mathbb{C}$. If φ and ψ are in $T^n(\mathbb{C})$ then the composition $\psi_\mathbb{C} \circ \varphi_\mathbb{C}^{-1}$ is in $\mathrm{Aut}(\pi(\mathcal{Y}_n))$ and $(\psi_\mathbb{C} \circ \varphi_\mathbb{C}^{-1}) \circ \varphi = \psi$. Hence the action is transitive and $T^n(\mathbb{C})$ is an $\mathrm{Aut}^*(\pi(\mathcal{Y}_n))$-torsor. The subgroup of inner automorphisms $\mathrm{Inn}(\pi(\mathcal{Y}_n))$ is a normal subgroup of $\mathrm{Aut}^*(\pi(\mathcal{Y}_n))$. Hence $t^n(\mathbb{C}) := T^n(\mathbb{C})/\mathrm{Inn}(\pi(\mathcal{Y}_n))$ is an $\mathrm{Out}^*(\pi(\mathcal{Y}_n)) := \mathrm{Aut}^*(\pi(\mathcal{Y}_n))/\mathrm{Inn}(\pi(\mathcal{Y}_n))$-torsor.

The following result is an analogue of the Ihara Injectivity Theorem (see [I1]).

Proposition 7.4. *The canonical map* $\mathrm{Out}^*(\pi(\mathcal{Y}_n)) \to \mathrm{Out}^*(\pi(\mathcal{Y}_{n-1}))$ *is injective for* $n \geq 5$.

Proof. Let $\mathrm{Out}_1^*(\pi(\mathcal{Y}_n)) := \mathrm{ker}(\mathrm{Out}^*(\pi(\mathcal{Y}_n)) \overset{N}{\to} \mathbb{C}^*)$, where $N(f) = \alpha_f$. The Lie algebra of $\mathrm{Out}^*(\pi(\mathcal{Y}_n))$ is the Lie algebra of special derivations of $L(\mathcal{Y}_n)$ modulo inner derivations. The Lie version is proved in [I1] section 4. The categories of complex unipotent algebraic groups and nilpotent Lie algebras of finite dimension over $\mathbb{C}$ are equivalent. The group $\mathrm{Out}_1^*(\pi(\mathcal{Y}_n))$ is pro-unipotent. Hence the Lie version implies the result for $\mathrm{Out}_1^*(\pi(\mathcal{Y}_n))$, and then also for $\mathrm{Out}^*(\pi(\mathcal{Y}_n))$.

The surjective homomorphisms $(p_{n+1})_* : \pi_1(\mathcal{Y}_{n+1}, y) \to \pi_1(\mathcal{Y}_n, y')$ and $(p_{n+1})_* : \pi(\mathcal{Y}_{n+1}) \to \pi(\mathcal{Y}_n)$ induce a morphism of torsors

$$t^{n+1}(\mathbb{C}) \to t^n(\mathbb{C})$$

compatible with $\mathrm{Out}^*(\pi(\mathcal{Y}_{n+1})) \to \mathrm{Out}^*(\pi(\mathcal{Y}_n))$.

Lemma 7.5. *The canonical morphism of torsors $t^{n+1}(\mathbb{C}) \to t^n(\mathbb{C})$ is injective for $n \geq 4$.*

This follows immediately from Proposition 7.4.

Corollary 7.6. *The monodromy homomorphism $\theta_{y,\mathcal{Y}_n} : \pi_1(\mathcal{Y}_n, y) \to \pi(\mathcal{Y}_n)$ is determined (up to conjugacy by an element of $\pi(\mathcal{Y}_n)$) by the homomorphism $\theta_{y',\mathcal{Y}_4} : \pi_1(\mathcal{Y}_4, y') \to \pi(\mathcal{Y}_4)$.*

Proof. Observe that $\theta_{y,\mathcal{Y}_n} \in t^n(\mathbb{C})$ and $\theta_{y',\mathcal{Y}_4}$ is the image of $\theta_{y,\mathcal{Y}_n}$ under the canonical morphism $t^n(\mathbb{C}) \to t^4(\mathbb{C})$.

Let $a := (a_1, \ldots, a_n, a_{n+1})$ be a sequence of $n+1$ different points in $\mathbb{P}^1(\mathbb{C})$ and let $V_a := \mathbb{P}^1(\mathbb{C}) \backslash \{a_1, \ldots, a_n, a_{n+1}\}$. The vector space $H(V_a)$ is spanned by $X_i := (\frac{dz}{z-a_i} - \frac{dz}{z-a_{n+1}})^*$ $i = 1, \ldots, n$. Let us set $X_{n+1} := -\sum_{i=1}^n X_i$. Let A_k denote a geometric generator of V_a, which is a loop around a_k. Let us set

$$T_a(\mathbb{C}) := \{f \in \mathrm{Hom}(\pi_1(V_a, x); \pi(V_a)) | \exists \alpha_f \in \mathbb{C}^*, \forall A_k \ f(A_k) \sim \alpha_f X_k\}.$$

Assume that $a = (a_i)_{i=1}^{n+1}$ is such that $a_1 = 0, a_2 = 1, a_3 = \infty$. The fibration

$$V_a \xrightarrow{k_{n+2}} \mathcal{Y}_{n+2} \xrightarrow{p_{n+2}} \mathcal{Y}_{n+1} \qquad (V_a = (p_{n+2})^{-1}(a_1, \ldots, a_{n+1}))$$

realizes $\pi(V_a)$ as a normal subgroup of $\pi(\mathcal{Y}_{n+2})$ $((k_{n+2})_*(X_i) = X_{i,n+2})$. Hence the group $\pi(\mathcal{Y}_{n+2})$ acts on $T_a(\mathbb{C})$; let

$$t_a(\mathbb{C}) := T_a(\mathbb{C})/\pi(\mathcal{Y}_{n+2}).$$

Observe that any $\pi(\mathcal{Y}_{n+2})$-conjugate of $X_{i,n+2}$ is in the image of $\pi(V_a)$. Hence the restriction map

$$(k_{n+2})^* : t^{n+2}(\mathbb{C}) \to t_a(\mathbb{C})$$

given by $f \to f_{|\pi_1(V_a,x)}$ is defined. We set

$$\tau_a(\mathbb{C}) := \mathrm{im}\ (t^{n+2}(\mathbb{C}) \to t_a(\mathbb{C})).$$

Observe that the diagram

$$
\begin{array}{ccc}
t^{n+2}(\mathbb{C}) & \xrightarrow{\ (k_{n+2})^*\ } & \tau_a(\mathbb{C}) \\
\downarrow pr & & \downarrow pr_1 \\
t^4(\mathbb{C}) & \xrightarrow[\approx]{\ (k_4)_*\ } & \tau_{0,1,\infty}(\mathbb{C})
\end{array}
$$

commutes where the map pr_1 is induced by the inclusion $V_a \hookrightarrow \mathbb{P}^1(\mathbb{C})\backslash\{0, 1, \infty\}$. The map $(k_4)_*$ is bijective because $\mathcal{Y}_4 = \mathbb{P}^1(\mathbb{C})\backslash\{0, 1, \infty\}$. Lemma 7.5 implies that the map pr is injective. The map $(k_{n+2})^*$ is surjective by the construction. Hence both maps, $(k_{n+2})^*$ and pr_1 are injective. Therefore we have proved the following result.

Proposition 7.7. *i) The $\pi(\mathcal{Y}_{n+2})$-conjugacy class of the monodromy homomorphism $\theta_{x,\mathcal{Y}_{n+2}} : \pi_1(\mathcal{Y}_{n+2}, x) \to \pi(\mathcal{Y}_{n+2})$ is determined by its restriction to $\pi_1(V_a, x')$.*
ii) The $\pi(\mathcal{Y}_{n+2})$-conjugacy class of the monodromy homomorphism $\theta_{x,V_a} : \pi_1(V_a, x) \to \pi(V_a)$ is determined by the monodromy homomorphism

$$
\theta_{x',\mathbb{P}^1(\mathbb{C})\backslash\{0,1,\infty\}} : \pi_1(\mathbb{P}^1(\mathbb{C})\backslash\{0, 1, \infty\}, x') \to \pi(\mathbb{P}^1(\mathbb{C})\backslash\{0, 1, \infty\}).
$$

§8. The Drinfeld-Ihara $\mathbb{Z}/5$-cycle relation.

In this section we show that the element which describes the monodromy of all iterated integrals on $\mathbb{P}^1(\mathbb{C}) \backslash \{0, 1, \infty\}$ satisfies the Drinfeld-Ihara relation.

8.1. Introduction

We recall that $Y_n = \left(\mathbb{P}^1(\mathbb{C})\right)^n_*$ and $\mathcal{Y}_n = \left(\mathbb{P}^1(\mathbb{C})\backslash\{0, 1, \infty\}\right)^{n-3}_*$. Let $a, b, c \in \mathbb{P}^1(\mathbb{C})$ be three different points and let $\varphi_{a,b,c}(z) = \dfrac{b-c}{b-a} \cdot \dfrac{z-a}{z-c}$. The map $\Phi_{4,5} : Y_5 \to \mathcal{Y}_5$ given by

$$
\Phi_{4,5}(x_1, x_2, x_3, x_4, x_5) = \left(\varphi_{x_1,x_2,x_3}(x_4), \varphi_{x_1,x_2,x_3}(x_5)\right)
$$

induces a bijection

$$
\varphi_{4,5} : {}^{Y_5}\!/\mathrm{PGL}_2(\mathbb{C}) \to \mathcal{Y}_5.
$$

The group Σ_5 acts on Y_5 by permutations. The action of Σ_5 on Y_5 induces an action of Σ_5 on $\mathcal{Y}_5$. The map $\sigma : \mathcal{Y}_5 \to \mathcal{Y}_5$, $\sigma(s, t) = \left(\dfrac{t-1}{t-s}, \dfrac{1}{s}\right)$ corresponds to the permutation $\tilde{\sigma}$ of Y_5 given by

$$
\tilde{\sigma}(x_1, x_2, x_3, x_4, x_5) = (x_2, x_3, x_4, x_5, x_1).
$$

Observe that the points

$$A = \left(\frac{\sqrt{5}-1}{2}, \frac{\sqrt{5}+1}{2}\right) \in \mathcal{Y}_5 \text{ and } B = \left(\frac{-\sqrt{5}-1}{2}, \frac{-\sqrt{5}+1}{2}\right) \in \mathcal{Y}_5$$

are fixed by σ.

The one-forms $\frac{ds}{s}, \frac{ds}{s-1}, \frac{dt}{t}, \frac{dt}{t-1}, \frac{ds-dt}{s-t}$ generate $A^1(\mathcal{Y}_5)$ and $H^1_{\mathrm{DR}}(\mathcal{Y}_5)$. Let S_0, S_1, T_0, T_1 and N be their formal duals. There are two non-trivial relations in $A^2(\mathcal{Y}_5)$:

$$\frac{ds}{s} \wedge \frac{dt}{t} + \frac{dt}{t} \wedge \frac{ds-dt}{s-t} + \frac{ds-dt}{s-t} \wedge \frac{ds}{s},$$

and

$$\frac{ds}{s-1} \wedge \frac{dt}{t-1} + \frac{dt}{t-1} \wedge \frac{ds-dt}{s-t} + \frac{ds-dt}{s-t} \wedge \frac{ds}{s-1}.$$

Elementary computations of linear algebra imply that the subspace $R(\mathcal{Y}_5)$ of $H(\mathcal{Y}_5)^{\otimes 2}$ is generated by

$$[S_i, N] + [T_i, N] \quad i = 0, 1;$$
$$[S_i, T_i] + [S_i, N] \quad i = 0, 1;$$
$$[T_i, S_i] + [T_i, N] \quad i = 0, 1;$$
$$[S_0, T_1] \quad \text{and} \quad [S_1, T_0]$$

where $[A, B] = A \otimes B - B \otimes A$.

We recall that $P(\mathcal{Y}_5)$ is a multiplicative group of the algebra of formal power series in non-commuting variables S_0, S_1, T_0, T_1 and N divided by the ideal generated by $R(\mathcal{Y}_5)$.

The principal fibre bundle

$$\mathcal{Y}_5 \times P(\mathcal{Y}_5) \to \mathcal{Y}_5$$

we equipped with the integrable connection given by the one form

$$\omega_{\mathcal{Y}_5} = \left(\frac{dt}{t-1} - \frac{dt}{t}\right) \otimes T_1 + \left(-\frac{dt}{t}\right) \otimes T_\infty$$
$$+ \left(\frac{ds-dt}{s-t} - \frac{dt}{t}\right) \otimes N + \frac{ds}{s} \otimes S_0 + \frac{ds}{s-1} \otimes S_1$$

where $T_\infty = -T_0 - T_1 - N$. More simply, we shall write ω instead of $\omega_{\mathcal{Y}_5}$.

8.2. Integration of ω

Recall that on $\mathbb{P}^1(\mathbb{C})\backslash\{0,1,\infty\}$ we have

8.2.0 $\qquad \Lambda_{\vec{\infty 1}}(z) \cdot a_{1\vec{\infty}}^{\vec{\infty 1}}(Y,Z) = \Lambda_{1\vec{\infty}}(z)$ (see Proposition 5.1).

Please notice that we are considering $a_{1\vec{\infty}}^{\infty 1}(Y,Z)$ as a power series in Y and Z. Thus we will be able to substitute Y and Z in $a_{1\vec{\infty}}^{\infty 1}(Y,Z)$ by other pairs of letters.

We shall denote the generators X, Y and Z from the beginning of section 6 by $T_0 + N$, T_1 and T_∞. The monodromy of $\Lambda_{\vec{\infty 1}}(z)$ is given by:

(around ∞) : $\Lambda_{\vec{\infty 1}}(z) \to \Lambda_{\vec{\infty 1}}(z) \cdot e^{-2\pi i T_\infty}$,

(around 1) : $\Lambda_{\vec{\infty 1}}(z) \to \Lambda_{\vec{\infty 1}}(z) \cdot a_{1\vec{\infty}}^{\infty 1}(T_1,T_\infty) \cdot e^{-2\pi i T_1} \cdot (a_{1\vec{\infty}}^{\infty 1}(T_1,T_\infty))^{-1}$

(see Theorem 5.3).

It follows from the definition of horizontal sections starting from a tangential base point, and also from the identity

$$\frac{dz}{z} \otimes (T_0 + N) + \frac{dz}{z-1} \otimes T_1 = (-\frac{dz}{z}) \otimes T_\infty + (\frac{dz}{z-1} - \frac{dz}{z}) \otimes T_1)$$

that asymptotically at ∞ and 1, we have respectively

8.2.1 $\qquad \Lambda_{\vec{\infty 1}}(z) \underset{z=\infty}{\sim} e^{\left(\int_1^z \frac{dt}{t}\right)T_\infty}$

i.e. $\underset{\substack{z\to\infty \\ z>1}}{\lim} \left(\Lambda_{\vec{\infty 1}}(z) \cdot e^{-\left(\int_1^z \frac{dt}{t}\right)T_\infty}\right) = 1$

and

$$\Lambda_{1\vec{\infty}}(z) \underset{z=1}{\sim} e^{-\left(\int_\infty^z (\frac{dt}{t-1} - \frac{dt}{t}))T_1\right)}$$

i.e. $\underset{\substack{z\to 1 \\ z>1}}{\lim} \left(\Lambda_{\vec{\infty 1}}(z) \cdot e^{-\left(\int_\infty^z (\frac{dt}{t-1} - \frac{dt}{t}))T_1\right)}\right) = 1$

Let $P_\varepsilon = (\varepsilon, 1+\varepsilon) \in \mathcal{Y}_5$ where ε is a small positive real number. Let $\Lambda_{P_\varepsilon}((s,t); \text{path})$ be a horizontal section of ω such that $\Lambda_{P_\varepsilon}(P_\varepsilon) = 1$. Let γ be a path in $\mathcal{Y}_5$ from P_ε to $\sigma(P_\varepsilon) = (\varepsilon, 1/\varepsilon)$ which is constant $(= \varepsilon)$ on the first coordinate. Assuming $s = \text{constant } (= \varepsilon)$ we have

$$\Lambda_{P_\varepsilon}(\varepsilon, z) \cdot a_{1\vec{\infty}}^{1+\varepsilon} = \Lambda_{1\vec{\infty}}(z).$$

Hence for small positive ε, we asymptotically have

8.2.2 $\qquad a_{1\vec{\infty}}^{1+\varepsilon} \underset{\varepsilon=0}{\sim} e^{\left(-\int_\infty^{1+\varepsilon} (\frac{dt}{t-1} - \frac{dt}{t}))T_1\right)}$.

It follows from 8.2.0, 8.2.1 and 8.2.2 and Proposition 5.1 that for small positive ε, we have

8.2.3 $\Lambda_{P_\varepsilon}\big(\sigma(P_\varepsilon);\gamma\big) \underset{\varepsilon=0}{\sim} e^{\left(\int_1^{1/\varepsilon} \frac{dz}{z}\right)T_\infty} \cdot a_{1\infty}^{\infty 1}(T_1,T_\infty) \cdot e^{\left(\int_\infty^{1+\varepsilon} \left(\frac{dt}{t-1}-\frac{dt}{t}\right)\right)T_1}.$

Let $p = \sigma^4(\gamma)\cdot\sigma^3(\gamma)\cdot\sigma^2(\gamma)\cdot\sigma(\gamma)\cdot\gamma$. Then $\Lambda_{P_\varepsilon}(P_\varepsilon;p)=1$ because the path p is contractible in $\mathcal{Y}_5$. On the other hand

$$1 = \Lambda_{P_\varepsilon}(P_\varepsilon,p) = \Lambda_{\sigma^4(P_\varepsilon)}\big(P_\varepsilon;\sigma^4(\gamma)\big)\cdot\Lambda_{\sigma^3(P_\varepsilon)}\big(\sigma^4(P_\varepsilon);\sigma^3(\gamma)\big)$$
$$\cdot\Lambda_{\sigma^2(P_\varepsilon)}\big(\sigma^3(P_\varepsilon);\sigma^2(\gamma)\big)\cdot\Lambda_{\sigma(P_\varepsilon)}\big(\sigma^2(P_\varepsilon);\sigma(\gamma)\big)\cdot\Lambda_{P_\varepsilon}\big(\sigma(P_\varepsilon);\gamma\big).$$

The formula

$$(\sigma^i)_*\big(\Lambda_{P_\varepsilon}(\sigma(P_\varepsilon);\gamma)\big) = \Lambda_{\sigma^i(P_\varepsilon)}\big(\sigma^{i+1}(P_\varepsilon);\sigma^i(\gamma)\big),$$

which follows from Corollary 1.7 implies that

$$1 = \sigma_*^4(L)\cdot\sigma_*^3(L)\cdot\sigma_*^2(L)\cdot\sigma_*(L)\cdot L$$

where $L = \Lambda_{P_\varepsilon}(\sigma(P_\varepsilon);\gamma)$. Let

$$\mathrm{L} = e^{\left(\int_1^{1/\varepsilon} \frac{dz}{z}\right)T_\infty} \cdot a_{1\infty}^{\infty 1}(T_1,T_\infty) \cdot e^{\left(\int_\infty^{1+\varepsilon} \left(\frac{dt}{t-1}-\frac{dt}{t}\right)\right)T_1}.$$

It follows from 8.2.3 that

$$1 \underset{\varepsilon=0}{\sim} \sigma_*^4(\mathrm{L})\cdot\sigma_*^3(\mathrm{L})\cdot\sigma_*^2(\mathrm{L})\cdot\sigma_*(\mathrm{L})\cdot\mathrm{L}.$$

The factors $e^{\left(\int_\infty^{1+\varepsilon} \left(\frac{dt}{t-1}-\frac{dt}{t}\right)\right)T_\infty \left(=\sigma_*^2(T_1)\right)}$ and $e^{\left(\int_1^{1/\varepsilon} \frac{dz}{z}\right)T_\infty}$ can be put together in the product $\sigma_*^4(\mathrm{L})\cdot\ldots\cdot\mathrm{L}$ because $T_\infty = \sigma_*^2(T_1)$ commutes with $\sigma_*(T_1) = S_0$ and $\sigma_*(T_\infty) = S_1$. After the calculations we get

$$\int_\infty^{1+\varepsilon}\left(\frac{dt}{t-1}-\frac{dt}{t}\right)+\int_1^{1/\varepsilon}\frac{dt}{t} = -\log(1+\varepsilon).$$

Repeating the same argument for S_1, S_1+T_1+N, T_1 and S_0 and passing to the limit $\varepsilon\to 0$ we get

$$\sigma_*^4(\mathfrak{a})\cdot\sigma_*^3(\mathfrak{a})\cdot\sigma_*^2(\mathfrak{a})\cdot\sigma_*(\mathfrak{a})\cdot\mathfrak{a} = 1$$

where $\mathfrak{a} = a_{1\infty}^{\infty 1}(T_1,T_\infty)$. The last formula can be written in the form

$$\mathfrak{a}(S_1+T_1+N,S_0)\cdot\mathfrak{a}(S_1,T_1)\cdot\mathfrak{a}(T_\infty,S_1+T_1+N)\cdot\mathfrak{a}(S_0,S_1)\cdot\mathfrak{a}(T_1,T_\infty) = 1$$

since $\sigma_*(S_0) = T_\infty$, $\sigma_*(S_1) = S_1 + T_1 + N$, $\sigma_*(T_0) = N$, $\sigma_*(T_1) = S_0$ and $\sigma_*(N) = -S_0 - S_1 - N$.

Let $\psi_5 : \mathbb{C}_*^4 \to \mathcal{Y}_5$ be given by $\psi_5(z_1, z_2, z_3, z_4) = \Phi_{4,5}(z_1, z_2, z_3, z_4, \infty)$. Let $(A_{ij})_{i,j}$ be formal duals of $\left(\dfrac{dz_i - dz_j}{z_i - z_j}\right)_{i,j}$. Then we have

$$\psi_{5*}(A_{12}) = -S_0 - S_1 - T_0 - T_1 - N,$$
$$\psi_{5*}(A_{13}) = S_1 + T_1 + N,$$
$$\psi_{5*}(A_{14}) = S_0,$$
$$\psi_{5*}(A_{23}) = S_0 + T_0 + N,$$
$$\psi_{5*}(A_{24}) = S_1,$$
$$\psi_{5*}(A_{34}) = -S_0 - S_1 - N.$$

Using $\psi_1 : \mathbb{C}_*^4 \to \mathcal{Y}_5$ given by $\psi_1(z_2, z_3, z_4, z_5) = \Phi_{4,5}(\infty, z_2, z_3, z_4, z_5)$ we get

$$\psi_{1*}(A_{23}) = S_0 + T_0 + N,$$
$$\psi_{1*}(A_{24}) = S_1,$$
$$\psi_{1*}(A_{25}) = T_1,$$
$$\psi_{1*}(A_{34}) = -S_0 - S_1 - N,$$
$$\psi_{1*}(A_{35}) = -T_0 - T_1 - N,$$
$$\psi_{1*}(A_{45}) = N.$$

Set $X_{ij} := \psi_{\varepsilon*}(A_{ij})$ $\varepsilon = 1, 5$; then $X_{15} = T_0$. Hence finally we get a formula

8.2.4 $\quad \mathfrak{a}(X_{13}, X_{14}) \cdot \mathfrak{a}(X_{24}, X_{25}) \cdot \mathfrak{a}(X_{35}, X_{13}) \cdot \mathfrak{a}(X_{14}, X_{24}) \cdot \mathfrak{a}(X_{25}, X_{35}) = 1.$

If we use $\Phi_{2,4} : \mathbb{C}_*^5 \to \mathcal{Y}_5$ given by $\Phi_{2,4}(0, s, 1, t, \infty) = (s, t)$ and repeat the calculations in $\mathcal{Y}_5$ we get the same formula as before, but the X_{ij}'s names of $S_0, S_1, \ldots$ are now different and the resulting formula is:

8.2.5 $\quad \mathfrak{a}(X_{15}, X_{12}) \cdot \mathfrak{a}(X_{23}, X_{34}) \cdot \mathfrak{a}(X_{45}, X_{15}) \cdot \mathfrak{a}(X_{12}, X_{23}) \cdot \mathfrak{a}(X_{34}, X_{45}) = 1.$

This is exactly the formula which appears in [I2, p. 106].

Proposition 8.3. *For any permutation σ of five letters we have*

i) $\quad \mathfrak{a}(X_{\sigma(13)}, X_{\sigma(14)}) \cdot \mathfrak{a}(X_{\sigma(24)}, X_{\sigma(25)}) \cdot \mathfrak{a}(X_{\sigma(35)}, X_{\sigma(13)})$
$\quad\quad\quad \cdot \mathfrak{a}(X_{\sigma(14)}, X_{\sigma(24)}) \cdot \mathfrak{a}(X_{\sigma(25)}, X_{\sigma(35)}) = 1,$

ii) $\quad \mathfrak{a}(X_{\sigma(15)}, X_{\sigma(12)}) \cdot \mathfrak{a}(X_{\sigma(23)}, X_{\sigma(34)}) \cdot \mathfrak{a}(X_{\sigma(45)}, X_{\sigma(15)})$
$\quad\quad\quad \cdot \mathfrak{a}(X_{\sigma(12)}, X_{\sigma(23)}) \cdot \mathfrak{a}(X_{\sigma(34)}, X_{\sigma(45)}) = 1,$

where $\sigma(ij) = \sigma(i)\sigma(j)$.

Proof. This follows from 8.2.4, 8.2.5 and Corollary 1.7.   ◇

Remark. The formulas of Proposition 8.3 take place in the group $P(\mathcal{Y}_5)$. If we apply log we obtain formulas in the group $\pi(\mathcal{Y}_5)$. In the sequel we shall work in the group $\pi(\mathcal{Y}_5)$.

We finish this section with a formula from which the Deligne $\mathbb{Z}/3$-cycle relation can be obtained. The proof is an imitation of Deligne's proof.

Proposition 8.4. *Let* $\alpha := \log \mathfrak{a}$. *In the group* $\pi(\mathcal{Y}_5)$ *we have*
i)
$$\alpha(X_{23}, X_{25})(-\pi i X_{23})\alpha(X_{35}, X_{23})(-\pi i X_{35})\alpha(X_{25}, X_{35})(-\pi i X_{25}) = -\pi i X_{14}$$
and
ii) $\alpha(X_{23}, X_{25})(\pi i X_{23})\alpha(X_{35}, X_{23})(\pi i X_{35})\alpha(X_{25}, X_{35})(\pi i X_{25}) = -\pi i X_{14}$.

Proof. We prove only i) since ii) is similar. Let $\tilde\sigma(x_1, x_2, x_3, x_4, x_5) = (x_1, x_5, x_2, x_4, x_3)$. Then the induced map $\sigma : \mathcal{Y}_5 \to \mathcal{Y}_5$ is given by $\sigma(s,t) = \left(\frac{t-1}{t} \cdot \frac{s}{s-1}, \frac{t-1}{t}\right)$ and $\sigma^2(s,t) = \left(\frac{s}{s-t}, \frac{1}{1-t}\right)$. Let $P_- = (r, 1-r)$ and $P_+ = (r, 1+r)$ where r is positive and small. Let $Q_- = (-r, 1-r)$ and $Q_+ = (-r, 1+r)$. Let γ be a path from $P_+ = (r, 1+r)$ to $\sigma^2(Q_-) = (r, 1/r)$, which is constant on the first coordinate. Let γ' be a path from Q_+ to $\sigma^2(P_-)$ passing through the point $\left(\frac{r}{2r-1}, 1+r\right)$ which is piecewise constant, first on the second coordinate, next on the first coordinate.

Let S be a path $[0, \pi] \ni \varphi \to (r, 1 + re^{i(\varphi+\pi)})$ and let S' be a path $[0, \pi] \ni \varphi \to (-r, 1 + re^{i(\varphi+\pi)})$. Let us consider the composition $p = \sigma(\gamma') \circ \sigma(S') \circ \sigma^2(\gamma) \circ \sigma^2(S) \circ \gamma' \circ S' \circ \sigma(\gamma) \circ \sigma(S) \circ \sigma^2(\gamma') \circ \sigma^2(S') \circ \gamma \circ S$. If we integrate the form ω along this path and pass to the limit as $r \to 0$ we obtain the square of the left hand side of the equality in i).

Let α be a loop in the opposite clockwise direction around $(0,0)$ in the plane $P = \{(s,t) \in \mathbb{C}^2 \mid \alpha s + \beta t = 0\}$. The integration of the form ω along α gives $(-2\pi i)(S_0 + N + T_0) = (-2\pi i)X_{23}$. In the model of $Y_*^5/PGL_2(\mathbb{C})$ in which the subspace $\{(x_1, x_2, x_3, x_4, x_5) \mid x_1 = x_4\}$ of $(\mathbb{P}^1(\mathbb{C}))^5$ degenerates to a point (for example for $\Phi_{2,5}(0, s, 1, \infty, t) = (s, t)$), the path p is homotopic to a loop around one of the points $(0,0)$, $(1,1)$ or (∞, ∞) in the plane passing through the corresponding point $(0,0)$, $(1,1)$ or (∞, ∞) (the point $(1,1)$ in the case of the model $\Phi_{2,5}$). Hence the square of the left hand side of the expression i) is also $(-2\pi i) \cdot X_{14}$.   ◇

Corollary 8.5. *For any permutation* σ *of five letters* $1, 2, 3, 4, 5$ *we have formulas* $i')$ *and* $ii')$, *which are obtained from formulas* $i)$ *and* $ii)$ *by replacing indices* $1, 2, 3, 4, 5$ *by* $\sigma(1), \sigma(2), \sigma(3), \sigma(4), \sigma(5)$.

Proof. Consider the map of Y_5 given by $(x_i)_{i=1,\ldots,5} \to (x_{\sigma(i)})_{i=1,\ldots,5}$. The induced map $\sigma : \mathcal{Y}_5 \to \mathcal{Y}_5$ satisfies $(\sigma^* \otimes id)\omega = (id \otimes \sigma_*)\omega$, which implies formulas i') and ii'). $\Diamond$

Remark 1. We have $X_{23} + X_{25} + X_{35} = X_{14}$ in the Lie algebra $\mathrm{Lie}(\mathcal{Y}_5)$. If we set $X_{14} = 0$ then the formulas i) and ii) reduce to the Deligne formula (see [D5]).

Remark 2. Let $f(z) = \frac{1}{1-z}$. Let us return to the notation from section 6. We have $f_*(X) = Y$ and $f_*(Y) = Z$. Hence we get

$$a_{01}^{10}(Y, Z) = f_*(a_{01}^{10}(X, Y)) = a_{1\infty}^{\infty 1}(Y, Z).$$

(We consider the element $a_{1\infty}^{\infty 1}$ as a power series in Y and Z.) Therefore we have

$$a_{01}^{10}(X, Y) = \mathfrak{a}(X, Y).$$

Let $p = [0, 1]$ be a path from $\vec{01}$ to $\vec{10}$. The element $a_{01}^{10}(X, Y)$ is the result of integration from $\vec{01}$ to $\vec{10}$. The element considered by Ihara is the result of the Galois action on the path p composed with the path p^{-1}. Both elements satisfy the $\mathbb{Z}/2$, $\mathbb{Z}/3$ and $\mathbb{Z}/5$ cycle relations. The analogy is more striking when we consider torsors corresponding to the homomorphism $\theta_{\vec{01}} : \pi_1(\mathbb{P}^1(\mathbb{C}) \setminus \{0, 1, \infty\}, \vec{01}) \to \pi(\mathbb{P}^1(\mathbb{C}) \setminus \{0, 1, \infty\})$ and the associated groups (see Appendix A2).

§9. Subgroups of the groups of automorphisms.

9.0. We summarize here the notation which will be used in the rest of the paper.

Let k be a field of characteristic zero. We say that X is an algebraic variety defined over k if X is an algebraic scheme over $\mathrm{Spec}\, k$. If A is a k-algebra, we set $X_A := X \underset{\mathrm{Spec}\, k}{\times} \mathrm{Spec}\, A$. We denote by $X(A)$ the set of A-points of X. Observe that $X(A) = X_A(A)$. We say that G is an affine algebraic (resp. pro-algebraic) group defined over k, if G is an affine algebraic (resp. pro-algebraic) group scheme over $\mathrm{Spec}\, k$.

Let L be a nilpotent (resp. pro-nilpotent) Lie algebra over a field k of characteristic zero. We equip the Lie algebra L with a multiplication given by the Baker-Campbell-Hausdorff formula. We denote by π the obtained group. The group π is the group of k-points in a connected affine algebraic unipotent (resp. pro-algebraic pro-unipotent) group scheme over k, which we denote by Π. Let R be a k-algebra. We denote by $\Pi(R)$ the group of R-points of Π. Let us set $\Pi_R := \Pi \underset{\mathrm{Spec}\, k}{\times} \mathrm{Spec}\, R$. Observe that $\Pi(R) = \Pi_R(R)$.

Let V be as in section 1.1. We shall assume that $A^1(V) \to H^1_{DR}(V)$ is an isomorphism. In 1.1 we defined the group $\pi(V)$. It is easy to see that $\pi(V)$ is the group of k-points of a connected affine pro-unipotent pro-algebraic group scheme over k. We denote by $\Pi(V)$ the corresponding group scheme. ($\Pi(V)$ is Spec of the algebra of k-valued functions on $\pi(V)$.)

Let $V = \mathbb{P}^1_{\mathbb{Q}} \setminus \{0, 1, \infty\}$. We recall that $\mathrm{Lie}(V)$ is the free Lie algebra over $\mathbb{Q}$ on $X = \left(\frac{dz}{z}\right)^*$ and $Y = \left(\frac{dz}{z-1}\right)^*$ and that $Z := -X - Y$. We recall that $L(V)$ is the completion of $\mathrm{Lie}(V)$ with respect to the filtration induced by the lower central series. Let us set $L' := [L(V), L(V)]$, $L'' := [L', L']$ and $L_2(V) := L(V)/_{L''}$. We denote by $\Pi_2(V)$ the group scheme over $\mathrm{Spec}\,\mathbb{Q}$ corresponding to $L_2(V)$. Let us set $\pi_2(V) = \Pi_2(V)(\mathbb{C})$.

Notation. Let $w \in L_2(V)$. We set $wX^iY^j := (\ldots(w, X)\ldots X)Y)\ldots Y)$. If $f(X, Y) = 1 + \sum\limits_{n=1, m=1} a_{nm}X^nY^m$ is a power series, then we set

$$(w, f(X, Y)) := w + \sum_{n=1, m=1} a_{nm}wX^nY^m.$$

9.0.1. Observe that $(wX^iY^j)X^aY^b = wX^{i+a}Y^{j+b}$ in $L_2(V)$. This implies that the elements X, Y and $(Y, X)X^{i-1}Y^{j-1}$ for $i = 1, 2, 3, \ldots$, $j = 1, 2, 3, \ldots$ form a linear topological basis of $L_2(V)$.

Let $F_2\big(\mathrm{Lie}(V)\big)$ be the Lie ideal of $L(V)$ topologically generated by Lie brackets in X and Y, which contain X at least twice. Let us set $L^2(V) := L(V)/F_2(L(V))$. We denote by Π^2 the corresponding group scheme over $\mathrm{Spec}\,Q$. To simplify notation let us set $\pi^2 := \Pi^2(\mathbb{C})$. If we replace X by Y in the above definition then the corresponding group of $\mathbb{C}$-points we denote by ϖ^2. The elements X, Y and $(YX)Y^i$ (resp. $(YX)X^i$) $i = 0, 1, 2, \ldots$ form a topological linear basis of π^2 (resp. ϖ^2). We denote by $\mathbf{G_m}$ the multiplicative group scheme $\mathrm{Spec}(Q[t, t^{-1}])$.

9.1. Let us set

$$\mathrm{Aut}^*(\pi_2(V)) := \{f \in \mathrm{Aut}(\pi_2(V)) \big| \exists\, \alpha_f \in \mathbb{C}^*, f(X) = \alpha_f X,$$

$$f(Y) \approx \alpha_f Y,\ f(Z) \sim \alpha_f Z\}$$

($\approx$ is a conjugation by an element of $(\pi_2(V), \pi_2(V))$ and $\sim$ is a conjugation by an element of $\pi_2(V)$). Let $p : \pi_2(V) \to \varpi^2$ be the natural projection. The map p induces $p_* : \mathrm{Aut}^*(\pi_2(V)) \to \mathrm{Aut}(\varpi^2)$. Let $\mathcal{C} : \mathbf{G_m}(\mathbb{C}) \to \mathrm{Aut}(\varpi^2)$ be given by $\mathcal{C}_t(X) = tX$, $\mathcal{C}_t(Y) = tY$. We shall investigate liftings of $\mathcal{C}$ to $\mathrm{Aut}^*(\pi_2(V))$.

Proposition 9.1. *i) Let* $\Phi : \mathbf{G_m}(\mathbb{C}) \to \mathrm{Aut}^*(\pi_2(V))$ *be given by*

$$\Phi_t(X) = tX$$

$$\Phi_t(Y) = (tY, \exp(\sum_{n=2}^{\infty} \sum_{\substack{i+j=n \\ i\geq 1, j\geq 1}} c_{i,j}(1-t^n)X^iY^j)), \quad \text{where } c_{i,j} \in \mathbb{C}.$$

Then Φ *is a homomorphism.*

ii)   All homomorphisms Φ *of* $\mathbf{G_m}(\mathbb{C})$ *into* $\mathrm{Aut}^*(\pi_2(V))$ *such that* $p_* \circ \Phi = C$ *and* $\Phi_t(X) = tX$ *are of this form.*

iii)   All one-dimensional subgroups G *of* $\mathrm{Aut}^*(\pi_2(V))$ *which are projected onto* $C(\mathbf{G_m}(\mathbb{C}))$ *by the map* p_* *and preserve the subgroup of* $\pi_2(V)$ *generated by* X *are of this form.*

iv)   The group G *is defined over a subfield* k *of* $\mathbb{C}$ *if and only if all coefficients* c_{ij} *are in* k.

Proof. The point i) is a straightforward verification if one uses 9.0.1. To show point ii) we can assume that $\Phi_t(Y) = (tY, \exp(\sum_{n=2}^{N-1} \sum_{\substack{i+j=n \\ i\geq 1, j\geq 1}} c_{i,j}(1 - t^n)X^iY^j + \sum_{\substack{i+j\geq N \\ i\geq 1, j\geq 1}} f_{i,j}(t)X^iY^j))$. Comparing coefficients of $\Phi_t(\Phi_s(Y))$ and $\Phi_s(\Phi_t(Y))$ we get $f_{i,j}(t) + t^N f_{i,j}(s) = f_{i,j}(s) + s^N f_{i,j}(t)$. This implies $f_{i,j}(t) = c_{i,j}(1 - t^N)$. Hence the point ii) is proved. If a subgroup of $\mathrm{Aut}^*(\pi_2(V))$ is one dimensional then the coefficients $f_{i,j}$ are algebraic functions of t. They cannot be multivalued functions because then the dimension of the subgroup would be greater than 1. Hence $f_{i,j}$ are Laurent polynomials of t. Now the point iii) follows from the proof of ii). The last point is obvious. $\diamond$

In §13 we shall also need results about the subgroups of $\mathrm{Aut}(\pi^2/\Gamma^{n+1}\pi^2)$ and $\mathrm{Aut}(\pi^2)$. Let $C' : \mathbf{G_m}(\mathbb{C}) \to \mathrm{Aut}(\pi^2/\Gamma^2\pi^2)$ be given by $C'_t(X) = tX$, $C'_t(Y) = tY$. Let $(p^n)_* : \mathrm{Aut}(\pi^2/\Gamma^{n+2}\pi^2) \to \mathrm{Aut}(\pi^2/\Gamma^2\pi^2)$ and $p_* : \mathrm{Aut}(\pi^2) \to \mathrm{Aut}(\pi^2/\Gamma^2\pi^2)$ be induced by projections of $\pi^2/\Gamma^{n+2}\pi^2$ and π^2 onto $\pi^2/\Gamma^2\pi^2$.

Corollary 9.2. *All one dimensional subgroups* G *of* $\mathrm{Aut}(\pi^2/\Gamma^{n+2}\pi^2)$ *(resp.* $\mathrm{Aut}(\pi^2)$*) which the map* $(p^n)_*$ *(resp.* p_**) projects onto* $C'(\mathbf{G_m}(\mathbb{C}))$ *are of the form*

$$G = \{f_t \in \mathrm{Aut}(\pi^2/\Gamma^{n+2}\pi^2) \ (resp. \mathrm{Aut}(\pi^2)) \big| f_t(X) = tX,$$

$$f_t(Y) = tY + \sum_{i=1}^{n(resp.\infty)} c_i(t - t^{i+1})((YX)Y^{i-1}) \text{ with } t \in \mathbb{C}^* \}.$$

262   Zdzisław Wojtkowiak

The group G is defined over a subfield k of $\mathbb{C}$ if and only if all c_i's are in k.

We shall denote the subgroup G of $\mathrm{Aut}(\pi^2/\Gamma^{n+2}\pi^2)$ (resp. $\mathrm{Aut}(\pi^2)$) considered in the corollary by $G(c_1, c_2, \ldots, c_n)$ (resp. $G((c_i)_{i=1}^{\infty}) = G(c_1, c_2, \ldots, c_n, \ldots))$.

Proposition 9.3. *Suppose that for each n we have a point $(c_{ij})_{\substack{i+j=n \\ i\geq 1, j\geq 1}} \in \mathbb{C}^{n-1}$ and a sequence $l_1^n, \ldots, l_{k_n}^n$ of linear forms on $\mathbb{C}^{n-1}$. Let V_n be a set of common zeros of $l_1^n, \ldots, l_{k_n}^n$. Let us set $L := \{(l_1^n, \ldots, l_{k_n}^n)_{n \in \mathbb{N}}\}$ and $c := \{((c_{ij})_{\substack{i+j=n \\ i\geq 1, j\geq 1}})_{n \in \mathbb{N}}\}$.*

i)   Let

$$G(c, L) := \{f \in \mathrm{Aut}^*(\pi_2(V)) \big| f(X) = tX,$$

$$f(Y) = (tY, \exp(\sum_{n=2}^{\infty} \sum_{\substack{i+j=n \\ i\geq 1, j\geq 1}} (c_{ij}(1 - t^n) + \beta_{ij})X^iY^j) \big| t \in \mathbb{C}^*,$$

$$\beta_{ij} \in \mathbb{C}, \forall_n \ \forall_{1 \leq i \leq k_n} l_i^n(\beta_{1,n-1}, \beta_{2,n-2}, \ldots, \beta_{n-1,1}) = 0\}.$$

Then $G(c, L)$ is a subgroup of $\mathrm{Aut}^*(\pi_2(V))$.

ii)   Any subgroup G of $\mathrm{Aut}^*(\pi_2(V))$, which preserves the subgroup of $\pi_2(V)$ generated by X and whose projection into $\mathrm{Aut}(\varpi^2)$ is $C(\mathbf{G_m}(\mathbb{C}))$ is of this form.

iii) Two subgroups $G(c, L)$ and $G(c', L')$ coincide if and only if for each n $(c_{ij})_{\substack{i+j=n \\ i\geq 1, j\geq 1}} + V_n = (c'_{ij})_{\substack{i+j=n \\ i\geq 1, j\geq 1}} + V'_n$.

iv)   The group $G(c, L)$ is defined over the subfield k of $\mathbb{C}$ if and only if for each n the affine space $(c_{ij})_{\substack{i+j=n \\ i\geq 1, j\geq 1}} + V_n$ is defined over k.

Proof. The point i) is a standard checking. Let G be as in the point ii). Let $G_1 := \ker(p_* : G \to \mathrm{Aut}(\varpi^2))$. Any element $f \in G_1$ is of the form $f(X) = X$, $f(Y) = (Y, \exp(\sum_{n=2}^{\infty} \sum_{\substack{i+j=n \\ i\geq 1, j\geq 1}} f_{i,j}X^iY^j))$. Observe that $f \to$

$((f_{i,j})_{\substack{i+j=n \\ i\geq 1, j\geq 1}})_{n=2,3,\ldots,N}$ defines a homomorphism $G_1 \to \sum_{n=2}^{N} \sum_{\substack{i+j=n \\ i\geq 1, j\geq 1}} \mathbf{G_a}(\mathbb{C})$ of G_1 into a direct sum of additive groups $\mathbf{G_a}(\mathbb{C})$. Hence there are only linear relations between various f_{ij}'s. The group G is an extension of $\mathbf{G_m}(\mathbb{C})$ by the pro-unipotent group G_1, hence there is a lifting $C_1 : \mathbf{G_m}(\mathbb{C}) \to G$ of C. If we calculate the coefficients of $C_{1t} \circ f \circ C_{1t}^{-1}$ we get that it can only be linear relations between the coefficients $(f_{i,n-i})_{i=1,\ldots,n-1}$. Hence for each n we have a finite number of linear forms $l_1^n, \ldots, l_{k_n}^n$ on $\mathbb{C}^{n-1}$ such that

$$G_1 = \{f | f(X) = X, f(Y) = (Y, \exp(\sum_{n=2}^{\infty} \sum_{\substack{i+j=n \\ i\geq 1, j\geq 1}} \beta_{i,j}X^iY^j)) \big|$$

$$\beta_{i,j} \in \mathbb{C}, \forall_n \ \forall_{1 \leq i \leq k_n} \ l_i^n(\beta_{1,n-1}, \beta_{2,n-2}, \dots) = 0\}.$$

Let $L = \{(l_1^n, \dots, l_{k_n}^n)_{n \in N}\}$. The group $C_1(\mathbf{G_m}(\mathbb{C}))$ is as in Proposition 9.1.i) for some sequences $(c_{ij})_{\substack{i+j=n \\ i \geq 1, j \geq 1}}$, $n = 2, 3, \dots$. Observe that $G_1 \circ C_1(\mathbf{G_m}(C)) = G(c, L)$. Hence for dimensional reasons $G = G(c, L)$. The points iii) and iv) are evident. $\diamond$

Let $(\varepsilon_i)_{i=1}^{\infty}$ be a sequence such that $\varepsilon_i \in \{0, 1\}$ for each i. Let $(c_i)_{i=1}^{\infty}$ be a sequence of complex numbers. For $i = 1, 2, \dots\}$, we set

$$G((c_i)_{i=1}^{\infty} \big| (\varepsilon_i)_{i=1}^{\infty}) = G(c_1, c_2, \dots | \varepsilon_1, \varepsilon_2, \dots) := \{f = f_{t,(\varepsilon_i \beta_i)_{i=1}^{\infty}} \in \mathrm{Aut}(\pi^2) \big|$$

$$f(X) = t \cdot X,$$

$$f(Y) = t \cdot Y + \sum_{i=1}^{\infty} ((\varepsilon_i \beta_i) + (1 - \varepsilon_i) c_i (t - t^{i+1}))((YX)Y^{i-1})$$

with $t \in \mathbb{C}^*, \beta_i \in \mathbb{C}$.

Replacing ∞ by n in the above definition gives a subgroup of $\mathrm{Aut}(\pi^2/\Gamma^{n+2}\pi^2)$, denoted by $G(c_1, \dots, c_n | \varepsilon_1, \dots, \varepsilon_n)$. Let $\delta((\varepsilon_i)_{i=1}^{n})$ be a number of ε_i equal to 1. Then $\dim G(c_1, \dots, c_n | \varepsilon_1, \dots, \varepsilon_n) = \delta((\varepsilon_i)_{i=1}^{n}) + 1$.

Corollary 9.4. *All subgroups of $\mathrm{Aut}(\pi^2/\Gamma^{n+2}\pi^2)$ (resp. $\mathrm{Aut}(\pi^2)$), which the map $(p^n)_*$ (resp. p_*) projects onto $C'(\mathbf{G_m}(C))$ and which preserve the subgroup of π^2 generated by X are of the form $G(c_1, \dots, c_n | \varepsilon_1, \dots, \varepsilon_n)$ (resp. $G((c_i)_{i=1}^{\infty} | (\varepsilon_i)_{i=1}^{\infty}))$. The group is defined over a subfield k of $\mathbb{C}$ if and only if all numbers $(1 - \varepsilon_i) c_i$ are in k.*

We shall denote by G_0 the subgroup $\{f_t \in \mathrm{Aut}(\pi^2/\Gamma^{n+2}\pi^2) \big| f_t(X) = tX, f_t(Y) = tY, t \in \mathbb{C}^*\}$ of $\mathrm{Aut}(\pi^2/\Gamma^{n+2}\pi^2)$. Let $\delta : \mathbb{C} \to \{0, 1\}$ be a map defined by $\delta(0) = 0, \delta(z) = 1$ if $z \neq 0$.

Lemma 9.5. *Let $f_0 \in \mathrm{Aut}(\pi^2/\Gamma^{n+2}\pi^2)$ be such that $f_0(X) = \alpha_0 X, f_0(Y) = \alpha_0 Y + \sum_{i=1}^{n} \beta_i^0((YX)Y^{i-1})$. Let G be the smallest closed algebraic subgroup of $\mathrm{Aut}(\pi^2/\Gamma^{n+2}\pi^2)$ such that $G_0 \subset G$ and $f_0 \in G$. Then*

$$G = G(0, 0, \dots, 0 \big| \delta(\beta_1^0), \dots, \delta(\beta_n^0)).$$

Proof. Let $\chi_t(X) = tX$ and $\chi_t(Y) = tY$ and let $f_1 = f_0 \circ (\chi_{\alpha_0})^{-1}$. Then $f_1(X) = X$ and $f_1(Y) = Y + \sum_{i=1}^{n} \beta_i^0 \alpha_0^{-1}((YX)Y^{i-1})$. Let $G_1 \subset G$ be a subgroup consisting of h such that $h(X) = X, h(Y) = Y + \sum_{i=1}^{n} \beta_i((YX)Y^{i-1})$.

Assume that for some i, $\beta_i^0 \neq 0$. Then $G_1 \neq \{Id\}$ because $f_1 \in G_1$. The subgroup G_1 is isomorphic to a subgroup of $(\mathbf{G_a}(C))^n$, hence it is given by a finite number of linear forms. Let $k = \chi_t \circ h \circ (\chi_t)^{-1}$. Then $k(X) = X$, $k(Y) = Y + \sum_{i=1}^{n} t^i \beta_i ((YX)Y^{i-1})$. Let $l(x_1, \ldots, x_n)$ be one of linear forms defining G_1. Then $l(t\beta_1, t^2\beta_2, \ldots, t^n\beta_n) = 0$ implies $l \equiv 0$. Hence the $\beta_1, \beta_2, \ldots, \beta_n$, if they are non-zero, they are linearly independent. Therefore $G = G(0, 0, \ldots, 0 | \delta(\beta_1^0), \ldots, \delta(\beta_n^0))$.   ◊

Corollary 9.6. *Let* $G \subset \mathrm{Aut}(\pi^2/_{\Gamma^{n+2}\pi^2})$ *be the smallest closed algebraic subgroup of* $\mathrm{Aut}(\pi^2/_{\Gamma^{n+2}\pi^2})$ *such that the subgroups* G_0 *and* $G(c_1, \ldots, c_n)$ *are contained in* G. *Then*

$$G = G(0, \ldots, 0 | \delta(c_1), \delta(c_2), \ldots, \delta(c_n)).$$

Proof. One takes any element f of $G(c_1, \ldots, c_n)$ such that $f(X) = \alpha X$ and $\alpha \neq 1$.   ◊

§10. Torsors.

Let G be a group. We say that a set T is a G-torsor if T is equipped with a free transitive action of G. We say that a subset $S \subset T$ is a subtorsor of T if there is a subgroup $H \subset G$ such that the natural action of H on S is free and transitive.

Main example. Let G_1 and G_2 be two groups. Assume that G_1 and G_2 are isomorphic. Then the set of isomorphisms from G_1 to G_2, which we denote by $\mathrm{Iso}(G_1, G_2)$, is an $\mathrm{Aut}(G_2)$-torsor.

For any non-empty subset $S \subset \mathrm{Iso}(G_1, G_2)$, the intersection of all subtorsors of $\mathrm{Iso}(G_1, G_2)$, which contain S, is a subtorsor of $\mathrm{Iso}(G_1, G_2)$, which we denote by $T(S)$.

10.1. Unipotent affine algebraic groups and torsors.

Lemma 10.1.1. *Let* G *be an affine unipotent algebraic group over a field* k *of characteristic zero. Then there is an affine algebraic group* $\mathrm{Aut}(G)$ *over* k *such that for any* k-*algebra* A *we have*

$$\mathrm{Aut}(G)(A) = \mathrm{Aut}(G_A).$$

Let G_1 *and* G_2 *be two affine unipotent algebraic groups over* k. *Then there is a smooth affine algebraic variety* $\mathrm{Iso}(G_1, G_2)$ *over* k, *such that for any* k-*algebra* A *we have*

$$\mathrm{Iso}(G_1, G_2)(A) = \mathrm{Iso}(G_{1A}, G_{2A}).$$

Proof. Let $\mathfrak{g}$ be the Lie algebra of G. Let us equip $\mathfrak{g}$ with the group law given by the Baker-Campbell-Hausdorff formula. The exponential map exp: $\mathfrak{g} \to G$ is an isomorphism of affine algebraic groups. The group automorphisms of $\mathfrak{g}$ coincide with the automorphisms of the Lie algebra $\mathfrak{g}$. One can easily give an ideal defining $\mathrm{Aut}_{\mathrm{Lie}}(\mathfrak{g})$ in $k[GL(\mathfrak{g})]$. One constructs $\mathrm{Iso}(G_1, G_2)$ in a similar way. $\diamond$

We say that an affine algebraic variety T over k is a G-torsor, if there is a morphism $T \times G \to T$ over $\mathrm{Spec}\, k$, which defines a free transitive action of G on T (see [S] page 149).

Let S be a closed subvariety of T and let H be a closed subgroup of G. We say that S is an H-subtorsor or a subtorsor of T if S is an H-torsor under the natural action of H.

Lemma 10.1.2. *Let T be a G-torsor. Let T_1 be an H_1-subtorsor of T and let T_2 be an H_2-subtorsor of T. Assume that $T_1 \cap T_2 \neq \emptyset$. Then the intersection $T_1 \cap T_2$ is an $H_1 \cap H_2$-subtorsor of T.*

10.1.3. Main example. Let G_1 and G_2 be two unipotent affine algebraic groups over k. Assume that there is an isomorphism $(G_1)_{\bar{k}} \to (G_2)_{\bar{k}}$ of algebraic groups. Then the algebraic variety $\mathrm{Iso}(G_1, G_2)$ is an $\mathrm{Aut}(G_2)$-torsor, if we equip $\mathrm{Iso}(G_1, G_2)$ with the obvious action of $\mathrm{Aut}(G_2)$.

Let $k \subset \mathbb{C}$ be a subfield of the field of complex numbers $\mathbb{C}$. Let $\Theta :$ $G_1(\mathbb{C}) \to G_2(\mathbb{C})$ be an isomorphism. Then Θ is a $\mathbb{C}$-point of $\mathrm{Iso}(G_1, G_2)$. We denote by $Z(\Theta)$ the k-Zariski closure of Θ in $\mathrm{Iso}(G_1, G_2)$ i.e. the smallest algebraic subset of $\mathrm{Iso}(G_1, G_2)$ defined over k, which contains Θ as a $\mathbb{C}$-point. The unipotent affine algebraic group G_i is isomorphic as an algebraic variety over k to the affine space A_k^m, hence Θ can be viewed as a $\mathbb{C}$-point $(\Theta_{ij})_{1 \leq i,j \leq m}$ of $A_k^{m^2}$. Let $k(\Theta)$ be the subfield of $\mathbb{C}$ generated over k by all $\Theta_{i,j}$.

Lemma 10.1.3. *The field $k(\Theta)$ does not depend on the choice of isomorphisms $G_i \approx A_k^m$ and the transcendental degree of the field $k(\Theta)$ over k is equal to the dimension of $Z(\Theta)$.*

Proof. This follows from Lemma 1.7 in [D2]. $\diamond$

Definition-Proposition 10.1.4. *Let $T(\Theta)$ be the intersection of all subtorsors T defined over k of $\mathrm{Iso}(G_1, G_2)$, which contain Θ as a $\mathbb{C}$-point. Then $T(\Theta)$ is a $G(\Theta)$-subtorsor of $\mathrm{Iso}(G_1, G_2)$ for some $G(\Theta) \subset \mathrm{Aut}(G_2)$.*

Proof. The intersection of a family of algebraic varieties coincides with an intersection of a finite number of them. Hence it follows from Lemma 10.1.2

that $T(\Theta)$ exists and it is unique. The group is also unique because it is an intersection of the corresponding subgroups of $\mathrm{Aut}(G_2)$.   ◇

Lemma 10.1.5. *Let G be a unipotent affine algebraic group over k. Then $\mathrm{Aut}(G)$ is an extension of an algebraic subgroup of $\mathrm{Aut}(G^{\mathrm{ab}})$ by a unipotent affine algebraic group. Hence the group $G(\Theta)$ is an extension of an algebraic subgroup of $\mathrm{Aut}(G^{\mathrm{ab}})$ by a unipotent affine algebraic group.*

Proof. Let $\mathfrak{g}$ be the Lie algebra of G and let $(\Gamma^i \mathfrak{g})_i$ be the filtration of $\mathfrak{g}$ by the lower central series. Any automorphism of the Lie algebra $\mathfrak{g}$ preserves the filtration and the induced automorphism of $\Gamma^i \mathfrak{g}/\Gamma^{i+1}\mathfrak{g}$ is determined by the automorphism of $\mathfrak{g}^{\mathrm{ab}} = \Gamma^1 \mathfrak{g}/\Gamma^2 \mathfrak{g}$. Hence $\mathrm{Aut}_{\mathrm{Lie}}(\mathfrak{g})$ is an extension of a closed subgroup of $GL(\mathfrak{g}^{\mathrm{ab}})$ by a unipotent group. The lemma follows from the identification of $\mathrm{Aut}(G)$ with $\mathrm{Aut}_{\mathrm{Lie}}(\mathfrak{g})$ by the exponential map $\exp : \mathfrak{g} \xrightarrow{\approx} G$.   ◇

Lemma 10.1.6. *Assume that $G(\Theta)$ is an extension of $\mathbf{G_m}$ (or G such that $H^1(\mathrm{Gal}(\bar{k}/k), G) = 0$) by a unipotent affine algebraic group N. Then $T(\Theta)$ has a k-point.*

Proof. It follows from [S] Proposition 4.1 that $H^1(\mathrm{Gal}(\bar{k}/k), N) = 0$. It follows from [S] Proposition 2.2 and the assumption of the lemma that $H^1(\mathrm{Gal}(\bar{k}/k), G(\Theta)) = 0$. Prop. 1.1 of [S] implies that $T(\Theta)(k) \neq \emptyset$.   ◇

Let $\Gamma^i(G)$ be a filtration of a group G by the lower central series. Let us set $G^{(i)} := G/\Gamma^{i+1}G$. The isomorphism $\Theta : G_1(\mathbb{C}) \to G_2(\mathbb{C})$ induces isomorphisms $\Theta^{(i)} : G_1^{(i)}(\mathbb{C}) \to G_2^{(i)}(\mathbb{C})$. Let $k < i$. The projections $G_j^{(i)} \to G_j^{(k)}$ for $j = 1, 2$ induce

$$\rho_k^i : \mathrm{Iso}(G_1^{(i)}, G_2^{(i)}) \to \mathrm{Iso}(G_1^{(k)}, G_2^{(k)}) \quad \text{and} \quad \rho(2)_k^i : \mathrm{Aut}(G_2^{(i)}) \to \mathrm{Aut}(G_2^{(k)}).$$

Lemma 10.1.7. *We have*
i) $\overline{\rho_k^i(Z(\Theta^{(i)}))} = Z(\Theta^{(k)}),$ *ii)* $\rho_k^i(T(\Theta^{(i)})) = T(\Theta^{(k)}),$
iii) $\rho(2)_k^i(G(\Theta^{(i)})) = G(\Theta^{(k)}).$

Proof. In the point i) $\overline{(\;\;)}$ means the k-Zariski closure and we omit its proof because we do not need this fact later. Let us set $p = \rho_k^i$ and $p' = \rho(2)_k^i$. Observe that the image $p'(G(\Theta^{(i)}))$ of the group $G(\Theta^{(i)})$ by the morphism p' is a closed subgroup of $\mathrm{Aut}(G_2^{(k)})$ defined over k. This implies that $p(T(\Theta^{(i)}))$ is a closed subvariety of $\mathrm{Iso}(G_1^{(k)}, G_2^{(k)})$ and a $p'(G(\Theta^{(i)}))$-torsor defined over k. This torsor contains $\Theta^{(k)}$ as a $\mathbb{C}$-point, so we have

$$T(\Theta^{(k)}) \subset p'(T(\Theta^{(i)})) \quad \text{and} \quad G(\Theta^{(k)}) \subset p'(G(\Theta^{(i)})).$$

Let P (resp. P') be the projection p (resp. p') restricted to $T(\Theta^{(i)})$ (resp. $G(\Theta^{(i)})$). Then $P^{-1}(T(\Theta^{(k)}))$ is a $P'^{-1}(G(\Theta^{(k)}))$-torsor defined over k, which contain $\Theta^{(i)}$ as a $\mathbb{C}$-point. Hence we get $P^{-1}(T(\Theta^{(k)})) = T(\Theta^{(i)})$ and $P'^{-1}(G(\Theta^{(k)})) = G(\Theta^{(i)})$ This implies that $p(T(\Theta^{(i)})) = T(\Theta^{(k)})$ and $p'(G(\Theta^{(i)})) = G(\Theta^{(k)})$. $\diamond$

10.2. Affine pro-algebraic pro-unipotent groups and torsors.

10.2.1 We assume that $G = \varprojlim_i G^{(i)}$, where the groups $G^{(i)}$ are affine unipotent algebraic groups over k and the morphisms $G^{(i)} \to G^{(j)}$ are also over k. We assume further that $G^{(i)} = G/\Gamma^{i+1}G$. Finally we assume that the Lie algebra $\mathfrak{g}$ of G is finitely presented, i.e. that for i big enough the number of relations defining $\mathfrak{g}/\Gamma^{i+1}\mathfrak{g}$, of degree less than $i + 1$ does not depend on i.

10.2.2. The condition that G is finitely presented implies that for i big enough the morphisms $\mathrm{Aut}(G^{(i+1)}) \to \mathrm{Aut}(G^{(i)})$ are surjective. We set

$$\mathrm{Aut}(G) := \varprojlim_i \mathrm{Aut}(G^{(i)}).$$

Similarly, if G_1 and G_2 satisfy 10.2.1 and if there is an isomorphism $(G_1)_{\bar{k}} \to (G_2)_{\bar{k}}$ of affine pro-algebraic pro-unipotent groups, then the morphisms $\mathrm{Iso}(G_1^{(i+1)}, G_2^{(i+1)}) \to \mathrm{Iso}(G_1^{(i)}, G_2^{(i)})$ are surjective for i big enough. We set

$$\mathrm{Iso}(G_1, G_2) := \varprojlim_i \mathrm{Iso}(G_1^{(i)}, G_2^{(i)}).$$

Observe that $\mathrm{Iso}(G_1, G_2)$ is an $\mathrm{Aut}(G_2)$-torsor defined over k.

10.2.3. Examples of groups satisfying 10.2.1.

1) Let V be a smooth algebraic variety over k. Let us set

$$G := \mathrm{Spec}(H^0_{\mathrm{DR}}(\rho^{\bullet -1}(x, x)))$$

(see §2 and [W1]). Then G satisfies 10.2.1.
2) Let V be as in 1.1. Then the group scheme $\Pi(V)$ satisfies 10.2.1.

Lemma 10.2.4. *Let G be as in 10.2.1. The affine group scheme $\mathrm{Aut}(G)$ is an extension of a closed subgroup of $GL(G^{\mathrm{ab}})$ by an affine pro-unipotent pro-algebraic group.*

This lemma follows from Lemma 10.1.5 and the definition of $\mathrm{Aut}(G)$.

10.2.5 Let G_1 and G_2 satisfy 10.2.1. Let $\Theta : G_1(\mathbb{C}) \to G_2(\mathbb{C})$ be an isomorphism. Then $\Theta = \varprojlim_i \Theta^{(i)}$, where $\Theta^{(i)} : G_1^{(i)}(\mathbb{C}) \to G_2^{(i)}(\mathbb{C})$ are isomorphisms for all i. Let us set

$$Z(\Theta) := \varprojlim_i Z(\Theta^{(i)}), \quad T(\Theta) := \varprojlim_i T(\Theta^{(i)}), \quad G(\Theta) := \varprojlim_i G(\Theta^{(i)}).$$

Then $T(\Theta)$ is a $G(\Theta)$-torsor.

§11. Torsors associated to non-abelian unipotent periods.

Let X be a smooth quasi-projective algebraic variety defined over a number field k. Assume that X has a k-point x. Let us fix an embedding $k \hookrightarrow \mathbb{C}$. In [W1] §7 (see also 2.0 in this paper) we have constructed affine pro-algebraic pro-unipotent finitely presented group schemes $\pi_1^B(X(\mathbb{C}), x)$ and $\pi_1^{DR}(X, x)$ over $\operatorname{Spec} \mathbb{Q}$ and $\operatorname{Spec} k$ respectively.

We set

$$\pi_1^B(X(\mathbb{C}, x))^{(n)} := \pi_1^B(X(\mathbb{C}), x)/\Gamma^{n+1}\pi_1^B(X(\mathbb{C}), x);$$
$$\pi_1^{DR}(X, x)^{(n)} := \pi_1^{DR}(X, x)/\Gamma^{n+1}\pi_1^{DR}(X, x).$$

Then we have:

$$\pi_1^B(X(\mathbb{C}), x) = \varprojlim_n \pi_1^B(X(\mathbb{C}), x)^{(n)} \quad \text{and} \quad \pi_1^{DR}(X, x) = \varprojlim_n \pi_1^{DR}(X, x)^{(n)}.$$

In [W1] Proposition 7.5 we have also constructed a homomorphism (called $B \circ \alpha$ in [W1])

$$\Phi_x : \pi_1^B(X(\mathbb{C}), x)(\mathbb{Q}) \to \pi_1^{DR}(X, x)(\mathbb{C})$$

such that the induced map on $\mathbb{C}$-points,

$$\varphi_x : \pi_1^B(X(\mathbb{C}), x)(\mathbb{C}) \to \pi_1^{DR}(X, x)(\mathbb{C})$$

is an isomorphism. We have $\varphi_x = \varprojlim_n \varphi_x^n$, where $\varphi_x^n : \pi_1^B(X(\mathbb{C}), x)^{(n)}(\mathbb{C}) \to \pi_1^{DR}(X, x)^{(n)}(\mathbb{C})$ is induced by φ_x. For each n we have an $\operatorname{Aut}(\pi_1^{DR}(X, x)^{(n)})$-torsor $\operatorname{Iso}^n := \operatorname{Iso}(\pi_1^B(X(\mathbb{C}), x)^{(n)} \times k, \pi_1^{DR}(X, x)^{(n)})$. Applying the construction from §10 to the isomorphism φ_x^n we get a $G(\varphi_x^n)$-torsor $T(\varphi_x^n)$ over $\operatorname{Spec} k$. The group $G(\varphi_x^n)$ is a subgroup of $\operatorname{Aut}(\pi_1^{DR}(X, x)^{(n)})$.

We set

$$G(\varphi_x) := \varprojlim_n G(\varphi_x^n) \quad \text{and} \quad T(\varphi_x) := \varprojlim_n T(\varphi_x^n).$$

We have projections

$$p_n : \operatorname{Aut}(\pi_1^{\mathrm{DR}}(X,x)^{(n)}) \to \operatorname{Out}(\pi_1^{\mathrm{DR}}(X,x)^{(n)}) :=$$
$$\operatorname{Aut}(\pi_1^{\mathrm{DR}}(X,x)^{(n)}))/\operatorname{Inn}(\pi_1^{\mathrm{DR}}(X,x)^{(n)}))$$

and

$$p : \operatorname{Aut}(\pi_1^{\mathrm{DR}}(X,x)) \to \operatorname{Out}(\pi_1^{\mathrm{DR}}(X,x)).$$

Definition 11.1. i)   The group $\mathcal{G}(X,x)^{(n)}$ is the image of $G(\varphi_x^n)$ in $\operatorname{Out}(\pi_1^{\mathrm{DR}}(X,x)^{(n)})$.
ii)   The group $\mathcal{G}(X,x)$ is the image of $G(\varphi_x)$ in $\operatorname{Out}(\pi_1^{\mathrm{DR}}(X,x))$.

Let x and y be two k-points of X. Let γ be a path in $X(\mathbb{C})$ from x to y. Then γ induces an isomorphism $c_\gamma : \pi_1^{\mathrm{DR}}(X_{\mathbb{C}},x) \to \pi_1^{\mathrm{DR}}(X_{\mathbb{C}},y)$. The induced isomorphism of outer automorphisms groups

$$(c_\gamma)_* : \operatorname{Out}(\pi_1^{\mathrm{DR}}(X_{\mathbb{C}},x)) \to \operatorname{Out}(\pi_1^{\mathrm{DR}}(X_{\mathbb{C}},y))$$

does not depend on the choice of γ and gives the canonical identification. We need this identification over $\operatorname{Spec} k$. Let us consider the morphism $\rho^\bullet : X^{\Delta[1]} \to X^{\partial\Delta[1]}$ of cosimplicial spaces. $\operatorname{Spec}\left(H_{\mathrm{DR}}^0(\rho^{\bullet-1}(x,y))\right)$ is a $\pi_1^{\mathrm{DR}}(X,y)$-torsor (see [W1] Section 3). It follows from [S] Proposition 4.1 that any $\pi_1^{\mathrm{DR}}(X,y)^{(n)}$-torsor is trivial, hence any $\pi_1^{\mathrm{DR}}(X,y)$-torsor is trivial (inverse limit of surjective maps of sets (k-points) is always non empty). Any k-point η of $\operatorname{Spec}\left(H_{\mathrm{DR}}^0(\rho^{\bullet-1}(x,y))\right)$ determines an isomorphism

$$c_\eta : \pi_1^{\mathrm{DR}}(X,x) \to \pi_1^{\mathrm{DR}}(X,y).$$

The isomorphism is unique up to conjugation by elements of $\pi_1^{\mathrm{DR}}(X,y)$. Hence the induced isomorphism

$$(c_\eta)_* : \operatorname{Out}(\pi_1^{\mathrm{DR}}(X,x)) \to \operatorname{Out}(\pi_1^{\mathrm{DR}}(X,y))$$

is canonical, in other words it does not depend on the choice of a k-point of $\operatorname{Spec}\left(H_{\mathrm{DR}}^0(\rho^{\bullet-1}(x,y))\right)$.

The isomorphisms φ_x and φ_y are related by the following commutative diagram

$$
\begin{array}{ccc}
\pi_1^{\mathrm{B}}(X(\mathbb{C}),x)(\mathbb{C}) & \xrightarrow{\varphi_x} & \pi_1^{\mathrm{DR}}(X,x)(\mathbb{C}) \\
\downarrow{\bar{c}_\gamma} & & \downarrow{c_\gamma} \\
\pi_1^{\mathrm{B}}(X(\mathbb{C}),y)(\mathbb{C}) & \xrightarrow{\varphi_y} & \pi_1^{\mathrm{DR}}(X,y)(\mathbb{C})
\end{array}
$$

where $\bar{c}_\gamma$ and c_γ are induced by the path γ. Observe that $c_\gamma = c_\eta \circ \operatorname{conj}(g)$, where η is a k-point of $\operatorname{Spec}\left(H_{\mathrm{DR}}^0((X;x,y))\right)$ and $\operatorname{conj}(g)$ is a conjugation by an element $g \in \pi_1^{\mathrm{DR}}(X,x)(\mathbb{C})$. This implies the following result.

Proposition 11.2. *The groups $\mathcal{G}(X,x)^{(n)}$ and $\mathcal{G}(X,y)^{(n)}$ coincide under the canonical isomorphism $(c_\eta)_*$. The groups $\mathcal{G}(X,x)$ and $\mathcal{G}(X,y)$ coincide under the canonical isomorphism $(c_\eta)_*$. We shall denote these groups by $\mathcal{G}_{\mathrm{DR}}(X)^{(n)}$ and $\mathcal{G}_{\mathrm{DR}}(X)$ respectively.*

§12. Torsors associated to the canonical unipotent connection with logarithmic singularities.

12.0. Let V be as in section 1.1. We shall assume that $A^1(V) \to H^1_{DR}(V)$ is an isomorphism. We assume that V has a k-point x.

We recall from Corollary 2.4 that we have an isomorphism $u : \pi_1^{DR}(V,x) \to \Pi(V)$ ($\Pi(V) := \mathrm{Spec}(\mathrm{Alg}\,\pi(V))$). Let $u(\mathbb{C}) : \pi_1^{DR}(V,x)(\mathbb{C}) \to \Pi(V)(\mathbb{C})$ be the induced isomorphism on $\mathbb{C}$-points.

We shall define homomorphisms

$$b^B_{\mathbb{Q}\mathrm{resp}.\mathbb{C}} : \pi_1(V(\mathbb{C}),x) \to \pi_1^B(V(\mathbb{C}),x)(\mathbb{Q})(\mathrm{resp}.\mathbb{C})$$

and

$$b^{DR} : \pi_1(V(\mathbb{C}),x) \to \pi_1^{DR}(V,x)(\mathbb{C}).$$

Let $\sum \omega_{i_1} \otimes \ldots \otimes \omega_{i_k}$ be a cocycle representing a class c in $H^0_{\mathrm{DR}}\big(\rho^{\bullet-1}(x,x)\big)$. (We recall from §2 that $\pi_1^{DR}(V,x) := \mathrm{Spec}\left(H^0_{\mathrm{DR}}\big(\rho^{\bullet-1}(x,x)\big)\right)$.) We can assume that all ω_{i_l} are one-forms (see 2.2.1). Let $\alpha \in \pi_1(V(\mathbb{C}),x)$. The value of c on $b^{DR}(\alpha)$ is given by the following formula:

12.1. $$c(b^{DR}(\alpha)) := \sum \int_{\alpha^{-1}} \omega_{i_1}, \ldots, \omega_{i_k} \quad (\text{see 2.8}).$$

The homomorphisms $b^B_{\mathbb{Q}}$ and $b^B_{\mathbb{C}}$ are defined in the same way. We want to calculate $G(\varphi_x)$-torsor $T(\varphi_x)$ associated to the comparison homomorphism

$$\varphi_x : \pi_1^B(V(\mathbb{C}),x)(\mathbb{C}) \to \pi_1^{DR}(V,x)(\mathbb{C}).$$

Lemma 12.2. *The $G(\varphi_x)$-torsor $T(\varphi_x)$ is isomorphic to $G(u(\mathbb{C}) \circ \varphi_x)$-torsor $T(u(\mathbb{C}) \circ \varphi_x)$.*

Proof. It follows from the fact that $u : \pi_1^{DR}(V,x) \to \Pi(V)$ is defined over k. ◇

We shall relate the $G(u(\mathbb{C}) \circ \varphi_x)$-torsor $T(u(\mathbb{C}) \circ \varphi_x)$ to the monodromy homomorphism $\theta_{x,V} : \pi_1(V(\mathbb{C}),x) \to \Pi(V)(\mathbb{C})$ of the form ω_V from §1.

Lemma 12.3. *We have*

$$u(\mathbb{C}) \circ b^{DR} = \theta_{x,V}.$$

Proof. The isomorphism u is induced by $\omega_{i_1} \otimes \ldots \otimes \omega_{i_k} \to (X_{i_1} \otimes \ldots \otimes X_{i_k})^*$ (see corollary 2.4.). The value of the class c on $b^{DR}(\alpha)$ is given by 12.1. The value of $\sum (X_{i_1} \otimes \ldots \otimes X_{i_k})^*$ on $\theta_{x,V}(\alpha)$ is $\sum (-1)^k \int_\alpha \omega_{i_k}, \ldots, \omega_{i_1}$ (see Definition 1.3 and Proposition 1.8). It follows from [Ch1] (1.6.2) that $\sum (-1)^k \int_\alpha \omega_{i_k}, \ldots, \omega_{i_1} = \sum \int_{\alpha^{-1}} \omega_{i_1}, \ldots, \omega_{i_k}$. $\Diamond$

It follows from the definition of the homomorphisms b^{DR} and $b^B_{\mathbb{C}}$ that

$$\varphi_x \circ b^B_{\mathbb{C}} = b^{DR}.$$

Let R be a k-algebra. Let $i_R : \pi^B_1(V(\mathbb{C}), x)(\mathbb{Q}) \to \pi^B_1(V(\mathbb{C}), x)(R)$ be the inclusion of $\mathbb{Q}$-points into R-points. We have $i_{\mathbb{C}} \circ b^B_{\mathbb{Q}} = b^B_{\mathbb{C}}$. We set $b^B_R = i_R \circ b^B_{\mathbb{Q}}$.

We define a functor $\mathcal{I}_V$ on k-algebras in the following way:

$$\mathcal{I}_V(R) := \{ f : \pi_1(V(\mathbb{C}), x) \to \Pi(V)(R) \mid \exists \text{ an isomorphism}$$
$$\bar{f} : \pi^B_1(V(\mathbb{C}), x)(R) \to \Pi(V)(R), \ \bar{f} \circ b^B_R = f \}.$$

Observe that $\bar{f}$ is uniquely determined by f. The functor $\mathcal{I}_V$ is represented by an affine pro-algebraic scheme over k, which we also denote by $\mathcal{I}_V$. Moreover $\mathcal{I}_V$ is an $\mathrm{Aut}\Pi(V)$-torsor. Observe that $\theta_{x,V} : \pi_1(V(\mathbb{C}), x) \to \Pi(V)(\mathbb{C})$ is a $\mathbb{C}$-point of $\mathcal{I}_V$ because $\theta_{x,V} = (u(\mathbb{C}) \circ \varphi_x) \circ b^B_{\mathbb{C}}$.

We denote by $T(\theta_{x,V})$ the smallest subtorsor defined over k of $\mathcal{I}_V$ such that $\theta_{x,V} \in T(\theta_{x,V})(\mathbb{C})$ (i.e. the intersection of all subtorsors of $\mathcal{I}_V$ defined over k, which contain $\theta_{x,V}$ as a $\mathbb{C}$-point). The corresponding group we denote by $G(\theta_{x,V})$.

Lemma 12.4. *The $G(\theta_{x,V})$-torsor $T(\theta_{x,V})$ is isomorphic to the $G(u(\mathbb{C}) \circ \varphi_x)$-torsor $T(u(\mathbb{C}) \circ \varphi_x)$.*

Proof. First we notice that the $\mathrm{Aut}(\Pi(V)$-torsor $\mathcal{I}_V$ is isomorphic to the $\mathrm{Aut}(\Pi(V)$ torsor $\mathrm{Iso}(\pi^B_1(V(\mathbb{C}), x) \times k, \Pi(V))$. The isomorphism is given by $f \to \bar{f}$. The equality $\theta_{x,V} = (u(\mathbb{C}) \circ \varphi_x) \circ b^B_{\mathbb{C}}$ implies the lemma. $\Diamond$

Corollary 12.5. *Let V be a projective line over k minus a finite number of k-points. Let S be a loop on $V(\mathbb{C})$ around a missing point. Then the element $b^{DR}(S) \in \pi^{DR}_1(V, v)(\mathbb{C})$ is conjugated to $s^{-2\pi i}$, where $s \in \pi^{DR}_1(V, v)(k)$.*

Proof. It follows from Lemma 12.3 and Corollary 5.4. $\Diamond$

12.6. Let V be a projective line $\mathbb{P}^1_k$ minus a finite number of k-points. If x is a tangential base k-point $\vec{v}$ then the groups π^B_1 and π^{DR}_1 are not defined in [W1]. We set $\pi^B_1(V(\mathbb{C}), \vec{v}) := (\pi_1(V(\mathbb{C}), \vec{v}))_{\mathbb{Q}}$ -Malcev rational completion of $\pi_1(V(\mathbb{C}), \vec{v})$. For the $\mathbb{Q}$-algebra R, the map $b^B_R : \pi_1(V(\mathbb{C}), \vec{v}) \to \pi^B_1(V(\mathbb{C}), \vec{v})(R)$ is the natural map $\pi_1(V(\mathbb{C}), \vec{v}) \to (\pi_1(V(\mathbb{C}), \vec{v}))_{\mathbb{Q}}(R)$ (see

§A.1). The monodromy homomorphism $\theta_{\vec{v},V} : \pi_1(V(\mathbb{C}), \vec{v}) \to \Pi(V)(\mathbb{C})$ is the Malcev $\mathbb{C}$-completion (see §A.1), hence there is an isomorphism $\varphi_{\vec{v}} :$ $\pi_1^B(V(\mathbb{C}), \vec{v})(\mathbb{C}) \to \Pi(V)(\mathbb{C})$ such that $\varphi_{\vec{v}} \circ b_{\mathbb{C}}^B = \theta_{\vec{v},V}$.

The $G(\theta_{\vec{v},V})$-torsor $T(\theta_{\vec{v},V})$ is defined as above. We set also $\pi_1^{DR}(V, \vec{v}) :=$ $\Pi(V)$. The isomorphism u is the identity and the $G(\theta_{\vec{v},V})$-torsor $T(\theta_{\vec{v},V})$ is isomorphic to $G(\varphi_{\vec{v}})$-torsor $T(\varphi_{\vec{v}})$.

§13. Partial information about $\mathcal{G}_{DR}(P_{\mathbb{Q}}^1 \setminus \{0, 1, \infty\})$.

13.1. Let $V = \mathbb{P}_{\mathbb{Q}}^1 \setminus \{0, 1, \infty\}$ and let v be a $\mathbb{Q}$-point of V or a tangential base $\mathbb{Q}$-point of V. For any $\mathbb{Q}$-algebra R we set:

$$\mathcal{T}(R) := \left\{ (x, y, z) \in \big(\Pi(V)(R)\big)^3 \mid \right.$$
$$\left. \exists \alpha \in R^*, \quad x = \alpha X, \quad y \approx \alpha Y, \quad z \sim \alpha Z, \quad x \cdot y \cdot z = 0 \right\}$$

and

$$\text{Aut}^* \big(\Pi(V)\big)(R) :=$$
$$\left\{ f \in \text{Aut}\big(\Pi(V)_R\big) \mid \exists \alpha \in R^*, \quad f(X) = \alpha X, \quad f(Y) \approx \alpha Y, \quad f(Z) \sim \alpha Z \right\}.$$

The functors $\mathcal{T}$ and $\text{Aut}^*(\Pi(V))$ are represented by affine pro-algebraic schemes over $\mathbb{Q}$, which we also denote by $\mathcal{T}$ and $\text{Aut}^*(\Pi(V))$. $\text{Aut}^*(\Pi(V))$ is an affine pro-algebraic group scheme, a subgroup of $\text{Aut}(\Pi(V))$. Moreover $\mathcal{T}$ is an $\text{Aut}^*(\Pi(V))$-torsor. (Let (x, y, z) and (x', y', z') be in $\mathcal{T}(\mathcal{R})$. The map $x \to x'$, $y \to y'$ can be extended to an automorphism of the Lie algebra, which is also an automorphism of the group $\Pi(V)(R)$. Hence the action is transitive.)

If we replace the group $\Pi(V)$ in the above definitions by $\Pi_2(V)$ (resp. Π^2), then we obtain an $\text{Aut}^*(\Pi_2(V))$-torsor $\mathcal{T}_2$ (resp. $\text{Aut}^*(\Pi^2)$-torsor $\mathcal{T}^2$).

Let us fix generators S_0', S_1' and S_∞' of $\pi_1(V(\mathbb{C}), v)$ which are loops around $0, 1$ and ∞ respectively, such that $S_0' \cdot S_1' \cdot S_\infty' = 1$.

Lemma 13.1.1. *The $\text{Aut}^*(\Pi(V))$-torsor $\mathcal{T}$ is a subtorsor of the $\text{Aut}(\Pi(V))$-torsor $\mathcal{I}_V$.*

Proof. Let $(x, y, z) \in \mathcal{T}(R)$. Let $f : \pi_1(V(\mathbb{C}), v) \to \Pi(V)(R)$ be given by $f(S_0') = x$, $f(S_1') = y$. The homomorphism $b_R^B : \pi_1(V(\mathbb{C}), v) \to$ $\pi_1^B(V(\mathbb{C}), v)(R)$ is the Malcev R-completion of $\pi_1(V(\mathbb{C}), v)$ (see §A.1), hence there is an isomorphism $\bar{f} : \pi_1^B(V(\mathbb{C}), v)(R) \to \Pi(V)(R)$ such that $\bar{f} \circ b_R^B = f$. Therefore $f \in \mathcal{I}_V(R)$. ◇

Let $\theta_v : \pi_1(V(\mathbb{C}), v) \to \Pi(V)(C)$ be the monodromy homomorphism of the one form ω_V. Observe that the triple $(\theta_v(S_0'), \theta_v(S_1'), \theta_v(S_\infty')) \in \mathcal{T}(\mathbb{C})$. Let $T'(\theta_v)$ be the smallest subtorsor defined over $\mathbb{Q}$ of $\mathcal{T}$ such that the triple

$(\theta_v(S'_0),\ \theta_v(S'_1),\ \theta_v(S'_\infty)) \in T(\mathbb{C})$ is a $\mathbb{C}$-point of $T'(\theta_v)$. Let $G'(\theta_v)$ be the corresponding group. We recall that in §12 we defined $G(\theta_v)$-torsor $T(\theta_v)$.

Lemma 13.1.2. *The $G(\theta_v)$-torsor $T(\theta_v)$ and the $G'(\theta_v)$-torsor $T'(\theta_v)$ are equal if we identify $\mathcal{T}$ with a subtorsor of $\mathcal{I}_V$ via the map from Lemma 13.1.1.*

We shall use the same notation $T(\theta_v)$ and $G(\theta_v)$ for both torsors and both groups.

13.2. We recall that the monodromy homomorphism $\theta_{\overrightarrow{01}} : \pi_1(V(\mathbb{C}), \overrightarrow{01}) \to \Pi(V)(\mathbb{C})$ is given by

$$\theta_{\overrightarrow{01}}(S_0) = (-2\pi i)X$$

$$\theta_{\overrightarrow{01}}(S_1) = (\alpha_{01}^{10}(X,Y))^{-1} \cdot (-2\pi i)Y \cdot \alpha_{01}^{10}(X,Y)$$

$$\theta_{\overrightarrow{01}}(S_\infty) = (-\pi i X) \cdot (\alpha_{01}^{10}(X,Z))^{-1} \cdot (-2\pi i)Z \cdot \alpha_{01}^{10}(X,Z) \cdot \pi i X$$

where $\alpha_{01}^{10}(X,Y) \in (\pi(V), \pi(V))$ and S_0, S_1, S_∞ are as in the figure in 0.2. The triple $(\theta_{\overrightarrow{01}}(S_0), \theta_{\overrightarrow{01}}(S_1), \theta_{\overrightarrow{01}}(S_\infty)) \in T(\mathbb{C})$. Hence we get the $G(\theta_{\overrightarrow{01}})$-torsor $T(\theta_{\overrightarrow{01}})$, a subtorsor of $\mathcal{T}$.

Proof of **Theorem B** from the introduction. Let v be a $\mathbb{Q}$-point of V. It follows from Lemmas 12.2 and 12.4 that the group schemes $G(\varphi_v)$ and $G(\theta_v)$ are isomorphic. The isomorphism is induced by the isomorphism $u : \pi_1^{DR}(V,v) \to \Pi(V)$. Hence the images of the groups $G(\varphi_v)$ and $G(\theta_v)$ in the groups of outer automorphisms are also isomorphic. The homomorphisms θ_v and $\theta_{\overrightarrow{01}}$ are conjugate; hence the images of groups $G(\theta_v)$ and $G(\theta_{\overrightarrow{01}})$ in $\mathrm{Out}(\Pi(V))$ are equal. Observe that the map $\mathrm{Aut}^*(\Pi(V)) \to \mathrm{Out}(\Pi(V))$ is injective. Hence $\mathcal{G}_{DR}(V)$ and $G(\theta_{\overrightarrow{01}})$ are isomorphic. $\Diamond$

13.3. Let $\theta : \pi_1(V(\mathbb{C}), \overrightarrow{01}) \to \Pi_2(V)(\mathbb{C})$ (resp. $\theta^2 : \pi_1(V(\mathbb{C}), \overrightarrow{01}) \to \Pi^2(\mathbb{C})$) be the composition of $\theta_{\overrightarrow{01}}$ with the projection on $\Pi_2(V)(\mathbb{C})$ (resp. $\Pi^2(\mathbb{C})$). The triple $(\theta(S_0), \theta(S_1), \theta(S_\infty)) \in \mathcal{T}_2(\mathbb{C})$ (resp. $(\theta^2(S_0), \theta^2(S_1), \theta^2(S_\infty)) \in \mathcal{T}^2(\mathbb{C})$). Repeating the construction from 13.1 we get a $G(\theta)$-torsor $T(\theta)$ (resp. $G(\theta^2)$-torsor $T(\theta^2)$), a subtorsor of $\mathrm{Aut}^*(\Pi_2(V))$-torsor $\mathcal{T}_2$ (resp. $\mathrm{Aut}^*(\Pi^2(\mathbb{C}))$-torsor $\mathcal{T}^2$).

The projections $p : \Pi(V) \to \Pi_2(V)$ and $p^2 : \Pi(V) \to \Pi^2$ induce homomorphisms of group schemes

$$p_* : \mathrm{Aut}^*(\Pi(V)) \to \mathrm{Aut}^*(\Pi_2(V)) \quad \text{and} \quad (p^2)_* : \mathrm{Aut}^*(\Pi(V)) \to \mathrm{Aut}^*(\Pi^2)$$

and morphisms of torsors compatible with homomorphisms of group schemes

$$p_* : \mathcal{T} \to \mathcal{T}_2 \quad \text{and} \quad (p^2)_* : \mathcal{T} \to \mathcal{T}^2.$$

Lemma 13.3.1. *We have*

$$p_*(T(\theta_{\overrightarrow{01}})) = T(\theta), \quad p_*(G(\theta_{\overrightarrow{01}})) = G(\theta)$$

and

$$(p^2)_*(T(\theta_{\overrightarrow{01}})) = T(\theta^2), \quad (p^2)_*(G(\theta_{\overrightarrow{01}})) = G(\theta^2).$$

This lemma follows from Lemma 10.1.7.

Observe that the group $G(\theta)$ is the image of the group $G(\theta_{\overrightarrow{01}})$. This implies **Theorem C** from the introduction.

Corollary 13.3.2. *The group $G(\theta^2)$ is isomorphic to a quotient of $\mathcal{G}_{DR}(P^1(\mathbb{C}) \setminus \{0,1,\infty\})$.*

13.4. Below we shall calculate the $G(\theta^2)$-torsor $T(\theta^2)$. We have $\Pi^2 = \varprojlim_n (\Pi^2/\Gamma^{n+2}\Pi^2)$. If in definitions of $\mathcal{T}$ and $\mathrm{Aut}^*(\Pi(V))$ in 13.1 we replace the group $\Pi(V)$ by $\Pi^2/\Gamma^{n+2}\Pi^2$, then we get an $\mathrm{Aut}^*(\Pi^2/\Gamma^{n+2}\Pi^2)$-torsor $\mathcal{T}_n^2$. Let Θ_n be the composition of θ^2 with the projection onto $\Pi^2/\Gamma^{n+2}\Pi^2(\mathbb{C}) = \Pi^2(\mathbb{C})/\Gamma^{n+2}\Pi^2(\mathbb{C})$. The triple $(\Theta_n(S_0), \Theta_n(S_1), \Theta_n(S_\infty)) \in \mathcal{T}_n^2(\mathbb{C})$. By definition $T_n := T(\Theta_n)$ is the smallest subtorsor of $\mathcal{T}_n^2$ defined over $\mathbb{Q}$ which contains the triple as a $\mathbb{C}$-point and $G_n = G(\Theta_n)$ is the corresponding group, a subgroup defined over $\mathbb{Q}$ of $\mathrm{Aut}^*(\Pi^2/\Gamma^{n+2}\Pi^2)$. Let $p_n^{n+1} : \mathcal{T}_{n+1}^2 \to \mathcal{T}_n^2$ and $p_n^{n+1} : \mathrm{Aut}^*(\Pi^2/\Gamma^{n+1+2}\Pi^2) \to \mathrm{Aut}^*(\Pi^2/\Gamma^{n+2}\Pi^2)$ be induced by the projection $\Pi^2/\Gamma^{n+1+2}\Pi^2 \to \Pi^2/\Gamma^{n+2}\Pi^2$. It follows from Lemma 10.1.7 that

13.4.1. $\qquad P_n^{n+1}(T_{n+1}) = T_n \quad \text{and} \quad P_n^{n+1}(G_{n+1}) = G_n,$

We recall that by definition

$$T(\theta^2) := \varprojlim_n T_n \quad \text{and} \quad G(\theta^2) := \varprojlim_n G_n.$$

We recall that the group of $\mathbb{Q}$-points $(\Pi^2/\Gamma^{n+2}\Pi^2)(Q)$ is a Lie algebra over $\mathbb{Q}$ equipped with the multiplication given by the Baker-Campbell-Hausdorff formula. A linear basis of this Lie algebra is given by X, Y, $(Y,X),\ldots,$ $(Y,X)Y^{n-1}$.

The homomorphism $\theta^2 : \pi_1(V(\mathbb{C}), \overrightarrow{01}) \to \Pi^2(\mathbb{C})$ is given by $\theta^2(S_0) = (-2\pi i)X$ and $\theta(S_1) = (-2\pi i)Y + \sum_{k=2}^{\infty} (2\pi i)\zeta(k)((YX)Y^{k-1})$ (see §6).

0. Calculations of T_0 and G_0.

We have $\Theta_0(S_0) = (-2\pi i)X$ and $\Theta_0(S_1) = (-2\pi i)Y$. Observe that $(2\pi i)$ is not a k-th root of a rational number for any $k = 1, 2, 3 \dots$. This implies that $T_0(\mathbb{C}) = \{\alpha X, \alpha Y \mid \alpha \in \mathbb{C}^*\}$ and $G_0(\mathbb{C}) = \{f_\alpha \mid f_\alpha(X) = \alpha X, \; f_\alpha(Y) = \alpha Y \mid \alpha \in \mathbb{C}^*\} \subset \mathrm{Aut}(\Pi^2(\mathbb{C})/\Gamma^2\Pi^2(\mathbb{C}))$.

1. Calculations of T_1 and G_1.

We have $\Theta_1(S_0) = (-2\pi i)X$ and $\Theta_1(S_1) = (-2\pi i)Y$. Hence we get

$$T_1(\mathbb{C}) = \{\alpha X, \alpha Y \mid \alpha \in \mathbb{C}^*\}$$

and

$$G_1(\mathbb{C}) = \{f_\alpha \mid f_\alpha(X) = \alpha X, \; f_\alpha(Y) = \alpha Y \mid \alpha \in \mathbb{C}^*\}.$$

2. Calculations of T_2 and G_2.

We have $\Theta_2(S_0) = (-2\pi i)X$, $\Theta_2(S_1) = (-2\pi i)Y + (2\pi i)\zeta(2)((YX)Y)$. Observe that $(2\pi i)\zeta(2) = \frac{1}{24}(-2\pi i)^3$. This implies that

$$T_2(\mathbb{C}) = \Big\{\alpha X, \alpha Y + \frac{1}{24}\alpha^3((YX)Y) \mid \alpha \in \mathbb{C}^*\Big\} \quad \text{and}$$
$$G_2(\mathbb{C}) = \{f_\alpha \mid f_\alpha(X) = \alpha X, \; f_\alpha(Y) = \alpha Y \mid \alpha \in \mathbb{C}^*\}.$$

3. Calculations of T_3 and G_3.

We have $\Theta_3(S_0) = (-2\pi i)X$, $\Theta_3(S_1) = (-2\pi i)Y + (2\pi i)\zeta(2)((YX)Y) + (2\pi i)\zeta(3)((YX)Y^2)$. Assume that $\dim G_3 = \dim T_3 = 1$. Then it follows from 13.4.1 and Corollary 9.2 that $G_3 = G(0, 0, c_3)$ where $c_3 \in \mathbb{Q}$. As T_3 is a one-dimensional variety we have $T_3(\mathbb{C}) = \{\alpha X, \; \alpha Y + \frac{1}{24}\alpha^3((YX)Y) + \beta_3((YX)Y^2) \mid \alpha \in \mathbb{C}^*, \; \beta_3 \in \mathbb{C}, \; p(\alpha, \beta_3) = 0\}$ for some Laurent polynomial $p(x, y) \in \mathbb{Q}[x, \frac{1}{x}, y]$. Observe that to each value of α corresponds exactly one value $\beta_3(\alpha)$, because T_3 is a G_3-torsor. Hence $\beta_3(\alpha) = p(\alpha)$ where $p(x) \in \mathbb{Q}[x, \frac{1}{x}]$. If $(\alpha X, \alpha Y + \frac{1}{24}\alpha^3((YX)Y) + p(\alpha)((YX)Y^2)) \in T_3(\mathbb{C})$, then

$$\Big((\alpha t)X, (\alpha t)Y + \frac{1}{24}(\alpha t)^3((YX)Y) + \big(p(\alpha)t^4 + c_3(t - t^4)\alpha\big)((YX)Y^2)\Big) \in T_3(\mathbb{C}).$$

Hence $p(\alpha t) = p(\alpha)t^4 + c_3(t - t^4)\alpha$. Therefore $p(x) = ax^4 + c_3 x$. The equality $(2\pi i)\zeta(3) = a(-2\pi i)^4 + c_3(-2\pi i)$ implies $a = 0$ and $c_3 = -\zeta(3)$. Therefore we get

$$\dim G_3 = 1 \quad \text{if and only if} \quad \zeta(3) \in \mathbb{Q}.$$

Therefore if $\zeta(3) \notin \mathbb{Q}$ (one knows that $\zeta(3)$ is irrational) then $\dim G_3 = 2$,

$$T_3(\mathbb{C}) = \Big\{\alpha X, \alpha Y + \frac{1}{24}\alpha^3((YX)Y) + \beta_3((YX)Y^2) \mid \alpha \in \mathbb{C}^*, \; \beta_3 \in \mathbb{C}\Big\}$$

and

$$G_3 = (0,0,0 \mid 0,0,1).$$

n. Calculations of T_n and G_n.

We have $\Theta_n(S_0) = (-2\pi i)X$,

$$\Theta_n(S_1) = (-2\pi i)Y + \sum_{k=2}^{n-1}(2\pi i)\zeta(k)\big((YX)Y^{k-1}\big) + (2\pi i)\zeta(n)\big((YX)Y^{n-1}\big).$$

Assume that

$$G_{n-1} = G(0,0,c_3,0,c_5,0,c_7,\ldots \mid 0,0,\varepsilon_3,0,\varepsilon_5,0,\varepsilon_7,\ldots)$$

(see Corollary 9.4), where $c_1 = c_2 = c_4 = \ldots = c_{2k} = \ldots = 0$, $c_{2k+1} = -\zeta(2k+1)$ and $\varepsilon_{2k+1} = 0$ if $\zeta(2k+1) \in \mathbb{Q}$, and $c_{2k+1} = 0$ and $\varepsilon_{2k+1} = 1$ if $\zeta(2k+1) \notin \mathbb{Q}$.

Then the torsor T_{n-1} corresponding to the group G_{n-1} is given by

$$T_{n-1}(\mathbb{C}) = \Big\{\alpha X,\ \alpha Y + \sum_{k=2,k\text{-pair}}^{n-1} r_k \alpha^{k+1}\big((YX)Y^{k-1}\big) +$$

$$\sum_{k=3,k\text{-impair}}^{n-1} \big((1-\varepsilon_k)c_k\alpha + \varepsilon_k\beta_k\big)\big((YX)Y^{k-1}\big) \mid \alpha \in \mathbb{C}^*, \forall k\ \beta_k \in \mathbb{C}\Big\}.$$

Assume that $n = 2p$. Then $\zeta(n)(2\pi i) = r_n(-2\pi i)^{n+1}$, $r_n \in \mathbb{Q}$. This equality, the fact that $(2\pi i)^n \notin \mathbb{Q}$ and the property 13.4.1 imply that

$$G_n = G(0,0,c_3,0,c_5,\ldots,c_{n-1},0 \mid 0,0,\varepsilon_3,0,\ldots,\varepsilon_{n-1},0)$$

and

$$T_n(\mathbb{C}) = \Big\{\alpha X,\ \alpha Y + \sum_{k=2,k\text{-even}}^{n} r_k \alpha^{k+1}\big((YX)Y^{k-1}\big) +$$

$$\sum_{k=3,k\text{-odd}}^{n} \big((1-\varepsilon_k)c_k\alpha + \varepsilon_k\beta_k\big)\big((YX)Y^{k-1}\big) \mid \alpha \in \mathbb{C}^*, \forall_k\ \beta_k \in \mathbb{C}\Big\}.$$

Assume that $n = 2p + 1$. Assume that $\dim G_n = \dim G_{n-1}$. Then

$$G_n = G(0,0,c_3,\ldots,0,c_n \mid 0,0,\varepsilon_3,0,\varepsilon_5,\ldots,0,0) \quad \text{(by Corollary 9.4)}$$

and

$$T_n(\mathbb{C}) = \{\alpha X, \; \alpha Y + \sum_{k=2, k\text{-even}}^{n-1} r_k \alpha^{k+1} ((YX)Y^{k-1}) +$$

$$\sum_{k=3, k\text{-odd}}^{n-1} ((1 - \varepsilon_k)c_k\alpha + \varepsilon_k\beta_k)((YX)Y^{k-1}) + \beta_n((YX)Y^{n-1}) \text{ with}$$

$$\alpha \in \mathbb{C}^*; \; \forall k, \beta_k \in \mathbb{C}; \; p(\alpha, \varepsilon_3 \cdot \beta_3, \varepsilon_5 \cdot \beta_5, \dots, \beta_n) = 0\}$$

for some polynomial $p(x, y_3, y_5, \dots, y_n) \in \mathbb{Q}[x, \frac{1}{x}, y_3, y_5, \dots, y_n]$.

Assume that for some α_0 there are two different β_n. Then there is g in $G_n(\mathbb{C})$ such that $g(X) = X$, $g(Y) = Y + \dots + b_n((YX)Y^{n-1})$ with $b_n \neq 0$. Then it follows from the proof of Lemma 9.5 that $G_n = G(0, 0, c_3, 0, \dots, 0, 0 \mid 0, 0, \varepsilon_3, 0, \dots, 0, 1)$ and $\dim G_n = \dim G_{n-1} + 1$. Hence for any $\alpha \in \mathbb{C}^*$ there is a unique β_n corresponding to that α. Observe that β_n is an algebraic function of $\alpha, \beta_3, \beta_5, \dots$, which depends only on α. Hence $\beta_n = p(\alpha)$ for some $p(x) \in \mathbb{Q}[x, \frac{1}{x}]$.

Choose any $f \in G_n(\mathbb{C})$. Then $f(X) = tX$ and

$$f(Y) = tY + \sum_{k=3, k\text{-odd}}^{n-1} ((1 - \varepsilon_k)c_k(t - t^{k+1}) + \varepsilon_k b_k)((YX)Y^{k-1})$$

$$+ c_n(t - t^{n+1})((YX)Y^{n-1})$$

for some $t \in \mathbb{C}^*$, $b_3, b_5, \dots, b_{n-2} \in \mathbb{C}$. Acting by f on any element of $T_n(\mathbb{C})$, we get

$$((\alpha t)X, \; (\alpha t)Y + \sum_{k=2, \; k\text{-even}}^{n-1} r_k(\alpha t)^{k+1}((YX)Y^{k-1})$$

$$+ \sum_{k=3, k\text{-odd}}^{n-1} (\alpha(1 - \varepsilon_k)c_k(t - t^{k+1}) + \alpha\varepsilon_k b_k + (1 - \varepsilon_k)c_k\alpha t^{k+1}$$

$$+ \varepsilon_k\beta_k t^{k+1})((YX)Y^{k-1}) + (\alpha c_n(t - t^{n+1}) + \beta_n t^{n+1})((YX)Y^{n-1}).$$

Therefore we get

$$p(\alpha t) = c_n\alpha(t - t^{n+1}) + p(\alpha)t^{n+1}.$$

This implies $p(x) = ax^{n+1} + c_n x$. The equality $\zeta(n)(-2\pi i) = a(2\pi i)^{n+1} + c_n(2\pi i)$ implies $p(x) = c_n x$ and $c_n = -\zeta(n)$. Therefore we get

$$\dim G_n = \dim G_{n-1} \quad \text{if and only if} \quad \zeta(n) \in \mathbb{Q}.$$

The final result is the following.

Proposition 13.5. *Let* $G = \varprojlim_{n} G_n$. *Then* $G(\theta^2) = G$ *and* $G =$

$$G(0, 0, c_3, 0, c_5, 0, c_7, \ldots, 0, c_{2k+1}, 0 \ldots \mid 0, 0, \varepsilon_3, 0, \varepsilon_5, 0 \ldots, 0, \varepsilon_{2k+1}, 0 \ldots),$$

where $\varepsilon_{2k+1} = 0$ *if and only if* $\zeta(2k+1) \in \mathbb{Q}$, *and then* $c_{2k+1} = -\zeta(2k+1)$. *The torsor* $T(\theta^2)(\mathbb{C})$ *is given by*

$$\Big\{ \alpha X, \alpha Y + \sum_{k=2, k\text{-even}}^{\infty} r_k \alpha^{k+1}\big((YX)Y^{k-1}\big) + \sum_{k=3, k\text{-odd}}^{\infty} \big((1 - \varepsilon_k)c_k\alpha +$$

$$+ \varepsilon_k \beta_k\big)\big((YX)Y^{k-1}\big) \mid \alpha \in \mathbb{C}^*, \text{for all } k, \ \beta_k \in \mathbb{C} \Big\},$$

where $\zeta(2k) = -r_{2k}(2\pi i)^{2k}$, *and* $c_{2k+1} = -\zeta(2k+1)$, $\varepsilon_{2k+1} = 0$ *if* $\zeta(2k+1) \in \mathbb{Q}$, $c_{2k+1} = 0$, $\varepsilon_{2k+1} = 1$ *if* $\zeta(2k+1) \notin \mathbb{Q}$.

Corollary 13.6. *The group* G *contains the group* $H := \{f_\alpha \mid f_\alpha(X) = \alpha X, \ f_\alpha(Y) = \alpha Y \mid \alpha \in \mathbb{C}^*\}$ *if and only if all numbers* $\zeta(2k + 1)$ *are irrational.*

Proof. It follows from Proposition 13.2 that all $\zeta(2k+1)$ are irrational if and only if $G = G(0, 0, 0, \ldots, 0, \ldots \mid 0, 0, 1, 0, 1, 0, \ldots 0, 1, 0, 1, 0, \ldots)$ (all $c_i = 0$, $\varepsilon_1 = 0$, all $\varepsilon_{2k} = 0$ and all $\varepsilon_{2k+1} = 1$ for $k = 1, \ldots$). Then we have $H \subset G$. Let $H \subset G = G(0, 0, c_3, 0, c_5, \ldots, 0, c_{2k+1}, 0 \ldots \mid 0, 0, \varepsilon_3, 0, \varepsilon_5, 0, \ldots, 0, \varepsilon_{2k+1}, 0, \ldots)$. Then there exists $f \in G$ such that $f(X) = \alpha X$ and

$$f(Y) = \alpha Y + \sum_{k=1}^{\infty} x_{2k+1}((YX)Y^{2k}),$$

where all $x_{2k+1} \neq 0$. Corollary 9.5 implies that $G = G(0, 0 \ldots, 0, \ldots \mid 0, 0, 1, 0, 1, \ldots 0, 1, 0, \ldots)$ (all $c_i = 0$, $\varepsilon_1 = \varepsilon_{2k} = 0$, $\varepsilon_{2k+1} = 1$ for $k = 1, 2, \ldots$). $\Diamond$

Corollary 13.7. *Let* $\mathcal{G}$ *be the smallest closed subgroup of* $\mathrm{Aut}(\pi^2)$ *defined over* $\mathbb{Q}$, *containing* G *and* H. *Then* $\mathcal{G} = G(0, 0 \ldots, 0, \ldots \mid 0, 0, 1, 0, 1, 0 \ldots)$ *(all* $c_i = 0$, $\varepsilon_1 = \varepsilon_{2k} = 0$, $\varepsilon_{2k+1} = 1$ *for* $k = 1, 2, \ldots$*).*

Proof. The group $\mathcal{G}$ contains f such that $f(X) = X$ and

$$f(Y) = Y + \sum_{k=1}^{\infty} x_{2k+1}((YX)Y^{2k})$$

where all $x_{2k+1} \neq 0$. Lemma 9.5 implies the corollary. $\Diamond$

13.8. Proof of **Theorem D** in the introduction. We recall from §6 that the monodromy homomorphism

$$\theta : \pi_1(V(\mathbb{C}), \overrightarrow{01}) \longrightarrow \Pi_2(V)(\mathbb{C})$$

is given by

$$S_0 \to (-2\pi i)X, \quad S_1 \to (-2\pi i)Y + \sum_{i=0,j=0}^{\infty} (2\pi i)\alpha_{i+1,j+1}\big((YX)X^iY^{j+1}\big)$$

(see §6, formula (7)). It is an observation of Drinfeld that the numbers $\alpha_{i,j}$ satisfy the equation

$$-1 + \sum_{n\geq 0, m\geq 1} \alpha_{n+1,m}u^{n+1} \cdot v^m = -\exp\Big(-\sum_{k=2}^{\infty} \sum_{\substack{i+j=k \\ i\geq 1, j\geq 1}} \frac{(k-1)!}{i!\,j!}\zeta(k)u^i \cdot v^j\Big)$$

(see [Dr]). Therefore we can describe the monodromy homomorphism θ in the following way

$$S_0 \to (-2\pi i)X, \quad S_1 \to \Big((-2\pi i)Y, \exp\Big(-\sum_{k=2}^{\infty} \sum_{\substack{i+j=k \\ i\geq 1, j\geq 1}} \frac{(k-1)!}{i!\,j!}\zeta(k)X^i \cdot Y^j\Big)\Big).$$

Let us temporarily write T' for the torsor given in Theorem D i) and G' for the corresponding group. Then $T(\theta)$ is a subtorsor of T' and the group $G(\theta)$ is a closed subgroup of G' because $(\theta(S_0), \theta(S_1)) \in T'(\mathbb{C})$. The image of the group G' in $\mathrm{Aut}(\Pi(V)/\Gamma^{n+2}\Pi(V))$ has the same dimension as the subgroup G_n of $\mathrm{Aut}(\Pi^2/\Gamma^{n+2}\Pi^2)$. The group $G(\theta)$ projects onto G_n, hence $G' = G(\theta)$ and $T' = T(\theta)$.

The point iii) of Theorem D and Corollary E follow from Corollary 13.7 and Corollary 13.6 respectively.

§14. The group $\mathcal{G}_{\mathrm{DR}}(V)$ for pointed projective lines and for configuration spaces.

In this section we shall translate the results about the monodromy representations from §§5-8 into results about the group $\mathcal{G}_{\mathrm{DR}}(V)$. Let $V = \mathbb{P}^1_k \setminus \{a_1, \ldots, a_{n+1}\}$ and let v be a k-point of V. We can assume that $a_1 = 0$, $a_2 = 1$ and $a_{n+1} = \infty$. We recall that $\mathrm{Lie}(V)$ is a free Lie algebra over k on $X_i := (\frac{dz}{z-a_i})^*$, $i = 1, 2, ..., n$. Let us set $X_{n+1} = -\sum_{i=1}^n X_i$. Let R be a k-algebra. Let us set

$$\mathrm{Aut}^*\big(\Pi(V)\big)(R) := \{f \in \mathrm{Aut}\big(\Pi(V)_R\big)\big| \exists \alpha_f \in R^*, \ f(X_1) = \alpha_f X_1,$$
$$f(X_2) \approx \alpha_f X_2, f(X_k) \sim \alpha_f X_k \ k = 3, ...n+1\}.$$

Here $\approx$ means a conjugation by an element of $\Pi(V)(R)$ with the zero coefficients at X_1 and X_2 and $\sim$ means conjugation by an element of $\Pi(V)(R)$. Similarly for $\mathcal{Y}_n = \left(P^1(\mathbb{C})\backslash\{0,1,\infty\}\right)_*^{n-3}$ we set

$$\mathrm{Aut}^*\left(\Pi(\mathcal{Y}_n)\right)(R) := \left\{ f \in \mathrm{Aut}\left(\Pi(\mathcal{Y}_n)_R\right) \big|\; \exists \alpha_f \in R^*, \;\; f(X_{i,j}) \sim \alpha_f X_{i,j} \right\}.$$

(The elements $X_{i,j}$ are as in §7.)

These functors are representable by affine pro-algebraic group schemes over k and over $\mathbb{Q}$, which we denote by $\mathrm{Aut}^*\left(\Pi(V)\right)$ and $\mathrm{Aut}^*\left(\Pi(\mathcal{Y}_n)\right)$ respectively.

Let us set

$$\mathrm{Out}^*\left(\Pi(V)\right) := \mathrm{image}\Big(\mathrm{Aut}^*(\Pi(V)) \to \mathrm{Aut}\big(\Pi(V)\big)/\mathrm{Inn}\big(\Pi(V)\big)\Big)$$

and

$$\mathrm{Out}^*\left(\Pi(\mathcal{Y}_n)\right) := \mathrm{Aut}^*\left(\Pi(\mathcal{Y}_n)\right)/\mathrm{Inn}\left(\Pi(\mathcal{Y}_n)\right).$$

We shall define an $\mathrm{Aut}^*(\Pi(V))$-torsor T and the $\mathrm{Aut}^*\left(\Pi(\mathcal{Y}_n)\right)$-torsor T^n as follows. Let $S_1,\ldots,S_{n+1}$ be geometric generators of $\pi_1(V(\mathbb{C}),v)$, i.e. loops around $a_1,\ldots,a_{n+1}$ respectively. For any k-algebra R we set

$$T(R) := \big\{ f \in \mathrm{Hom}(\pi_1(V(\mathbb{C}),v); \Pi(V)(R)) \;\big|\; \exists \alpha_f \in R^*, \; f(S_1) = \alpha_f X_1,$$
$$f(S_2) \approx \alpha_f X_2, \; f(S_k) \sim \alpha_f X_k \;\; k = 3,\ldots,n+1 \big\}$$

and

$$T^n(R) = \big\{ \varphi \in \mathrm{Hom}(\pi_1(\mathcal{Y}_n(\mathbb{C}),y); \Pi(\mathcal{Y}_n)(R)) \;\big|\; \exists \alpha_\varphi \in R^* \text{ such that}$$
$$\forall A_{ij}, \; \varphi(A_{ij}) \sim \alpha_\varphi X_{ij} \big\}.$$

(The elements A_{ij} are as in §7.)

Proposition 14.2. *We have*

$$\mathcal{G}_{DR}(V) \subset \mathrm{Out}^*(\Pi(V)).$$

Proof. Let us choose the vector $\overrightarrow{01}$ as a tangential base point. Let S_1 be a loop around 0 as in the picture of §4. Let us choose a tangent vector v_k at each point a_k. Assume that $v_2 = \overrightarrow{10}$. Let Γ be a family of paths from the base point to $v_2,\ldots,v_{n+1}$ such that γ_2 is the interval $[0,1]$. Let $S_1, S_2,\ldots,S_{n+1}$ be a sequence of geometric generators associated to Γ. Let $\theta_{\overrightarrow{01}} : \pi_1(V(\mathbb{C}),\overrightarrow{01}) \to \Pi(V)(\mathbb{C})$ be the monodromy homomorphism. It follows from Corollary 5.4 and Theorem 6.1 that

$$\left(\theta_{\overrightarrow{01}}(S_1), \theta_{\overrightarrow{01}}(S_2),\ldots,\theta_{\overrightarrow{01}}(S_{n+1})\right) \in T(\mathbb{C}).$$

Hence $G(\theta_{\overrightarrow{01}}) \subset \mathrm{Aut}^*(\Pi(V))$. This implies that $\mathcal{G}_{DR}(V) \subset \mathrm{Out}^*(\Pi(V))$ if we identify $\pi_1^{DR}(V, v)$ and $\Pi(V)$ by the homomorphism u from Corollary 2.4. ◇

Observe that T^n is a $\mathrm{Aut}^*(\Pi(\mathcal{Y}_n))$-torsor. Hence $T^n/\mathrm{Inn}(\Pi(\mathcal{Y}_n))$ is an $\mathrm{Out}^*(\Pi(\mathcal{Y}_n))$-torsor.

Proposition 14.3.
$$\mathcal{G}_{DR}(\mathcal{Y}_n) \subset \mathrm{Out}^*(\Pi(\mathcal{Y}_n)).$$

Proof. Corollary 7.3 implies that $\theta_{y,\mathcal{Y}_n} \in T^n(\mathbb{C})$. Hence the proposition follows.

Proposition 14.4. *For $n \geq 4$ we have isomorphisms*

$$\mathcal{G}_{DR}(\mathcal{Y}_{n+1}) \approx \mathcal{G}_{DR}(\mathcal{Y}_n)$$

induced by the projection $\mathcal{Y}_{n+1} \to \mathcal{Y}_n$.

Proof. The proposition follows from Proposition 7.4, Lemma 7.5 and Corollary 7.6. ◇

Observe that $\mathcal{Y}_4 = P_{\mathbb{Q}}^1 \setminus \{0, 1, \infty\}$. Hence we have the following corollary.

Corollary 14.5. *For $n \geq 4$ we have*

$$\mathcal{G}_{DR}(\mathcal{Y}_n) \approx \mathcal{G}_{DR}(\mathbb{P}_{\mathbb{Q}}^1 \setminus \{0, 1, \infty\}).$$

The group $\Pi(V)$ is a normal subgroup of $\Pi(\mathcal{Y}_{n+2})$. Hence $\Pi(\mathcal{Y}_{n+2}) \subset \mathrm{Aut}(\Pi(V))$. Proposition 7.7 implies the following result.

Proposition 14.6. *We have*

$$\mathcal{G}_{DR}(V) \cdot \Pi(\mathcal{Y}_{n+2})/\Pi(\mathcal{Y}_{n+2}) \approx \mathcal{G}_{DR}(\mathbb{P}_{\mathbb{Q}}^1 \setminus \{0, 1, \infty\}).$$

§15. Conjectures.

Let X be a smooth quasi-projective algebraic variety defined over a number field k and let x be a k-point or a tangential base k-point. Denote by $\pi_1^{et}(X \times_k \bar{k}, x)_l$ the pro-l completion of the étale fundamental group of $X \times_k \bar{k}$. Let $\Phi_X : \mathrm{Gal}(\bar{k}/k) \to \mathrm{Out}(\pi_1^{et}(X \times_k \bar{k}, x)_l)$ be the natural representation of the Galois group into the outer automorphisms of the pro-l completion of the étale fundamental group.

The monodromy representations of iterated integrals, i.e. the homomorphisms

$$\varphi_x : \pi_1^B(X(\mathbb{C}), x)(\mathbb{C}) \to \pi_1^{DR}(X, x)(\mathbb{C})$$

have many properties analogous to the action of $\mathrm{Gal}(\bar{k}/k)$ on étale fundamental groups. We give some examples.

Example 1. Let X be the projective line minus a finite number of k-points. Let S be a loop on X around a missing point. If $\sigma \in \mathrm{Gal}(\bar{k}/k)$ then $\sigma(S)$ is conjugate to $S^{\chi(\sigma)}$, where χ is the cyclotomic character (see [I3]). It follows from Corollary 12.5 that $\varphi_x(S)$ is conjugate to an element $s^{2\pi i}$, where $s \in \pi_1^{DR}(X, x)(k)$.

Example 2. Assume that $\bar{X}$ is a smooth projective variety and D is a divisor with normal crossings in $\bar{X}$. Assume that $X := \bar{X} \setminus D$ is defined over a field k. Let S be a loop in X around an irreducible component of D. As before $\sigma(S) \sim S^{\chi(\sigma)}$ in the case of a Galois action. It follows from Corollary 7.2 that $\varphi_x(S) \sim s^{2\pi i}$, where $s \in \pi_1^{DR}(X, x)(k)$ and X is a configuration space of n points in $\mathbb{C}$ (see §7).

Example 3. The element $a_{01}^{10}(X, Y)$ satisfies $\mathbb{Z}/2$, $\mathbb{Z}/3$ and $\mathbb{Z}/5$ cycle relations. The element considered by Ihara in [I2] satisfies analogous relations. (see also Remark 2 at the end of §8.)

Example 4. Chudnowsky has shown that the transcendence degree of the field generated by abelian periods is 2 if X is an elliptic curve with complex multiplication. The existence of a non-trivial endomorphism of X implies that the image of $\mathrm{Gal}(\bar{k}/k)$ is contained in a two dimensional subgroup of $\mathrm{GL}(H^1_{et}(X \times_k \bar{k})_l)$.

In the second part of this paper we investigated the group $\mathcal{G}_{DR}(X)$. Our results show that this group remains very close to the image of $\mathrm{Gal}(\bar{k}/k)$ in the group of outer automorphisms of the étale fundamental group of X. We shall formulate some conjectures to make this relation precise. Let $\pi_1^{et}(X \times_k \bar{k}, x)_l^{(n)} \otimes \mathbb{Q}$ be the rational Malcev completion of

$$\pi_1^{et}(X \times_k \bar{k}, x)_l / \Gamma^{n+1} \pi_1^{et}(X \times_k \bar{k}, x)_l.$$

Let $\Phi_X^n : \mathrm{Gal}(\bar{k}/k) \to \mathrm{Out}\big(\pi_1^{et}(X \times_k \bar{k}, x)_l^{(n)} \otimes \mathbb{Q}\big)$ be the natural homomorphism.

Conjecture 15.1. *There exists an affine algebraic subgroup G_n of the group* $\mathrm{Out}\big(\pi_1^{et}(X \times_k \bar{k}, x)_l^{(n)} \otimes \mathbb{Q}\big)$ *such that $\Phi_X^n(\mathrm{Gal}(\bar{k}/k))$ is open in $G_n(\mathbb{Q}_l)$ for the non-archimedean metric of $\mathbb{Q}_l$. Let $\mathrm{Lie}(G_n)$ be a Lie algebra of the group G_n.*

We conjecture furthermore that the Lie algebras $\mathrm{Lie}(G_n)$ form a compatible system. We denote the resulting pro-Lie algebra by $\mathrm{Lie}(\Phi_X(\mathrm{Gal}(\bar{k}/k)))$.

(We hope that such a conjecture has already been made by someone else. We need it to compare the group $\mathcal{G}_{DR}(X)$ with the image of $\mathrm{Gal}(\bar{k}/k)$. We have formulated the conjecture in analogy with [D4, 8.14].)

Let us assume that X is defined over $\mathbb{Q}$.

Conjecture 15.2. *There is an isomorphism*

$$\mathrm{Lie}(\Phi_X(\mathrm{Gal}(\bar{\mathbb{Q}}/\mathbb{Q}))) \approx \mathrm{Lie}(\mathcal{G}_{DR}(X)) \otimes \mathbb{Q}_l .$$

We recall that we have the isomorphism

$$\varphi_x^{(n)} : \pi_1^{\mathrm{B}}(X(\mathbb{C}), x)^{(n)}(\mathbb{C}) \to \pi_1^{\mathrm{DR}}(X, x)^{(n)}(\mathbb{C}).$$

Let $Z(\varphi_x^{(n)})$ be the k-Zariski closure of $\varphi_x^{(n)}$ in

$$\mathrm{Iso}\big(\pi_1^{\mathrm{B}}(X(\mathbb{C}), x)^{(n)} \times k; \pi_1^{\mathrm{DR}}(X, x)^{(n)}\big).$$

We recall that $T(\varphi_x^{(n)})$ is the corresponding torsor (see §11).

Question 15.3. *Is it true that*

$$Z(\varphi_x^{(n)}) = T(\varphi_x^{(n)})?$$

Assume that X is a complement of a divisor with normal crossings in a smooth projective scheme of finite type over a number field k and that $H_{DR}^1(V) = A^1(V)$. In this case the group $\mathcal{G}_{DR}(X)$ can be considered as a subgroup of $\mathrm{Out}(\Pi(X))$. Let

$$\mathcal{C}_X : \mathbf{G_m} \to \mathrm{Aut}_{\mathrm{group}}(\Pi(X)) = \mathrm{Aut}_{\mathrm{Lie\ algebra}}(L(X))$$

be a homomorphism such that $t \in \mathbf{G_m}(k)$ acts on $H(X)$ as a multiplication by t.

Question 15.4. *Does the group $\mathcal{G}_{\mathrm{DR}}(V)$ contain the image of $\mathcal{C}_X$?*

Corollary 15.5. *Let $V = \mathbb{P}_{\mathbb{Q}}^1 \setminus \{0, 1, \infty\}$ and let $v = \overrightarrow{01}$ be a tangential base point. If Questions 15.3 and 15.4 have affirmative answers for V and v then the transcendence degree of the field $\mathbb{Q}(2\pi i)(\zeta(3), \zeta(5), \ldots \zeta(2k+1))$ is $k + 1$.*

The corollary follows from Theorems A, C and D and Lemma 10.1.3.

§A.1. Malcev completion.

A1.0. Let G be a unipotent algebraic group over a field k of characteristic zero. The exponential map $\exp : \mathrm{Lie}(G) \to G$ is an isomorphism of $\mathrm{Lie}(G)$ equipped with the multiplication given by the Baker-Campbell-Hausdorff formula with G. For a nilpotent Lie algebra L over k equipped with the multiplication given by the B.-C.-H. formula we have

$$\mathrm{Aut}_{\mathrm{group}}(L) = \mathrm{Aut}_{\mathrm{Lie\ algebra}}(L).$$

Let R be a k-algebra. Then the same holds for a unipotent algebraic group scheme over R.

A1.1. Let π be a finitely generated nilpotent group. Then there exists an affine unipotent algebraic group scheme $\pi_{\mathbb{Q}}$ over $\mathrm{Spec}\,\mathbb{Q}$ and a homomorphism $b_{\mathbb{Q}} : \pi \to \pi_{\mathbb{Q}}(\mathbb{Q})$ functorial with respect to homomorphisms $\pi \to \pi'$, which has the following universal property: Let $G_{\mathbb{Q}}$ be a unipotent algebraic group over $\mathrm{Spec}\,\mathbb{Q}$. Let $\varphi : \pi \to G_{\mathbb{Q}}(\mathbb{Q})$ be a homomorphism. Then there is a unique morphism of unipotent algebraic groups $\varphi_{\mathbb{Q}} : \pi_{\mathbb{Q}} \to G_{\mathbb{Q}}$ such that on $\mathbb{Q}$-points we have $\varphi_{\mathbb{Q}}(\mathbb{Q}) \circ b_{\mathbb{Q}} = \varphi$. The pair $(\pi_{\mathbb{Q}},\ b_{\mathbb{Q}})$ is unique up to a unique isomorphism.

A1.2. Let K be a $\mathbb{Q}$-algebra and let $\pi_K := \pi_{\mathbb{Q}} \times \mathrm{Spec}\,K$. The homomorphism $b_K : \pi \to \pi_{\mathbb{Q}}(\mathbb{Q}) \hookrightarrow \pi_K(K)$, which is the composition of $b_{\mathbb{Q}}$ and the inclusion has the same universal property with respect to homomorphisms $\varphi : \pi \to G_K(K)$, where G_K is a unipotent algebraic group scheme over K.

Proof. It follows from A1.1 that there is a unique $f : \pi_{\mathbb{Q}}(\mathbb{Q}) \to G_K(K)$ such that $f \circ b_Q = \varphi$. We can assume that f is a morphism of Lie algebras (see A1.0). We extend f to $\pi_K(K)$ by linearity.

A1.3. If π is a finitely generated group then we set $\pi_K := \varprojlim_n (\pi/\Gamma^n \pi)_K$.

The group π_K together with the homomorphism $b_K : \pi \to \pi_K(K)$ has the same universal property with respect to homomorphisms $\varphi : \pi \to G_K(K)$, where G_K is a pro-unipotent pro-algebraic group scheme over K. We call the pair $(\pi_K,\ b_K : \pi \to \pi_K(K))$ the Malcev K-completion of π. It is unique up to a unique isomorphism.

A1.4. It follows from [Ch2] (Theorems 2.1.1 and 2.7.2) and [Q] (Lemma 3.3 and (1.4)) that the pair $(\pi_1^{DR}(V, v),\ b^{DR} : \pi_1(V(\mathbb{C}), v) \to \pi_1^{DR}(V, v)(\mathbb{C}))$ (resp. $(\pi_1^{B}(V(\mathbb{C}), v),\ b_{\mathbb{Q}}^{B} : \pi_1(V(\mathbb{C}), v) \to \pi_1^{B}(V(\mathbb{C}), v)(\mathbb{Q})))$ is the Malcev $\mathbb{C}$-completion (resp. rational completion) of $\pi_1(V(\mathbb{C}), v)$.

§A.2. The torsor and the group corresponding to the Drinfeld-Ihara relation.

A.2.0. In this section we describe the torsor and the group which correspond to $\mathbb{Z}/2$, $\mathbb{Z}/3$ and $\mathbb{Z}/5$-relations. It is interesting to observe that the $\mathbb{Z}/3$-relation for the group is different from the $\mathbb{Z}/3$-relation in the Galois theory.

Let us set $X = P^1(\mathbb{C}) \setminus \{0, 1, \infty\}$. Let S_0, S_1 and S_∞ be the generators of $\pi_1(X, \overrightarrow{01})$ (see the introduction, and more particularly the picture on p. 5). Observe that $S_\infty \cdot S_1 \cdot S_0 = 1$.

Notation. If a and b are elements of the group G we set $a^b := b \cdot a \cdot b^{-1}$.

A.2.1. For any $\mathbb{Q}$-algebra k, let us set:

$$TG(k) := \Big\{ f : \pi_1(X, \overrightarrow{01}) \to \pi(X)(k) | \exists t \in k^* \text{ and } a(X, Y) \in \pi'(X)(k) \text{ with}$$

$$f(S_0) = -tX$$

$$f(S_1) = a(X, Y) \cdot (-tY) \cdot a(X, Y)^{-1}$$

$$*_\infty) \; f(S_\infty) = (-\tfrac{1}{2}tX)^{-1} \cdot a(X, Z) \cdot (-tZ) \cdot a(X, Z)^{-1} \cdot (\tfrac{1}{2}tX)$$

$$*_2) \; a(X, Y) \cdot a(Y, X) = 0$$

$$*_3) \; (-\tfrac{1}{2}tY) \cdot a(Y, Z) \cdot (-\tfrac{1}{2}tZ) \cdot a(Z, X) \cdot (-\tfrac{1}{2}tX) \cdot a(X, Y) = 0,$$

$$*_5) \; a(X_{12}, X_{15}) \cdot a(X_{34}, X_{23}) \cdot a(X_{15}, X_{45}) \cdot a(X_{23}, X_{12}) \cdot a(X_{45}, X_{34}) = 0 \Big\}$$

where the last equality takes place in the group $\pi(\mathcal{Y}_5)(k)$.

The condition $*_3$ is equivalent to the condition $*_3'$) obtained by replacing $-\tfrac{1}{2}$ by $\tfrac{1}{2}$ in $*_3$.

A.2.2. We shall define an affine group scheme G over $\mathbb{Q}$ in the following way. For a $\mathbb{Q}$-algebra k we set:

$$G(k) := \Big\{ g \in \mathrm{Aut}\big(\pi(X)(k)\big) | \exists s \in k^* \text{ and } \alpha(X, Y) \in \pi'(X)(k) \text{ with}$$

$$g(X) = sX$$

$$g(Y) = (sY)^{\alpha(X,Y)}$$

$$**_\infty) \; g(Z) = (sZ)^{\alpha(X,Z)}$$

$$**_2) \; \alpha(X, Y) \cdot \alpha(Y, X) = 0$$

$$**_3) \; \alpha(Y, Z) \cdot \alpha(Z, X) \cdot \alpha(X, Y) = 0$$

$$**_5) \; \alpha(X_{12}, X_{15}) \cdot \alpha(X_{34}, X_{23}) \cdot \alpha(X_{15}, X_{45}) \cdot \alpha(X_{23}, X_{12}) \cdot \alpha(X_{45}, X_{34}) = 0.$$

The last equality takes place in the group $\pi(\mathcal{Y}_5)(k)$.

Lemma A.2.2.1. *Let us suppose that $\alpha(X,Y)$ and $A(X,Y)$ are in $\pi'(X)(k)$ and $s \in k^*$. Then there is a unique element $\beta(X,Y) \in \pi'(X)(k)$ such that*

$$\alpha\big(sX, (sY)^{\beta(X,Y)}\big) \cdot \beta(X,Y) = A(X,Y).$$

Proof. From the equation one gets induction formulas for the coefficients of $\beta(X,Y)$.   ◇

Proposition A.2.2.2. *The set $G(k)$ is a group.*

Proof. Let $g(Y) = (sY)^{\alpha(X,Y)}$ and $g_1(Y) = (tY)^{\beta(X,Y)}$. Then $\alpha_{g_1 \circ g}(X,Y) = \alpha\big(tX, (tY)^{\beta(X,Y)}\big)\cdot\beta(X,Y)$. It is immediate to check the conditions $**_\infty$, $**_2$ and $**_3$ for $\alpha_{g_1 \circ g}(X,Y)$. To show that relation $**_5$ holds, we must use the group $\pi(\mathcal{Y}_5)$. Let $g \in G(k)$. We shall extend g to an automorphism of $\pi(\mathcal{Y}_5(k))$ in the following way. We set

$$\hat{g}(X_{12}) = sX_{12}, \quad \hat{g}(X_{34}) = sX_{34}, \quad \hat{g}(X_{45}) = (sX_{45})^{\alpha(X_{34},X_{45})},$$

$$\hat{g}(X_{23}) = (sX_{23})^{\alpha(X_{34},X_{45})\cdot\alpha(X_{12},X_{23})}, \quad \hat{g}(X_{51}) = (sX_{51})^{\alpha(X_{12},X_{51})}.$$

Applying the automorphism $\hat{g}$ of $\pi(\mathcal{Y}_5)(k)$ to the relation $**_5$ for g we get the relation $**_5$ for $g_1 \circ g$.

Assume that $g_1 \circ g = \mathrm{id}$. Then the element $\beta(X,Y)$ is uniquely determined by the equation

$$\alpha\big(\tfrac{1}{s}X, (\tfrac{1}{s}Y)^{\beta(X,Y)}\big) \cdot \beta(X,Y) = 0$$

(see Lemma A.2.2.1). Applying g_1 to the equality $\alpha(X,Y) \cdot \alpha(Y,X) = 0$, we get

$$\alpha\big(\tfrac{1}{s}Y, (\tfrac{1}{s}X)^{\beta(X,Y)^{-1}}\big) \cdot \beta(X,Y)^{-1} = 0.$$

It follows from Lemma A.2.2.1 that

$$\beta(X,Y)^{-1} = \beta(Y,X).$$

One proves the conditions $**_\infty$, $**_3$ and $**_5$ for $\beta(X,Y)$ similarly.   ◇

A.2.3. We shall show that TG is a G-torsor.

Proposition A.2.3.1. *The set $TG(k)$ is a $G(k)$-torsor, i.e. the group $G(k)$ acts freely and transitively on $TG(k)$.*

If $f \in TG(k)$ and $g \in G(k)$ then the verification that $g \circ f \in TG(k)$ is instantaneous. The action of $G(k)$ on $TG(k)$ is obviously free.

Suppose that $f \in TG(k)$, $f(S_1) = (-tY)^{a(X,Y)}$ and $h \in TG(k)$, $h(S_1) = (-tY)^{b(X,Y)}$. We are looking for $g \in G(k)$ such that $g \circ f = h$. If $g(Y) = (\phi Y)^{\alpha(X,Y)}$ then $\alpha(X,Y)$ is determined uniquely by the equation

A.2.3.2 $$a\big(\phi X, (\phi Y)^{\alpha(X,Y)}\big) \cdot \alpha(X,Y) = b(X,Y)$$

where $\phi = \frac{s}{t}$.

Suppose that we have already shown the conditions $**_\infty$ and $**_2$ for $\alpha(X,Y)$. Applying the automorphism g to the formula $*_3$ for f we get

$$a\big(\phi Y, (\phi Z)^{\beta(Y,X) \cdot \beta(X,Z)}\big) \cdot \beta(Y,X) \cdot \beta(X,Z) = b(Y,Z).$$

Replacing X, Y by Y and Z in A.2.3.2 and using Lemma A.2.2.1 we get $\beta(Y,Z) \cdot \beta(Z,X) \cdot \beta(X,Y) = 0$. The proofs of $*_\infty$, $*_2$ and $*_5$ are similar. $\Diamond$

Proposition A.2.3.3. *The functors*

$$TG : \mathbb{Q} - \text{algebras} \to \text{Sets} \ and \ G : \mathbb{Q} - \text{algebras} \to \text{Groups}$$

are representable by an affine pro-scheme over $\mathbb{Q}$ and an affine pro-group scheme over $\mathbb{Q}$. We denote these pro-schemes by TG and G respectively. The pro-scheme TG is a G-torsor.

Proof. This follows from the definitions of the functors TG and G, Proposition A.2.2.2 and Proposition A.2.3.1. $\Diamond$

One can show that the monodromy representation $\theta_{\overrightarrow{01}} : \pi_1(X, \overrightarrow{01}) \to \pi(X)(\mathbb{C})$ is an element of $TG(\mathbb{C})$. One takes $t = 2\pi i$ and $a(X,Y) = \log a_{\overrightarrow{10}}^{\overrightarrow{01}}(X,Y)$. This implies the following result.

Corollary A.2.3.4.

$$\mathcal{G}_{\mathrm{DR}}(P_{\mathbb{Q}}^1 \setminus \{0,1,\infty\}) \subset G.$$

References

[A]   K. Aomoto, Special Values of Hyperlogarithms and Linear Difference Schemes, *Illinois J. of Math.* **34** (2) (1990), 191-216.

[An]   Y.Andre, *G-Functions and Geometry*, Aspects of Mathematics, F. Vieweg & Sohn Publ., 1989.

[C]   P.Cartier, Développements récents sur les groupes de tresses, applications à la topologie et à l'algèbre, *Séminaire Bourbaki*, 1989-90, no. 716, Astérisque 189-190 (1990), SMF Publ., 17-67.

[Ch1] K. T. Chen, Algebra of iterated path integrals and fundamental groups, *Transactions of the AMS* **156**, (1971), 359 – 379.

[Ch2] K. T. Chen, Extention of C^∞ function algebra by integrals and Malcev completion of π_1, *Adv. in Math.* **23** (1977), 181 –210.

[D1] P. Deligne, Théorie de Hodge II, *Publ. Math. IHES* **40**, (1971), 5 – 58.

[D2] P. Deligne, Hodge Cycles on Abelian Varieties, Lecture Notes in Math. **900**, Springer-Verlag 1982, 9 – 100.

[D3] P. Deligne, Equations différentielles à points singuliers Réguliers, Lecture Notes in Math. **163**, Springer Verlag, 1970.

[D4] P. Deligne, Le groupe fondamental de la droite projective moins trois points, in *Galois Groups over* $\mathbb{Q}$, Math. Sc. Res. Ins. Publ. **16**, 1989, 79 – 297.

[D5] P. Deligne, letter to Bloch, 2.02.1984.

[Dr] W. G. Drinfeld, On quasitriangular quasi-Hopf algebras and a group closely connected with $Gal(\bar{\mathbb{Q}}/\mathbb{Q})$, *Leningrad Math. J.* **2 (4)**, (1991), 829 –860.

[G] A.Grothendieck, On the De Rham cohomology of algebraic varieties, Math. Publ. I.H.E.S. **29** (1966), 95-103.

[H] R.M.Hain, On a generalization of Hilbert's 21st problem, *Ann. Scient. Ec. Norm. Sup.* 4e série **19** (1986), 609-627.

[HZ] R.M.Hain, S.Zucker, A guide to unipotent variations of mixed Hodge structure, Lecture Notes in Math. **1246**, Springer Verlag, 92 – 106.

[KO] N. Katz and T. Oda, On the differentiation of De Rham cohomology classes with respect to parameters, *J. Math. Kyoto Univ.* **8** (1968), 199 – 213.

[I1] Y. Ihara, Automorphisms of pure sphere braid groups and Galois representations, in *The Grothendieck Festschrift*, Volume II, Birkhäuser, 1990, 353 – 373.

[I2] Y. Ihara, Braids, Galois groups, and some arithmetic functions, Proc. of the ICM 1990 (vol. I) (1991), 99 – 120.

[I3] Y. Ihara, Profinite braid groups, Galois representations and complex multiplications, *Annals of Math.* **123** (1986), 43 – 106.

[L] A.Lichnerowicz, *Théorie globale des connexions et des groupes d'holonomie*, Edizioni Cremonese Roma, 1955.

[MKS] W. Magnus, A. Karrass, and D. Solitar, *Combinatorial Group Theory*, Pure and Applied Mathematics, **XIII**, Interscience Publ., 1966.

[R]   R. Ree, Lie elements and an algebra associated with shuffles, *Annals of Mathematics* **68**(2) (1958), 210 – 220.

[S]   T. A. Springer, Galois cohomology of linear algebraic groups, in Algebraic Groups and Discontinuous Subgroups, Proc. of Symp. in Pure Math. **IX**, AMS (1966), 149 – 158.

[W1]   Z. Wojtkowiak, Cosimplicial objects in algebraic geometry, in *Algebraic K-theory and Algebraic Topology,* Kluwer Academic Publishers, 1993, pp. 287 – 327.

[W2]   Z. Wojtkowiak, Monodromy of polylogarithms and cosimplicial spaces, preprint I.H.E.S., 1991.

[W3]   Z. Wojtkowiak, A note on the monodromy representation of the canonical unipotent connection on $\mathbb{P}^1(\mathbb{C})\backslash\{a_1,\ldots,a_n\}$, preprint MPI, Bonn, 1990.

[W4]   Z. Wojtkowiak, Functional equations of iterated integrals with regular singularities, *Nagoya Math. J.* **142** (1996), 145-159.

[W5]   Z. Wojtkowiak, Non-abelian unipotent periods, monodromy of iterated integrals, RIMS-969, February 1994 preprint.

Université de Nice-Sophia Antipolis   Research Institute
Département de Mathématiques   for Mathematical Sciences
Laboratoire Jean Alexandre Dieudonné   Kyoto University
U.R.A. au C.N.R.S., No 168   Kitashirakawa, Sakyo-ku,
Parc Valrose - B.P.N° 71   Kyoto 606, Japan
06108 Nice Cedex 2, France

Part IV. Universal Teichmüller theory

The Universal Ptolemy Group and Its Completions

Robert C. Penner*

§0. Introduction

A new model of a universal Teichmüller space was introduced in [P2], and a universal analogue of the mapping class groups, called the "universal Ptolemy group", was defined and studied. In our concentration on the geometry in [P2], we perhaps obfuscated the essentially easy algebraic and combinatorial arguments underlying this definition, and one goal here is to give a gentle survey of these aspects and a complete definition of the universal Ptolemy group. To be sure, the geometric side of the story provides the main calculational tools of the theory (and explains the terminology "Ptolemy group"), but we do not discuss these aspects here.

To hopefully gain some perspective on the results in this paper, we next review the "universal (decorated) Teichmüller theory" from [P2]. In this context, the appellate "universal" means that we seek certain infinite-dimensional spaces (actually, we shall find infinite-dimensional symplectic Fréchet manifolds supporting a group action) together with maps of each of the "classical (decorated) Teichmüller spaces" into our universal object in such a way that classical combinatorial, topological, and geometric structures arise as the pull-back or restriction of corresponding (group invariant) structures on the universal objects.

The classical "(decorated) Teichmüller theory" may be described as follows. Fix a smooth oriented surface F_g^s of genus $g \geq 0$ with $s \geq 0$ punctures or distinguished points, and consider the corresponding *Teichmüller space* $\mathcal{T}_g^s$ consisting of all complete finite-area metrics of constant Gauss curvature -1 on F_g^s modulo push-forward by diffeomorphisms of F_g^s which are homotopic to the identity. The *moduli space* $\mathcal{M}_g^s$ is likewise defined to be the collection of all such metrics modulo push-forward by all orientation-preserving diffeomorphisms of F_g^s. The *mapping class group* MC_g^s consists of all homotopy classes of orientation-preserving diffeomorphisms of F_g^s, and MC_g^s thus acts on $\mathcal{T}_g^s$ with quotient $\mathcal{M}_g^s$.

In case $s \geq 1$, we may consider the *decorated Teichmüller space* $\tilde{\mathcal{T}}_g^s$ which is by definition the total space of the $\mathbb{R}_+^s$-bundle $\tilde{\mathcal{T}}_g^s \to \mathcal{T}_g^s$, where the fiber over a point of $\mathcal{T}_g^s$ is the set of all s-tuples of horocycles, one horocycle about each puncture of F_g^s. MC_g^s evidently also acts on $\tilde{\mathcal{T}}_g^s$ by simultaneously

* Partially supported by the National Science Foundation

pushing-forward metrics and permuting horocycles, and the quotient of $\tilde{\mathcal{T}}_g^s$
by this MC_g^s-action is the *decorated moduli space* $\tilde{\mathcal{M}}_g^s$.

The sequel does not actually require any real mastery of these classical
constructions, and we have included the precise definitions here only for
completeness (see [P1] for further details and references) and to describe
the following *classical square of* F_g^s

$$\tilde{\mathcal{T}}_g^s \;\longrightarrow\; \mathcal{T}_g^s$$

$$\downarrow \qquad\qquad \downarrow$$

$$\tilde{\mathcal{M}}_g^s \;\longrightarrow\; \mathcal{M}_g^s$$

where each morphism in the square is just a corresponding forgetful map,
so MC_g^s acts transitively on the fibers of the vertical maps. We shall study
here a corresponding *universal square*

$$\widetilde{\mathcal{T}ess} \;\longrightarrow\; \mathcal{T}ess$$

$$\downarrow \qquad\qquad \downarrow$$

$$\widetilde{\mathcal{M}od} \;\longrightarrow\; \mathcal{M}od$$

relating our corresponding universal spaces and (equally to the point) define a countable group G, the "universal Ptolemy group", one of whose
completions acts transitively on the fibers of the vertical maps.

There are in fact several reasonable completions of G as we shall discuss.
Indeed, it seems clear (cf. [LS] in this volume) that the Galois group of
the algebraic closure of $\mathbb{Q}$ over $\mathbb{Q}$ (to be denoted here simply *Galois*) arises
as a group of automorphisms of a suitable completion of G explaining our
exposition of this material here.

To abide by our stated goal to gently survey, we shall only highlight
various aspects of the theory restricting attention to the righthand side
$\mathcal{T}ess \to \mathcal{M}od$ of the universal square for simplicity. This is already sufficient
to describe the group G, which is our primary focus in this paper. We define
the spaces $\mathcal{T}ess$ and $\mathcal{M}od$ in §1 and then define and discuss the group
G in §2 taking this opportunity to describe a few recent related results
of Imbert-Kontsevich and Lochak-Schneps. §3 and §4 contain complete
proofs of several new results; specifically, §3 describes a natural completion
of G which acts transitively on the fibers of the map $\mathcal{T}ess \to \mathcal{M}od$ and
improves our discussion of this in [P2;§4], while §4 describes some related
transitivity results which give a new structure theorem for normalizers of
Fuchsian groups and suggest still other completions of G as we shall discuss.

It is a pleasure to acknowledge helpful conversations on various aspects of this work with Bob Guralnick, Pierre Lochak, Leila Schneps, Vlad Sergiescu, and Dennis Sullivan.

§1. Tesselations

A *tesselation* $\tau = \{\gamma_i\}$ is a countable and locally-finite family of geodesics γ_i, $i \geq 1$, in the Poincaré disk $\mathbf{D}$ so that each component of $\mathbf{D} - \cup\tau$ is an ideal triangle in $\mathbf{D}$. More explicitly, each γ_i is a bi-inifinite geodesic with well-defined endpoints on the circle S_∞^1 at infinity, the γ_i are pairwise disjoint and decompose $\mathbf{D}$ into triangles whose vertices lie at infinity, and this decomposition of $\mathbf{D}$ is furthermore locally-finite in $\mathbf{D}$. We refer to a choice of edge γ_i together with a specification of orientation on it as a "distinguished oriented edge" or a "doe" for short, and we typically denote a tesselation τ together with a doe by τ'. As a further point of notation, we let $\tau^0 \subseteq S_\infty^1$ denote the collection of vertices at infinity of the triangles complementary to τ, so τ^0 is some countable dense subset of the circle at infinity.

There is a standard classical tesselation τ_*, called the *Farey tesselation* and illustrated in Figure 1, defined by reflections about the sides of a fixed ideal triangle in $\mathbf{D}$. Explicitly, we have labeled three of our vertices $0 = \frac{0}{1}$, $1 = \frac{1}{1}$, $\infty = \frac{1}{0}$, and these determine a suitable corresponding triangle complementary to τ_*. We specify that the doe is the oriented edge running from $\frac{0}{1}$ to $\frac{1}{0}$ (i.e., a diameter of $\mathbf{D}$) to determine the standard Farey tesselation τ_*' with doe.

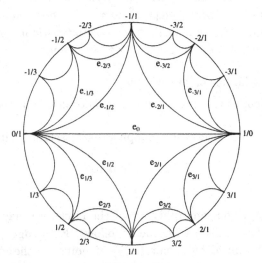

Figure 1

As is well-known [Ra], iteratively labeling the ideal vertices τ^0_* of τ'_* by $\frac{p}{q}, \frac{r}{s} \mapsto \frac{p+r}{q+s}$ as illustrated in Figure 1 gives a bijective enumeration of the rationals. (Indeed, the problem of giving a one-to-one enumeration of the rationals was first solved in this manner by the mineralogist Farey without proofs, which were essentially immediately supplied by Cauchy.) Indeed, we say that $\frac{0}{1}, \frac{1}{1}, \frac{1}{0}, -\frac{1}{1}$ are the "first generation" of the enumeration, and each application of the addition law above increases the "generation" by one. For instance, $\pm\frac{1}{2}$ and $\pm\frac{2}{1}$ are the generation two points. Thus, τ^0_* is canonically identified with the set of rational points in the circle at infinity which comes equipped with its Farey enumeration.

Furthermore, the Farey enumeration of τ^0_* also determines a one-to-one enumeration of τ_* itself by $\mathbb{Q} - \{+1, -1\}$ as illustrated in Figure 1; namely, given a triangle t complementary to τ_*, one edge of t, call this edge e, separates the other two edges of t from the doe, and we label e by the Farey number of the vertex opposite e in t (letting 0 denote the doe of Farey by convention). We shall require this enumeration in §3 below.

Now, if τ' is any tesselation with doe, then there is a unique mapping $f_{\tau'} : \tau^0_* \to \tau^0$ defined iteratively as follows. The respective initial and terminal points of the doe of τ'_* (that is, the points $\frac{0}{1}$ and $\frac{1}{0}$) are mapped by $f_{\tau'}$ to the corresponding endpoints of the doe of τ'. Next, map the vertex of the triangle to the right of the doe in τ'_* to the vertex of the corresponding triangle to the right of the doe in τ'; continue in this way a generation at a time to iteratively define the function $f_{\tau'}$. Not only is $f_{\tau'}$ evidently a bijection between the dense sets τ^0_* and τ^0, but it is also order-preserving (for the counter-clockwise ordering on S^1_∞) by construction. It follows from elementary topology and one-dimensional dynamics that $f_{\tau'}$ interpolates a homeomorphism of S^1_∞ which maps τ^0_* to τ^0; this homeomorphism of the circle is also denoted $f_{\tau'}$ and is called the *characteristic mapping* of τ'.

Formalizing part of the discussion above, we define

$$\mathcal{T}ess' = \{\text{tesselations of } \mathbf{D} \text{ with doe}\},$$

$$Homeo_+ = \{\text{orientation} - \text{preserving homeomorphisms of the circle } S^1_\infty\},$$

so the assignment of characteristic mapping to a tesselation with doe induces a function

$$\mathcal{T}ess' \to Homeo_+$$

$$\tau' \mapsto f_{\tau'}$$

Conversely, given a homeomorphism f of the circle, we may define a corresponding tesselation $f(\tau')$ with doe as follows. Each edge γ of the Farey tesselation is determined by a pair $\{x, y\}$ of points in the circle, there is an associated "image" geodesic $f(\gamma)$ in $\mathbf{D}$ spanned by $\{f(x), f(y)\}$, and we take the "image"

$$f(\tau') = \{f(\gamma) : \gamma \in \tau_*\},$$

with doe given by the edge from $f(\frac{0}{1})$ to $f(\frac{1}{0})$. It is straight-forward to check that $f(\tau')$ is actually a tesselation and that the assignment $f \mapsto f(\tau')$ is the inverse of the characteristic mapping. We have proved

Theorem 1. *The characteristic mapping* $\tau' \mapsto f_{\tau'}$ *induces a bijection between* $Tess'$ *and* $Homeo_+$.

There are immediately a myriad open questions, for instance, specify some family of orientation-preserving homeomorphisms of the circle and characterize the corresponding family of tesselations.

Now, the Möbius group $M\ddot{o}b$ (i.e., the group of orientation-preserving isometries of $\mathbf{D}$, a group isomorphic to $PSL(2,\mathbb{R})$) acts on the left on both $Tess'$ and $Homeo_+$. We finally define the *universal Teichmüller space*

$$Tess = Tess'/M\ddot{o}b \equiv Homeo_+/M\ddot{o}b,$$

where the topology is induced as the quotient of the pull-back by the characteristic mapping of the compact-open topology on $Homeo_+$.

Since the Möbius group acts transitively on ordered triples of points in the circle, we may simply consider "normalized" representatives of Möbius-orbits of tesselations, namely, tesselations where the doe is the edge from $\frac{0}{1}$ to $\frac{1}{0}$ and the triangle to the right of it is the one spanned by $\frac{0}{1}, \frac{1}{1}, \frac{1}{0}$; the corresponding "normalized" homeomorphisms of the circle evidently fix each of these points.

In fact, it is easy to give global coordinates on $Tess$ as follows. Given a tesselation τ', let us specify an edge γ of the Farey tesselation, so there is then a corresponding edge of τ', namely the image $\delta = f_{\tau'}(\gamma)$. Insofar as τ' is a tesselation, δ separates two ideal triangles complementary to τ, which together determine an ideal quadrilateral. One of the vertices of this quadrilateral, call it x, is distinguished in that γ separates it from the triangle to the right of the doe, so we may enumerate the vertices starting from the distinguished one in counter-clockwise order around the circle as x, a, b, c. Recall that the cross ratio of these points is the value

$$\frac{x-a}{x-c}\frac{b-c}{b-a}$$

of x under the fractional linear transformation mapping $a, b, c \mapsto 0, 1, \infty$, and this is a complete $\mathbb{R}_-$-valued invariant of Möbius-orbits of ordered four-tuples of points in the circle.

Assigning in this way a cross ratio of points in τ^0 to each edge of Farey descends to a mapping

$$E : \mathcal{T}ess \to (\mathbb{R}_-)^{\tau_\bullet}$$

by Möbius invariance of cross ratios. Furthermore, since cross ratios are a complete invariant, we conclude that E is an injection. In fact, we have

Theorem 2. *Giving* $(\mathbb{R}_-)^{\tau_\bullet}$ *the product topology, E is an embedding onto a path connected open set, so $\mathcal{T}ess$ has naturally the structure of a Fréchet space.*

In fact, it turns out that $\mathcal{T}ess$ also has a canonical formal symplectic structure as we shall amplify momentarily.

Insofar as $\mathcal{T}ess$ has been identified with $Homeo_+/M\ddot{o}b$, we can compare it with other such models, namely, with *Bers' universal Teichmüller space* $Homeo_{qs}/M\ddot{o}b$, where $Homeo_{qs}$ denotes the space of quasi-symmetric homeomorphisms of the circle (that is, boundary values of quasi-conformal homeomorphisms of the disk $\mathbf{D}$), and with the L^2 *universal Teichmüller space* $Diff_+/M\ddot{o}b$, where $Diff_+$ denotes the space of orientation-preserving diffeomorphisms of the circle (and the quotients are again by the left action of $M\ddot{o}b$). We therefore have the inclusions

$$Diff_+/M\ddot{o}b \to Homeo_{qs}/M\ddot{o}b \to Homeo_+/M\ddot{o}b \equiv \mathcal{T}ess,$$

so we are naturally generalizing the other well-known universal Teichmüller spaces; we mention that the spaces in this chain of inclusions are respectively a Hilbert, Banach, and Fréchet space. In fact, these inclusions are also geometrically natural, for instance in the sense that our formal symplectic form on $\mathcal{T}ess$ pulls back under these inclusions to the Kirillov-Kostant two-form on $Diff_+/M\ddot{o}b$. It follows that our formal symplectic form gives an honest (i.e., convergent) symplectic form on the subspace of all $C^{\frac{3}{2}+\epsilon}$ homeomorphisms of the circle.

To complete our description of the righthand side of the universal square, we must still define our model $\mathcal{M}od$ of a universal moduli space, and we shall be content here to describe this simply as a set. As before, we first define the set $\mathcal{M}od'$ to be the collection of all countable dense subsets of the circle together with the specification of three labeled points a, b, c among the countable set. The Möbius group acts on the left on $\mathcal{M}od'$, and we let $\mathcal{M}od = \mathcal{M}od'/M\ddot{o}b$ denote the set of $M\ddot{o}b$ orbits. As before, by taking the points a, b, c to be $\frac{0}{1}, \frac{1}{1}, \frac{1}{0}$, we may consider "normalized" countable dense subsets.

The mapping $\tau' \mapsto \tau^0$ which furthermore takes a, b, respectively, to the initial, terminal point of the doe and c to be the other vertex of the triangle in τ' to the right of the doe descends to a map

$$\mathcal{T}ess \to \mathcal{M}od,$$

which defines the morphism on the righthand side of the universal square. To see that this morphism is a surjection, fix some normalized countable dense set, and enumerate it $x_1, x_2, x_3, \ldots$, where we assume that x_1, x_2, x_3 agree with $\frac{0}{1}, \frac{1}{1}, \frac{1}{0}$. At each generation of the Farey enumeration, map the new vertices to the least index possible. Specifically, at a given generation of the Farey enumeration, the circle has been decomposed into a finite collection of open intervals. In each interval, choose the x_i with i least among points lying in the interval and map the next generation of vertices of Farey to these points in each interval. This determines a tesselation τ with τ^0 the given countable dense set, as desired.

In fact, the quotient topology on $\mathcal{M}od$ is not Hausdorff. Indeed, let us take the rationals τ_*^0 and choose one irrational point s in the circle. Define a normalized sequence of tesselations τ_i' which agree with the Farey tesselation up to some generation i, include an extra vertex at s, and then continue exhausting all the rationals. Each of these tesselations has $\tau_i^0 = \tau_*^0 \cup \{s\}$ by construction yet the limit in i converges to the Farey tesselation, so the two normalized countable dense sets τ_*^0 and $\tau_* \cup \{s\}$ cannot be separated. With a little more work, one similarly sees that the largest Hausdorff quotient of $\mathcal{M}od$ is a singleton.

§2. The Ptolemy Group

We shall define the group G discussed in the introduction in three stages: groupoid, monoid, and finally group.

To define the groupoid, suppose that τ' is a tesselation with doe and that $\gamma \in \tau$ is an edge of τ. γ separates two triangles in $\mathbf{D} - \tau$, which together with γ comprise an ideal quadrilateral of which γ is a diagonal; let γ^* denote the other diagonal of this quadrilateral. We may alter τ to produce a new tesselation

$$\tau_\gamma = (\tau \cup \{\gamma^*\}) - \{\gamma\}$$

as in Figure 2a. We say that τ_γ arises from τ by applying an "elementary move" along $\gamma \in \tau$. Notice that there is a canonical identification of $\tau - \{\gamma\}$ with $\tau_\gamma - \{\gamma^*\}$, so it makes sense to stipulate that if γ is not the doe of τ', then τ_γ inherits the given doe of τ' as its doe to determine τ_γ'. On the other hand, if γ is the doe of τ', then we specify γ^* as the distinguished edge of τ_γ taken with an orientation determined by the condition that γ, γ^* (in this order) meet with a positive orientation in $\mathbf{D}$. Thus, the elementary move on the doe has order four; see Figure 2b. For later use, two further sequences of moves are given in Figures 2c and 2d.

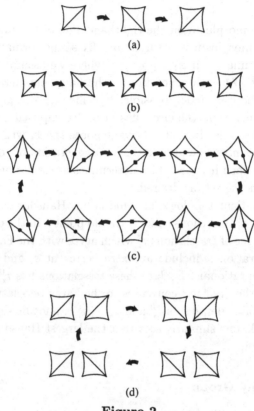

(a)

(b)

(c)

(d)

Figure 2

Define the groupoid Pt' as a category, where the objects are elements of $\mathcal{T}ess'$ and a morphism from σ' to τ' is a finite sequence $\sigma' = \tau'_0, \tau'_1, \ldots, \tau'_{n+1} = \tau'$ of tesselations with doe, where τ'_{i+1} arises from τ'_i by an elementary move along some edge of τ_i, for $i = 0, \ldots, n$. Notice that if there is a morphism between σ' and τ', then $\sigma^0 = \tau^0$. Passing to $M\ddot{o}b$-orbits, we find a category, called the *universal Ptolemy groupoid* Pt, whose objects are given by elements of $\mathcal{T}ess$ and whose morphisms are given by $M\ddot{o}b$-orbits of sequences of tesselations with doe differing by elementary moves as above.

To pass to a monoid, observe that given any $\tau' \in \mathcal{T}ess'$, the characteristic mapping establishes a bijection between the edges τ_* of Farey and the edges of τ. To specify an elementary move along an edge of τ', we might instead simply specify the corresponding edge of τ'_*. Thus, if $q \in \tau_*$, then we let $q \cdot \tau'$ denote the tesselation with doe gotten from τ' by applying the elementary move along the edge $f_{\tau'}(q)$ of τ' corresponding to $q \in \tau_*$ via the characteristic mapping.

Define the free monoid M generated by τ_*, and inductively extend the action in the previous paragraph by setting

$$(q_n \cdots q_2 q_1) \cdot \tau' = q_n \cdot (\ldots \cdot q_2 \cdot (q_1 \cdot \tau') \ldots),$$

thereby defining an action of M on $\mathcal{T}ess'$. Passing to $M\ddot{o}b$-orbits, we find an action of M on $\mathcal{T}ess$ as well. This replacement of a groupoid by a monoid is a general categorical construction applicable whenever the morphisms from any object are a priori identified with some fixed set.

To finally define the group, consider the sub-monoid K of M consisting of those words which act identically on τ'_*, that is, whose corresponding groupoid elements begin and end at the Farey tesselation τ'_* with its fixed doe. For instance, the fourth power of the doe of Farey lies in K as in Figure 2b.

The *universal Ptolemy group* is the quotient $G = M/K$, where we take the quotient of M generated by insertion and deletion of words in K. To see that G is actually a group, it remains only to verify that there are inverses, namely, that K is big enough to allow inverses in G. For instance, the third power of the doe corresponds to the inverse of the doe in G. In the remaining case of an edge other than the doe, return to the earlier notation of a quadrilateral with diagonals γ and γ^* as in Figure 2a; the word in M corresponding to the composition of the move on γ with the move on γ^* lies in K and provides the required inverse in G.

This defines the group G, which we next identify with two other known groups. To this end, suppose that $g \in G$ and consider applying the corresponding elementary moves to the Farey tesselation τ'_* with its doe to produce another tesselation $g \cdot \tau'_*$ with doe. There is a corresponding characteristic mapping $f_g = f_{g \cdot \tau'_*}$, giving a representation

$$G \to Homeo_+$$
$$g \mapsto f_g$$

of G which is faithful by definition (of the relations K). We mention here that there are also several other interesting essentially geometric representations of G (partly discussed in [P2;p. 180-181] and to be taken up further elsewhere).

One can compute directly that in fact the homeomorphism f_g actually lies in the subgroup $PPSL(2, \mathbb{Z})$ of *piecewise* $PSL(2, \mathbb{Z})$ homeomorphisms of the circle. Specifically, if $A \in PPSL(2, \mathbb{Z})$, then there is a finite decomposition of the circle into intervals with rational endpoints, and on each such interval

I, the function A restricts to an element $A|_I$ of $PSL(2, \mathbb{Z})$. Of course, there are conditions on the restrictions $A|_I$ to guarantee that they combine to give a homeomorphism, and in fact, this homeomorphism is then *automatically* once continuously differentiable (cf. [P2; p. 206-207] and [P2;§3], where various properties of $PPSL(2, \mathbb{Z})$ homeomorphisms are discussed).

In a recent DEA memoire [I1] and also in Imbert's contribution [I2] to this volume, we find the following result:

Theorem 3. [Imbert-Kontsevich] *The assignment* $g \mapsto f_g$ *is an isomorphism between the groups* G *and* $PPSL(2, \mathbb{Z})$.

In fact, there is yet another well-studied group (see [I2] and the references therein for a survey) isomorphic to $G \equiv PPSL(2, \mathbb{Z})$, namely, the *Thompson group* Th defined to be the group of piecewise linear homeomorphisms of the circle where the "breakpoints" (in the piecewise structure) are required to be rational numbers of the form $p/2^n$.

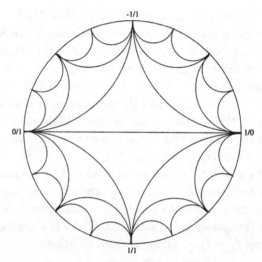

Figure 3

The connection between Th and our current discussion is as follows. Consider the dyadic tesselation τ_2 defined a generation at a time as with Farey, but where at each generation, we take the (Euclidean) midpoint of the interval as the next vertex; see Figure 3. With the usual doe (from $\frac{0}{1}$ to $\frac{1}{0}$) on τ_2, there is a characteristic mapping $f : \tau_2' \to \tau_*'$. This homeomorphism evidently conjugates Th to $PPSL(2, \mathbb{Z})$ in $Homeo_+$, and so we find that furthermore

$$G \equiv PPSL(2, \mathbb{Z}) \equiv Th.$$

Using Theorem 3 and a known presentation of Th, [LS] have given a new essentially geometric presentation of these groups as follows.

Corollary 4. [Lochak-Schneps] *The groups G, $PPSL(2, \mathbb{Z})$ and Th admit a presentation with relations*

$$\alpha^4, \beta^3, (\alpha\beta)^5, [\beta\alpha\beta, \alpha^2\beta\alpha\beta\alpha^2], [\beta\alpha\beta, \alpha^2\beta\alpha^2\beta\alpha\beta\alpha^2\beta^2\alpha^2],$$

where $[\cdot, \cdot]$ denotes commutator and

   $\alpha =$ the elementary move on the doe

   $\beta =$ the first six consecutive frames of Figure 2c where $\bullet$ is the doe.

In other words, the stated relations generate the sub-monoid K discussed above.

To explain the presentation, observe that β just cyclically permutes the doe around the base triangle hence has order 3, while α as the move on the doe has order 4. The word $\alpha\beta$ generates the well-known pentagon relation and has order 5. The remaining two relations correspond respectively to the commutativity of elementary moves on disjoint but contiguous quadrilaterals, and quadrilaterals which are separated by one triangle (cf. Figure 2d).

Another consequence of Theorem 3 is that the group G is simple (since Th is known to be so), and we therefore cannot simply take the profinite completion of G (there is none!) as a candidate completion to support the desired action of Galois. [LS] describes a certain extension of G by braid groups which does indeed support an action of Galois (where the pentagon relation in G corresponds to the pentagon relation in Galois).

In the remaining sections of this paper, we briefly discuss other sensible completions of G which would seem suitable in various contexts but which are not yet known to support any action of Galois.

§3. The Transitive Completion

In analogy to the mapping class groups, we seek a completion of G which acts transitively on the fibers of the map $\mathcal{T}ess \to \mathcal{M}od$. In fact, in [P2] we gave a a sort of completion (to be recalled below) of G, but it was a set-theoretic completion, not a group-theoretic one. We also proposed there a group-theoretic completion $\bar{G}$, which was a "huge" group based on "nets of countable ordinals".

We find here that the construction in [P2] is already essentially enough to give a suitable group-theoretic completion $\tilde{G}$ of G based on a much smaller index set. We refer to $\tilde{G}$ as the "transitive completion" since it has the required transitivity property. Furthermore, we go on to identify $\tilde{G}$ with a natural group of homeomorphisms of the circle.

To begin, we recall some material from [P2;§4.2]. We consider semi-infinite words $w = \cdots q_2 q_1$ where $q_i \in \tau_*$, and we read these words from right to left as a corresponding sequence of elementary moves as before. Furthermore, using the identification of τ_* with $\mathbb{Q} - \{+1, -1\}$ discussed in §1, we regard w as a sequence of rationals.

We must impose two conditions, and the first one is:

> $w = \cdots q_2 q_1$ is *stable* if for each $n \geq 1$, there is some $L \geq 1$ so that if $\ell \geq L$, then the Farey generation of the rational q_ℓ is at least n; that is, the generation of q_ℓ tends uniformly to infinity in ℓ.

Given a stable semi-infinite word $w = \cdots q_2 q_1$ and $\tau' \in \mathcal{T}ess'$, recursively define $\tau'_{j+1} = q_{j+1} \cdot \tau'_j$, where $\tau'_0 = \tau'$, and let f_j denote the characteristic mapping of τ'_j, for $j \geq 0$. Let $P_n \subseteq \mathbf{D}$ denote the ideal polygon of 2^{n+1} sides whose vertices comprise the set P_n^0 of Farey points of generation at most n and let P_n^1 denote the set of geodesics in τ_* lying in P_n. Since w is stable, for each $n \geq 1$, there is some j_n so that $f_j(e) = f_{j_n}(e)$ for each $j \geq j_n$ and each $e \in P_n^1$. Define

$$Q_n^0 = \{f_{j_n}(x) : x \in P_n^0\},$$
$$Q_n^1 = \{f_{j_n}(e) : e \in P_n^1\},$$

and let $Q_n \subseteq \mathbf{D}$ denote the closed convex hull of Q_n^0. By construction, Q_n^1 determines an ideal triangulation of Q_n which agrees with the restriction of τ_j to Q_n for each $j \geq j_n$, and $Q_m \subseteq Q_n$ for $m \leq n$.

Our second technical condition on a semi-infinite word $w = \cdots q_2 q_1$ is:

> w is *convergent* provided that $\bigcup \{Q_n^0 : n \geq 1\} = \tau^0$; that is, the finite sets $\{Q_n^0\}$ exhaust the countable set τ^0.

We proved in [P2;§4] that stable and convergent semi-infinite words have a well-defined action on $\mathcal{T}ess'$ and $\mathcal{T}ess$, and this action is transitive on the fibers of $\mathcal{T}ess \to \mathcal{M}od$. This basic action and transitivity result will be simply assumed here. (Transitivity essentially follows from the Axiom of Choice and a finite transitivity result to the effect that finite sequences of elementary moves act transitively on triangulations with doe of a finite polygon).

To define the completion $\tilde{G}$, first consider the monoid $\tilde{M}$ consisting of all *finite* concatenations $w = w_n \cdots w_2 w_1$, where each w_i is either a finite word in τ_* or a stable and convergent semi-infinite word in τ_*. In case vu is a subword of w, where u is semi-infinite, we may think of the first letter of v (and notice that both finite and semi-infinite words have a well-defined first letter) as a limit ordinal for the sequence u. Notice that $\tilde{M}$ is the smallest monoid containing both M and the set of stable convergent semi-infinite sequences.

It follows from results already mentioned that there is a well-defined action of $\tilde{M}$ on $\mathcal{T}ess'$ and $\mathcal{T}ess$, and we as usual denote the action of $w \in \tilde{M}$ on $\tau' \in \mathcal{T}ess'$ by $w \cdot \tau'$. Define

$$\tilde{K} = \{w \in \tilde{M} : w \cdot \tau'_* = \tau'_*\},$$

so $\tilde{K}$ is a sub-monoid of $\tilde{M}$. Finally, define the *transitive completion* $\tilde{G}$ of G to be the quotient

$$\tilde{G} = \tilde{M}/\tilde{K},$$

which is defined to be the monoid of equivalence classes of elements of $\tilde{M}$, where the equivalence relation is generated by insertion and deletion of (finitely many) elements of $\tilde{K}$.

Theorem 5. *The transitive completion $\tilde{G}$ is a group which acts transitively on the fibers of the map $\mathcal{T}ess \to \mathcal{M}od$.*

Proof. To see that $\tilde{G}$ is a group, it remains only to verify that $\tilde{K}$ is large enough so that every class in $\tilde{M}$ is invertible, and to this end, it suffices to show separately that the class of each element of G and of each stable and convergent semi-infinite word has an inverse. We have already argued above that elements of G have inverses relative to $K \subseteq \tilde{K}$, so suppose finally that w is a stable and convergent semi-infinite word.

Since $w \cdot \tau'_*$ is a well-defined tesselation τ' with doe and $\tau^0 = \tau^0_*$ by convergence, we may apply the transitivity result from [P2;§4] to produce another stable and convergent semi-infinite word u so that $u \cdot \tau' = \tau'_*$; thus, the class of u is inverse to the class of w, as desired. Finally, transitivity of $\tilde{G}$ follows again from [P2;§4].   q.e.d.

It is worth pointing out that it follows from the proof the every equivalence class in $\tilde{G}$ actually admits a representative w which is a single-letter word in $\tilde{M}$, that is, either $w \in G$ or w is a stable and convergent semi-infinite word (so, in a sense, [P2;§4] was the whole story). It is also worth pointing out that this representative may not always be the most desirable; for instance, if we wish to work equivariantly for a Fuchsian group Γ, then a useful representative $w = w_n \cdots w_2 w_1$ takes each w_i to be a Γ-orbit of elementary moves on the universal cover. We shall discuss these possibilities further in the next section.

A reasonable paradigm for the construction of $\tilde{G}$ is to consider the monoid m consisting of all finite concatenations of words which are themselves either a finite sequence of natural numbers or a "stable" semi-infinite sequence $\cdots n_2 n_1$ of naturals, where a sequence is "stable" if each natural number occurs in it only finitely often. We construct a sub-monoid k of m generated by n^2 for each natural number n and all commutators. The quotient group $g = m/k$ is isomorphic to the operation of symmetric difference on the power set of the natural numbers.

Insofar as $\tilde{G}$ acts on $\mathcal{T}ess'$, given $w \in \tilde{G}$, we might consider the characteristic map $f_w = f_{w \cdot \tau'_*}$ of $w \cdot \tau'_*$. We also define

$$V = \{f \in Homeo_+ : f(\mathbb{Q}) = \mathbb{Q}\},$$

certainly a natural subgroup of $Homeo_+$.

Theorem 6. *The assignment $w \mapsto f_w$ establishes an isomorphism between the groups $\tilde{G}$ and V.*

Proof. First consider an element $w \in \tilde{G}$, so w is represented by $w_n \cdots w_2 w_1$, where each w_i is either finite or stable and convergent semi-infinite. Convergence is used to verify that the characteristic map of each w_i indeed sends $\mathbb{Q}$ to $\mathbb{Q}$, hence functoriality of characteristic mappings finally proves that $f_w \in V$.

On the other hand, suppose that $f \in V$. There is a well-defined "image" $f(\tau'_*) \in \mathcal{T}ess'$ defined as before, so by transitivity of $\tilde{G}$, there is some word w_f with $w_f \cdot \tau'_* = f(\tau'_*)$.

The assignment $f \mapsto w_f$ induces the inverse homomorphism, as required.

$$q.e.d.$$

§4. Normalizers of Fuchsian Groups

Though parts of what we discuss in this section apply in the more general setting of an arbitrary Fuchsian group, we shall restrict ourselves once and for all for simplicity to the case of a torsion-free finite-index subgroup $\Gamma < PSL(2, \mathbb{Z})$.

The surface $F_\Gamma = \mathbf{D}/\Gamma$ is a finite branched cover of the "modular curve" $\mathbf{D}/PSL(2, \mathbb{Z})$, which has the ramification points $0, 1, \infty$ of respective ramification indices $2, 3, \infty$, and the points of F_Γ lying over $\infty \in \mathbf{D}/PSL(2, \mathbb{Z})$ are necessarily punctures of F_Γ since $\Gamma < PSL(2, \mathbb{Z})$ is finite-index. In contrast, the points lying over $0, 1 \in \mathbf{D}/PSL(2, \mathbb{Z})$ are smooth points of F_Γ since Γ is torsion-free.

Furthermore, $PSL(2, \mathbb{Z})$ leaves the Farey tesselation τ_* invariant (cf. [P1; §6]), so τ_* descends to an *ideal triangulation* Δ_Γ of F_Γ, that is, a decomposition of F_Γ into ideal triangles. Indeed, the edges of Δ_Γ are naturally identified with the set of Γ-orbits of edges of τ_*. More generally, if Δ is any ideal triangulation of F_Γ, then we may lift Δ to a tesselation $\tilde{\Delta}$ of $\mathbf{D}$; Γ again acts on $\tilde{\Delta}$, and the edges of Δ are naturally identified with the Γ-orbits of edges $\tilde{\Delta}$. Moreover, for any ideal triangulation Δ of F_Γ, we have $\tilde{\Delta}^0 = \tau_*^0$.

An edge e of an ideal triangulation Δ of F_Γ is of one of the two following types. We may choose a lift $\tilde{e}$ of e to $\mathbf{D}$, and this lift is the common frontier of two triangles t_1, t_2 complementary to $\tilde{\Delta}$ in $\mathbf{D}$; let q denote the quadrilateral spanned by these two triangles. It may be that $t_1 \neq t_2$ (so the projection is an embedding on the interior of q), and in this case, we say that e is *locally separating*. In the remaining case that $t_1 = t_2$, q projects to a once-punctured monogon in F_Γ, and e is the geodesic connecting the puncture to the cusp on the boundary; in this case, we say that e is *locally non-separating*.

Given a locally separating edge e of Δ, we may sensibly perform an elementary move as before on e in F_Γ to produce another triangulation of F_Γ. In analogy to the construction of the universal Ptolemy group, let us specify some doe e of Δ in order to enumerate the edges of Δ by τ_*: Given a specified doe $e \in \Delta$, choose some lift $\tilde{e} \in \tilde{\Delta}$ to get $\tilde{\Delta}' \in \mathcal{T}ess'$, so the characteristic map $\tau_* \to \tilde{\Delta}$ induces a surjection $\tau_* \to \Delta$. Γ-invariance of $\tilde{\Delta}$ shows that this enumeration is independent of the choice of $\tilde{e}$ lying over e. (This enumeration depends only on the combinatorics of edges lying to the left/right of the doe in F_Γ.) In particular, the doe of Farey τ_*' gives rise to a canonical doe of Δ_Γ, so every F_Γ comes equipped with a canonical ideal triangulation Δ_Γ' with doe.

We may specify, as before, a sequence of elementary moves on a triangulation Δ' with doe of F_Γ instead as a sequence of elements of τ_*. We must include also letters corresponding to elementary moves on locally non-separating edges to achieve this *a priori* enumeration of Δ, and we simply think of these letters here and throughout as acting identically on triangulations with doe. Thus, the free monoid M on τ_* acts naturally on the set of triangulations with doe of F_Γ, and we may build the kernel K_Γ corresponding to all sequences of moves which act identically on Δ_Γ' to define the *Ptolemy group* $G_\Gamma = M/K_\Gamma$ of Γ.

Lemma 7. G_Γ *is a group.*

Proof. Again we must simply prove the existence of inverses. To this end, recall from [P1;Proposition 7.1] that finite sequences of elementary moves in F_Γ act transitively on ideal triangulations of F_Γ. Thus, given $w \in M$, we may find $u \in M$ so that $(uw) \cdot \Delta_\Gamma'$ agrees with the tesselation (without doe) underlying Δ_Γ'. We can finally find a word $v \in M$ in the letters α^2 and β

of Corollary 4 above to arrange that $(vuw) \cdot (\Delta'_\Gamma) = \Delta'_\Gamma$, so the class of vu is inverse to the class of w.   q.e.d.

In particular, if $e \in \tau_*$ corresponds to a locally non-separating edge in F_Γ, then e lies in K_Γ, and if $e_1, e_2 \in \tau_*$ lie over a common edge of F_Γ, then $e_1 = e_2$ in G_Γ. In fact, we are more keenly interested here in another group, a certain extension of G_Γ by Γ to be discussed below.

In case $\tilde{e}_1$ and $\tilde{e}_2$ are edges of $\tilde{\Delta}'$ lying over a locally separating edge e of Δ, then the elementary moves on $\tilde{e}_1$ and $\tilde{e}_2$ commute as in Figure 2d. Explicitly, we have

$$f_{\tilde{e}_1}(\tilde{e}_2) \ \tilde{e}_1 \ = \ f_{\tilde{e}_2}(\tilde{e}_1) \ \tilde{e}_2 \ \text{ in } G_\Gamma,$$

where, as before, $f_{\tilde{e}_i}(\tilde{e}_j)$ is the "image" of $\tilde{e}_j$ under the characteristic map $f_{\tilde{e}_i}$ of $\tilde{e}_i \cdot \tau'_*$.

Now, suppose that e is a locally separating edge in a triangulation Δ' of F_Γ with doe. Enumerate the edges $\tilde{e}_i$, $i \geq 1$, of $\tilde{\Delta}$ lying over e regarded as elements of τ_*, and consider the formal semi-infinite word

$$\cdots f_{\tilde{e}_3\tilde{e}_2\tilde{e}_1}(\tilde{e}_4) \ f_{\tilde{e}_2\tilde{e}_1}(\tilde{e}_3) \ f_{\tilde{e}_1}(\tilde{e}_2) \ \tilde{e}_1.$$

Our immediate goal is to show that this is actually a stable and convergent semi-infinite word. Assuming this, the order $\tilde{e}_1, \tilde{e}_2, \ldots$ on the edges lying over e is immaterial according to the commutativity discussed in the previous paragraph; that is, different orders give rise to the same action on tesselations with doe.

On the other hand, it is convenient in the sequel to actually specify a particular order as follows. Choose a fundamental domain $E \in \mathbf{D}$ for Γ which is a connected ideal polygon triangulated by $\tilde{\Delta}$, where we may assume that the doe lies in E and some fixed lift $\tilde{e}_1$ of e lies in the interior of E. We say that E itself is "first generation" and also that the vertices of E are "first generation". Of course, the frontier edges of E are identified in pairs by Γ, and across each frontier edge of E there is therefore some other contiguous fundamental domain; these are the "second generation" fundamental domain, and their vertices taken together form the "second generation" of points in the circle. One continues in this way to associate a generation to each ideal point of $\tilde{\Delta}$ and to each Γ-orbit of E. Let us choose an enumeration $E = E_1, E_2, \ldots$ of the Γ-orbits of E so that the generation of E_{i+1} is at least as great as the generation of E_i, for each $i \geq 1$. In the interior of each E_i, there is a unique lift $\tilde{e}_i$, and we define the Γ-move to be the semi-infinite word

$$w_q^\Gamma = \cdots f_{\tilde{e}_3\tilde{e}_2\tilde{e}_1}(\tilde{e}_4) \ f_{\tilde{e}_2\tilde{e}_1}(\tilde{e}_3) \ f_{\tilde{e}_1}(\tilde{e}_2) \ \tilde{e}_1,$$

where $q \in \tau_*$ corresponds to $e \in \Delta'$ as above.

Lemma 8. *For any edge $q \in \tau_*$ and any Γ, w_q^Γ is a stable and convergent semi-infinite word.*

Proof. If the edge q corresponds to a locally non-separating edge, then w_q^Γ acts identically and is thus stable and convergent by convention. In case the edge q corresponds to a locally separating edge, then adopt the notation of the previous paragraph. Stability is clear just from the existence of the ordering above on fundamental domains by non-decreasing generation. Since $\tilde{e}_i$ lies in the interior of E_i by construction, there is always at least one triangle in E_i with $\tilde{e}_i$ in its frontier and separated from the doe by $\tilde{e}_i$. Thus, the set of points in the circle of generation at most n contains the set of Farey points of generation at most n by induction, so the word is convergent as well. $\hspace{4cm} q.e.d.$

It follows from [P2;§4] and Lemma 8 that for any edge $q \in \tau_*$ and any Γ-invariant tesselation τ' with doe of **D**, there is a well-defined corresponding tesselation with doe $w_q^\Gamma \cdot \tau'$, and we furthermore have its characteristic map $f_{q,\tau'}^\Gamma$.

Consider the normalizer

$$N_{Homeo_+}(\Gamma) = \{f \in Homeo_+ : f\Gamma f^{-1} = \Gamma\}$$

of Γ in $Homeo_+$. It follows from the defining equation $f\Gamma f^{-1} = \Gamma$ that f necessarily maps the set of parabolic fixed points of Γ to itself, that is, f necessarily satisfies $f(\mathbb{Q}) = \mathbb{Q}$. It follows that $f \in V$, so $N_V(\Gamma) = N_{Homeo_+}(\Gamma)$, and we let

$$N(\Gamma) = N_V(\Gamma) = N_{Homeo_+}(\Gamma)$$

denote the normalizer of Γ in either group.

Lemma 9. *For any edge $q \in \tau_*$, any Γ, and any Γ-invariant tesselation τ' with doe, we have $f_{q,\tau'}^\Gamma \in N(\Gamma)$.*

Proof. To see that $f_{q,\tau'}^\Gamma$ normalizes Γ, consider two copies of the universal cover of F_Γ, where Γ acts on each copy, but we take τ' in the first copy and $w_q^\Gamma \cdot \tau'$ in the second copy. $f_{q,\tau'}^\Gamma$ induces a topological mapping on the universal covers which sends τ' to $w_q^\Gamma \cdot \tau'$, and so $f_{q,\tau'}^\Gamma$ conjugates Γ to itself. $\hspace{4cm} q.e.d.$

We next show that conversely any $f \in N(\Gamma)$ can be expressed as the characteristic map of a finite composition $w_n \cdots w_2 w_1$, where each w_i is a

Γ-move. The proof again depends upon [P1; Proposition 7.1] as in Lemma 7.
One may think of this result as a kind of structure theorem for normalizers
of Fuchsian groups. At the same time, these results together may be seen
as a sort of converse to Nielsen's famous theorem that a homeomorphism
of a surface F_Γ lifts to a mapping on **D** which extends continuously to a
homeomorphism of S^1_∞ normalizing Γ.

Given a Γ-move w_q^Γ, where $q \in \tau_*$, we may regard the stable and convergent semi-infinite word w_q^Γ as acting on τ'_* itself. Letting M as usual denote
the free monoid over τ_*, an element $w = q_n \cdots q_2 q_1 \in M$ now acts on τ'_* by

$$ w \cdot \tau'_* = w_{q_n}^\Gamma \cdot (\ldots \cdot w_{q_2}^\Gamma \cdot (w_{q_1}^\Gamma \cdot \tau'_*) \ldots). $$

Define $\tilde{K}_\Gamma = \{w \in M : w \cdot \tau'_* = \tau'_*\}$ and the Γ *covering Ptolemy group*

$$ \tilde{G}_\Gamma = M/\tilde{K}_\Gamma. $$

One sees that $\tilde{G}_\Gamma$ is indeed a group arguing just as in the proof of Lemma 7.
Furthermore, $\tilde{G}_\Gamma$ acts on the set of Γ-invariant tesselations with doe as
before, and we have an exact sequence

$$ 1 \to \Gamma \to \tilde{G}_\Gamma \to G_\Gamma \to 1. $$

If $w \in \tilde{G}_\Gamma$, then we let f_w denote the characteristic map of $w \cdot \tau'_*$.

Theorem 10. *The assignment $w \mapsto f_w$ establishes an isomorphism between
the groups $\tilde{G}_\Gamma$ and $N(\Gamma)$.*

Proof. Using Lemma 9, the mapping is evidently an injective homomorphism. To see that it is surjective, choose a fundamental domain for Γ, and
take its "image" under f as before, for any fixed f in $N(\Gamma)$. This is again
a fundamental domain for Γ, as one can check, and it comes equipped with
an induced triangulation with doe.

Now, as in the proof of Lemma 7, there is a sequence of Γ-moves taking
Farey to its "image" under f and mapping the doe correctly, so surjectivity
follows.

Finally, use that f normalizes Γ to prove that this construction is independent of the choice of fundamental domain.   *q.e.d.*

Relating the constructions of this section to the transitive completion $\tilde{G}$
discussed in §3, we have

Corollary 11. *For any Γ, the assignment $M \to \tilde{M}$ induced by $q \mapsto w_q^\Gamma$
descends to a canonical injection $N(\Gamma) \equiv \tilde{G}_\Gamma \to \tilde{G}$.*

As to completions, there is a homomorphism

$$N(\Gamma_1) \equiv \tilde{G}_{\Gamma_1} \to \tilde{G}_{\Gamma_2} \equiv N(\Gamma_2)$$

whenever $\Gamma_2 < \Gamma_1$ defined as follows. Given $q \in \tau'_*$ and any Γ_1-invariant tesselation τ' with doe, there are induced triangulations with doe Δ'_1 and Δ'_2, respectively, of F_{Γ_1} and F_{Γ_2}. Let $e \in \Delta_1$ correspond to $q \in \tau_*$ and consider the full pre-image of e under the projection $F_{\Gamma_2} \to F_{\Gamma_1}$; choose labels $q_i \in \tau_*$, for $i = 1, \ldots, n$ corresponding to this family of arcs in F_{Γ_2}, and define

$$q \mapsto f_{q_{n-1}\cdots q_2 q_1}(q_n) \ \cdots \ f_{q_2 q_1}(q_3) \ f_{q_1}(q_2) \ q_1$$

to induce the mapping $\tilde{G}_{\Gamma_1} \to \tilde{G}_{\Gamma_2}$. As before, this is well-defined (i.e., independent of the choice of ordering and lifts) provided that e is locally separating in Δ_1.

One interesting family of possibilities for completions of G involves various inverse limit type constructions for the $\tilde{G}_\Gamma$, where, given $\Gamma_2 < \Gamma_1$ and an edge in F_{Γ_1} (i.e., a Γ_1-move), we choose some non-empty collection of edges lying over it in F_{Γ_2} (i.e., a corresponding family of Γ_2-moves); we furthermore require these choices to be compatible (and we might impose various further constraints to get various further completions).

Finally and most naturally, one might take the directed set of torsion-free finite-index groups $\Gamma < PSL(2,\mathbb{Z})$ ordered by reverse inclusion as usual, and consider the direct limit of the $N(\Gamma) \equiv \tilde{G}_\Gamma$ under the maps discussed above; this is another completion of G which Subhashis Nag and I have recently proved to be isomorphic to the commensuarability mapping class group of [Su] (where the base surface required in [Su] is taken to be the modular curve). Furthermore, there are reasons to believe that this direct limit indeed admits a faithful representation of *Galois* as a group of automorphisms.

References

[I1] M. Imbert, Constructions universelles dans la théorie des espaces de Teichmüller, Mémoire de DEA, Grenoble (1994).

[I2] M. Imbert, this volume

[LS] Lochak and Schneps, The universal Ptolemy-Teichmüller groupoid, this volume.

[P1] R. C. Penner, The decorated Teichmüller space of punctured surfaces, *Comm. Math. Phys.* **113** (1987), 299-339.

[P2] —, Universal constructions in Teichmüller theory, *Adv. Math.* **98** (1993), 143-215.

[Ra] H. Rademacher, *Lectures on Elementary Number Theory*, Blaisdell, New York, 1964.

[Su] D. P. Sullivan, "Relating the universalities of Milnor-Thurston, Feigenbaum and Ahlfors-Bers", in the Milnor Festschrift *Topological Methods in Modern Mathematics*, eds. L. Goldberg and A. Phillips, Publish or Perish (1993), 543-563.

Departments of Mathematics and Physics/Astronomy
University of Southern California
Los Angeles, CA 90089

Sur l'isomorphisme du groupe de Richard Thompson

avec le groupe de Ptolémée

Michel Imbert

§0. Introduction.

Le but de cet article est de donner une démonstration de l'observation suivante de M. Kontsevitch: *le groupe universel de Ptolémée G défini par R. Penner dans [Pe1] est isomorphe au groupe de R. Thompson* $\mathrm{PL}_2(S^1)$.

Le groupe $\mathrm{PL}_2(S^1)$ a été découvert par R.Thompson dans un contexte algébrique et ce sont ses notes manuscrites non publiées qui constituent la référence de base. Le groupe G est lui un ingrédient important dans la théorie de Penner des espaces universels de Teichmüller, il possède par exemple des complétions intéressantes (voir l'article de Penner dans ce volume); d'autre part P. Lochak et L. Schneps ont étendu le groupoïde universel de Ptolémée (à partir duquel R. Penner a construit G) en un nouveau groupoïde qui possède une complétion profinie sur laquelle le groupe de Galois absolu agit en tant que groupe d'automorphismes (voir leur article dans ce volume). Nous suivrons le plan suivant: dans une première partie nous rappellerons la définition de $\mathrm{PL}_2(S^1)$ et quelques formes équivalentes de ce groupe, ce qui nous permettra de démontrer dans une deuxième partie qu'il est isomorphe à G, et ceci de deux manières.

Dans tout l'article nous noterons $Homeo_+$ le groupe des homéomorphismes du cercle préservant l'orientation, et $\mathbb{Q}(2)$ l'ensemble des nombres dyadiques de l'intervalle $[0,1]$, ie les nombres de la forme $\frac{q}{2^n}$ compris entre 0 et 1 avec q et n entiers naturels. Enfin nous identifierons souvent S^1 avec $\mathbb{R} \cup \{\infty\}$ et avec $[0,1]$ où 0 et 1 sont identifiés.

Je suis redevable à M. Kontsevitch pour la communication de ce résultat, et je remercie V. Sergiescu, par lequel j'en ai eu connaissance, de m'avoir encouragé à le démontrer.

§1. Le groupe de R.Thompson

§1.1. Définition, survey.

Définition 1.1. $PL_2(S^1)$ est le sous-groupe de *Homeo*$_+$ formé des éléments qui sont affines par morceaux et tels que:

1. les points de coupure appartiennent à $\mathbb{Q}(2)$.
2. l'image de 0=1 appartient à $\mathbb{Q}(2)$.
3. les dérivés de chaque application affine constituant un élément du groupe sont des puissances de 2.

Ce groupe, défini en 1965 par R. Thompson, ainsi que F le groupe des homéomorphismes de [0,1] linéaires par morceaux possédant les propriétés 1 et 3, fut introduit au début par son auteur pour construire des groupes de présentation finie avec un problème de mots non résoluble (voir [M,Th]). Il s'est avéré ensuite posséder des propriétés remarquables et il est apparu dans de nombreux contextes mathématiques. Voici une liste non exhaustive de ces propriétés:

$PL_2(S^1)$ est un groupe infini, simple, de présentation finie. C'est un résultat de R. Thompson qui est démontré dans [CFP], article où l'on trouvera les propriétés basiques de ce groupe.

W. Thurston a trouvé une interprétation de $PL_2(S^1)$ liée aux structures projectives entières par morceaux (voir plus bas).

Dans [GhS] les auteurs classifient les actions de $PL_2(S^1)$ sur le cercle suivant la dynamique topologique de cette action, et donnent une description de ses représentations dans $\mathrm{Diff}_r(S^1)$ pour r≥ 2; d'autre part ils donnent une série de résultats sur l'homologie de ce groupe (par exemple $H_n(PL_2(S^1), \mathbb{Z})$ est de type fini pour tout n) ainsi que diverses relations avec les classes caractéristiques.

Dans les années 80, F est également apparu en théorie de l'homotopie et en homologie des groupes. En particulier, K.S. Brown et R. Geoghegan ont montré que les groupes F et $PL_2(S^1)$ sont de type FP_∞ (voir par exemple [Br]).

§1.2. Le groupe modulaire par morceaux.

Définition 1.2. $PPSL_2(\mathbb{Z})$ est le groupe des homéomorphismes préservant l'orientation de $\mathbb{R} \cup \{\infty\}$ qui sont $PSL_2(\mathbb{Z})$ par morceaux, avec des points de coupure rationnels.

Théorème 1.1. $PPSL_2(\mathbb{Z})$ *est isomorphe à* $PL_2(S^1)$.

Esquisse de la démonstration. Elle repose sur une bijection canonique que nous noterons i entre $\mathbb{Q} \cup \{\infty\}$ et $\mathbb{Q}(2)$ et qui est construite ainsi. On envoie d'abord $0/1$ sur $0 = 1$, $1/1$ sur $1/4$, $\infty = 1/0$ sur $1/2$ et $-1/1$ sur $3/4$. Pour une fraction réduite a/b, appelons $\max(|a|, |b|)$ l'ordre de a/b. Alors si $q \in \mathbb{Q}^+$ est d'ordre n, q s'écrit $\frac{a+c}{b+d}$ où a/b et c/d sont d'ordre au plus $n - 1$ (voir [Ch] pages 6 à 8). Soit $d_1 = i(a/b)$ et $d_2 = i(c/d)$; alors on pose $i(q) = \frac{d_1 + d_2}{2}$, et on procède de même si q est négatif.

Par densités respectives de $\mathbb{Q} \cup \{\infty\}$ et $\mathbb{Q}(2)$ dans $\mathbb{R} \cup \{\infty\}$ et $[0,1]$, et comme i préserve l'ordre, il existe un unique homéomorphisme de $\mathbb{R} \cup \{\infty\}$ dans S^1 dont la restriction à $\mathbb{Q} \cup \{\infty\}$ est i. On démontre alors que

$$I : \mathrm{PPSL}_2(\mathbb{Z}) \to \mathrm{PL}_2(S^1)$$
$$f \mapsto i \circ f \circ i^{-1}$$

est un isomorphisme de groupe. ◇

§1.3. Couples d'arbres.

Les éléments de $\mathrm{PL}_2(S^1)$ peuvent être paramétrés par des couples d'arbres binaires ordonnés munis d'une permutation cyclique des feuilles. Pour ceci la référence est [Br].

Définition 1.3. Un arbre binaire ordonné est un arbre S tel que:

1. S ait un sommet distingué v_0 (appelé la racine) tel que si S n'est pas réduit à v_0 alors v_0 est de valence deux.

2. si v est un sommet de S de valence ≥ 2, alors il y a exactement deux arêtes $e_{v,l}$ et $e_{v,r}$, appelées respectivement arête gauche et arête droite, qui contiennent v et qui ne sont pas contenues dans le chemin d'arêtes allant de v_0 à v.

Les sommets de valence un sont appelés des feuilles et peuvent donc être ordonnés de gauche à droite.

Définition 1.4. Un intervalle standard dyadique (i.s.d) est un intervalle fermé inclus dans $[0,1]$ de la forme $[\frac{a}{2^n}, \frac{a+1}{2^n}]$ où a et n sont des entiers positifs ou nuls. Une partition de $[0,1]$ est dyadique si et seulement si les intervalles de la partition sont des i.s.d.

Il existe un arbre infini dyadique noté A_∞ dont les sommets sont les i.s.d, et une arête est une paire (I, J) d'i.s.d telle que si J est la moitié gauche (resp. droite) de I alors (I, J) est une arête gauche (resp. droite). Nous appelons ici un arbuste un arbre binaire ordonné fini identifié à un sous-arbre binaire ordonné fini de A_∞ (i.e. un sous-arbre fini de racine v_0 dont les arêtes gauches (resp. droites) sont des arêtes gauches (resp. droites) de A_∞). Il y a une bijection canonique entre les arbustes et les partitions dyadiques de [0,1], les feuilles correspondants aux intervalles de la partition.

Exemple.

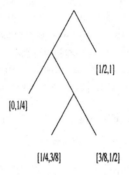

[0,1/4]

[1/2,1]

[1/4,3/8]   [3/8,1/2]

Définition 1.5. Un diagramme arboricole est une paire ordonnée d'arbustes (R, S) qui ont le même nombre de feuilles, munie d'une permutation cyclique des feuilles, σ, modulo la relation d'équivalence engendrée de la manière suivante. Soit (R, S, σ) un tel diagramme, I la n-ième feuille de R et $\sigma(I)$ la feuille correspondante de S. Si on rajoute à R n'importe quel arbre binaire B ordonné à la racine I et si on rajoute à S le même arbre à la racine $\sigma(I)$, alors on dira que (R, S, σ) est équivalent à $(R \cup B, S \cup B, \sigma')$ où σ' se déduit de σ et de l'identité sur B.

Si $f \in \mathrm{PL}_2(S^1)$, alors il existe des partitions dyadiques P et Q telles que f soit affine sur les intervalles de P, et les envoient sur les intervalles de Q (voir lemme 2.2 de [CFP]). On peut donc associer à f un diagramme arboricole (R, S, σ) où R est l'arbuste associé à P et S celui qui est associé à Q. Si $f(0) = 0$ alors σ=id.

Notons DA l'ensemble des diagrammes arboricoles. Si $(Q, R, \sigma) \in$DA et $(R, T, \tau) \in$DA correspondent respectivement à f et g alors $(Q, T, \tau\sigma)$ correspond à $g \circ f$.

Théorème 1.2. DA *est un groupe isomorphe à* $\mathrm{PL}_2(S^1)$.

On en trouve la preuve dans [CFP], où les auteurs donnent aussi une présentation de DA en terme de trois générateurs et six relations, trouvée par R. Thompson (voir la figure suivante où ces trois générateurs sont exprimés dans $\mathrm{PPSL}_2(\mathbb{Z})$, $\mathrm{PL}_2(S^1)$ et DA).

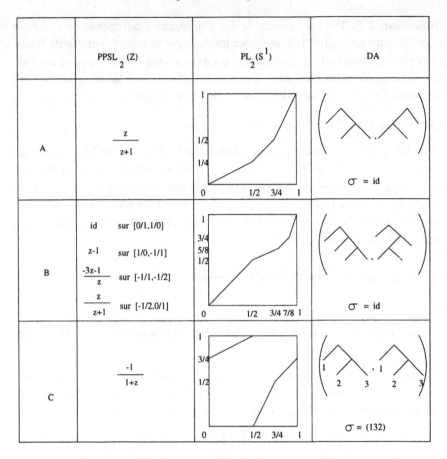

§1.4. Couples de polygônes triangulés.

Le groupe DA est isomorphe à un groupe que nous noterons DP qui consiste en des couples de polygônes triangulés ordonnés munis d'une permutation. Il s'agit d'une construction de dualité classique, voir par exemple [STT].

Soit donc R un arbuste. A chaque sommet on fait correspondre un triangle; en particulier on fait correspondre à la racine de l'arbuste v_0 un triangle de base T_0. A une arête de R entre deux sommets on fait correspondre une instruction de collage des deux triangles correspondants le long d'une arête. On forme ainsi un polygône triangulé $P(R)$. Grâce à la propriété 2 de R, on observe que si T est un triangle de $P(R)$ de valence ≥ 2, alors il y a exactement deux triangles $S_{T,l}$ et $S_{T,r}$ adjacents à T et non contenus dans le chemin de triangles allant de T_0 à T. Les triangles de valence un (i.e. avec deux cotés libres) correspondent aux feuilles de l'arbuste; ils sont dits externes. Ils sont ordonnés de gauche à droite.

Définition 1.6. DP est l'ensemble des diagrammes polygonaux constitués de couples de polygônes triangulés ordonnés avec le même nombre de triangles externes, munis d'une permutation de ces triangles externes, et modulo une relation d'équivalence analogue à celle définie pour les diagrammes arboricoles. On obtient sans difficulté le théorème suivant:

Théorème 1.3. *DP est un groupe isomorphe à DA.*

Un fait important pour l'isomorphisme entre $\mathrm{PL}_2(S^1)$ et le groupe universel de Ptolémée est le résultat suivant de transitivité : étant donnés deux polygônes triangulés à n cotés, il existe une série finie de mouvements élémentaires permettant de passer de l'un à l'autre, où un mouvement élémentaire est un changement de diagonale à l'intérieur d'un quadrilatère (voir [STT] et lemme 4.4 de [Pe1]).

§2. Le groupe universel de Ptolémée.

§2.1. Tesselations.

Le paragraphe 2.1 reprend quelques notions des deux articles de R. Penner cités en références.

Définition 2.1. Une tesselation τ est une collection dénombrable et localement finie de géodésiques du demi-plan de Poincaré $\mathcal{H}$ dont les régions complémentaires sont des triangles idéaux de sommets appartenants à $\mathbb{R} \cup \{\infty\}$.

$\tau^{(0)}$ dénotera les sommets de τ. Ils sont denses dans $\mathbb{R} \cup \{\infty\}$. $\tau^{(2)}$ désigne l'ensemble des triangles complémentaires de τ dans $\mathcal{H}$. Il faut considérer des couples (τ, e) où τ est une tesselation et e est une géodésique distinguée des autres que l'on oriente. L'ensemble de ces couples est noté TESS′ par R. Penner. L'avantage de ces couples réside dans le fait qu'ils sont combinatoirement rigides, ie : si (τ_i, e_i) pour $i = 1, 2$ sont deux tels couples et x_i, y_i sont les sommets de e_i, alors il existe une unique bijection

$$f : \tau_1^{(0)} \to \tau_2^{(0)}$$
$$x_1 \mapsto x_2$$
$$y_1 \mapsto y_2$$

et telle que si (x, y, z) engendre un triangle de $\tau_1^{(2)}$ alors $\big(f(x), f(y), f(z)\big)$ engendre un triangle de $\tau_2^{(2)}$. f respecte l'ordre et détermine une application caractéristique

$$F : \tau_1 \to \tau_2$$
$$e = (x, y) \mapsto F(e) = \big(f(x), f(y)\big).$$

Définition 2.2. L'espace universel de Teichmüller, TESS, dû à R. Penner est défini comme les classes d'équivalence des orbites des couples (τ, e) de TESS$'$ modulo l'action de $\mathrm{PSL}_2(\mathbb{R})$.

Il existe une tesselation particulière, dite de Farey, qui joue un rôle important dans la théorie de Penner. On considère le triangle géodésique de sommets 0, 1, ∞. La tesselation de Farey consiste en les images de ce triangle par $\mathrm{PSL}_2(\mathbb{Z})$.

Le groupe modulaire est alors le sous-groupe de $\mathrm{PSL}_2(\mathbb{R})$ qui préserve cette tesselation. On note τ_* cette tesselation, et son arête orientée notée e_0 est celle qui va de 0 vers ∞. Ce qui est remarquable, c'est qu'il y a une bijection entre $\tau_*^{(0)}$ et $\mathbb{Q} \cup \{\infty\}$: si $T \in \tau_*^{(2)}$ possède deux sommets assignés respectivement à a/b et c/d alors le troisième sommet est assigné à $\frac{a+c}{b+d}$.

Soit maintenant $\mathbb{Q}' = \mathbb{Q} \setminus \{-1, 1\}$. Alors on a la bijection suivante:

$$\mathbb{Q}' \to \tau_*$$
$$0 \mapsto e_0$$
$$q \mapsto e_q,$$

où e_q est définie ainsi: soit $e \in \tau_*$ ($e \neq e_0$), U la composante connexe de $\mathcal{H} \setminus \{e\}$ qui ne contient pas e_0, et T l'unique triangle de $\tau_*^{(2)} \cap U$ ayant e dans sa frontière. Un des sommets de T n'appartient pas à e; il est assigné à un rationnel q, et alors $e = e_q$. Voir la figure suivante représentant la tesselation de Farey.

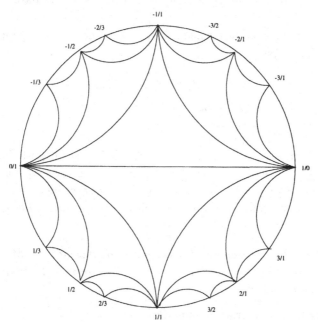

Lemme 2.1. *$Homeo_+/\mathrm{PSL}_2(\mathbb{R})$ et TESS sont des espaces topologiques homéomorphes.*

Esquisse de la démonstration. Si $g \in Homeo_+$, on lui associe la tesselation $(g(e) : e \in \tau_*)$ où si x et y sont les sommets de e alors $g(e)$ est l'unique géodésique joignant $g(x)$ à $g(y)$, munie de l'arête orientée distinguée joignant $g(0)$ à $g\{\infty\}$. D'autre part si $(\tau, e) \in$ TESS', on lui associe l'unique élément de $Homeo_+$ qui prolonge $f : \tau_*^{(0)} \to \tau^{(0)}$. $\diamond$

§2.2. Le groupe universel de Ptolémée et l'isomorphisme remarquable.

Soit $(\tau_1, e) \in$ TESS' et a une arête de τ_1; soit a' l'autre diagonale du quadrilatère formé par les deux triangles ayant a dans leur frontière; enfin soit $F : \tau_* \to \tau_1$ l'application caractéristique et $a = F(e_{q_1})$. Alors $\tau_2 = (\tau_1 \backslash (a)) \cup (a')$ est une autre tesselation. On écrit $\tau_2 = q_1 \cdot \tau_1$ et on dit que l'on a obtenu τ_2 par un q_1-mouvement. Il faut remarquer que $\tau_1^{(0)} = \tau_2^{(0)}$. D'autre part on possède une bijection

$$\Phi_1 : \tau_1 \to \tau_2$$
$$a \mapsto a',$$

et qui vaut l'identité ailleurs. Si l'arête orientée distinguée de τ_1 est différente de a, alors elle sera aussi celle de τ_2; sinon a' devient l'arête distinguée, et on l'oriente de telle façon que (a, a') soit une base directe du plan.

Soit M le monoïde libre formé sur $\mathbb{Q}'$ et soit une action à gauche de M sur TESS donnée par:

$$M \times \mathrm{TESS} \to \mathrm{TESS}$$
$$(q_1 \cdots q_n = m, [\tau_1]) \mapsto [m \cdot \tau_1]$$

où $m \cdot \tau_1 = q_1(\cdots (q_{n-1}(q_n \cdot \tau_1)) \cdots)$.

Ensuite, si $m = q_1 \cdots q_l$ et $n = q'_1 \cdots q'_p \in M$ et si on pose $\Phi_m = \Phi_1 \circ \Phi_2 \cdots \circ \Phi_l$ et $\Phi_n = \Phi'_1 \circ \Phi'_2 \cdots \circ \Phi'_p$, alors m et n sont dits équivalents si et seulement si pour tout $[\tau] \in$ TESS, on a $[m \cdot \tau] = [n \cdot \tau]$ et les correspondances induites Φ_m et Φ_n sont identiques. La classe d'équivalence de $m \in M$ est notée $[m]$.

Lemme 2.2. *G, l'ensemble des classes d'équivalence de M, muni de l'opération $[m][n] = [mn]$ est un groupe, dit universel de Ptolémée, qui agit à gauche sur TESS.*

On trouve la démonstration dans [Pe1].

Voici maintenant le théorème qui motive cet article, initialement démontré dans [Im].

Théorème 2.3. *G est isomorphe à* $\mathrm{PPSL}_2(\mathbb{Z})$, *et par conséquent à* $\mathrm{PL}_2(S^1)$.

Démonstration. Démontrons d'abord l'existence d'un homomorphisme injectif de groupes entre G et $\mathrm{PPSL}_2(\mathbb{Z})$. Dans [Pel], R. Penner donne une représentation fidèle de G dans l'ensemble des applications de τ_* dans τ_*. J'ai remarqué dans [Im] qu'en passant à l'ensemble des sommets $\tau_*^{(0)}$ cela fournissait un plongement de G dans $Homeo_+$. Enfin dans son article dans ce volume, R. Penner résume cela ainsi: soit $g \in G$, et $f_g : \tau_*^{(0)} \to (g \cdot \tau_*)^{(0)} = \tau_*^{(0)}$ l'application correspondante. Elle se prolonge de manière unique en un élément de $Homeo_+$ qui préserve l'ensemble des nombres rationnels et que l'on note encore par f_g. Alors

$$h : G \to Homeo_+$$

$$g \mapsto f_g$$

est une représentation fidèle de G. Je démontre alors que l'image de G est incluse dans le sous-groupe $\mathrm{PPSL}_2(\mathbb{Z})$ de $Homeo_+$. Soit donc $q \in \mathbb{Q}'$, qu'on suppose strictement positif, et Q le quadrilatère formé par les deux triangles possédant e_q dans leur frontière. Il y a une unique géodésique $e_{q'}$ appartenant à la frontière de Q qui est dans la composante de $\mathcal{H} \setminus Q$ contenant e_0. Si $q = c/c'$ est strictement supérieur à $q' = b/b'$, alors on voit facilement que les sommets de Q sont dans l'ordre a/a', b/b', c/c', d/d' avec $c = b + d, c' = b' + d'$ et $b = a + d, b' = a' + d'$. Dans la composante de $\mathcal{H} \setminus Q$ qui contient e_0, y compris sur $e_{q'}$, on a $F_q =$id. Dans les trois autres composantes de $\mathcal{H} \setminus Q$, F_q est une bijection respectant la structure triangulaire d'une partie infinie de τ_* dans une autre partie infinie de τ_*. Comme $\mathrm{Aut}(\tau_*)$ est le groupe $\mathrm{PSL}_2(\mathbb{Z})$, sur chacune de ces trois composantes, y compris sur leur frontière avec Q, F_q est égal à un élément de $\mathrm{PSL}_2(\mathbb{Z})$. On en déduit que f_q est une bijection de $\mathbb{R} \cup \{\infty\}$ qui se comporte comme un élément de $\mathrm{PSL}_2(\mathbb{Z})$ sur chacun des intervalles $[a/a', b/b']$, $[b/b', q]$, $[q, d/d']$, $[d/d', a/a']$, i.e. $f_q \in \mathrm{PPSL}_2(\mathbb{Z})$. On procède de manière analogue si q est strictement inférieur à q'. Enfin si q est négatif la démonstration est également analogue.

Pour prouver que notre homomorphisme est en fait un isomorphisme de groupes, il suffit de trouver $h' : \mathrm{PPSL}_2(\mathbb{Z}) \to G$ tel que $h \circ h' =$id sur $\mathrm{PPSL}_2(\mathbb{Z})$.

Les éléments de $\mathrm{PPSL}_2(\mathbb{Z})$ sans points de coupure sont ceux de $\mathrm{PSL}_2(\mathbb{Z})$; par contre il n'existe pas d'éléments avec un ou deux points de coupure. Soit $g \in \mathrm{PPSL}_2(\mathbb{Z})$ avec au moins trois points de coupure. On lui associe $\tau(g^{-1}) = (g^{-1}(e), e \in \tau_*)$. Soient $r_1, r_2, \ldots, r_p (p \geq 3)$ les p points de coupure de g^{-1}, et posons $s_i = g^{-1}(r_i)$ et $n = \max(\text{ordre}(s_i))$ pour $i \in \{1, \ldots, p\}$; soit P_n le polygône géodésique régulier ouvert dont les sommets sont les rationnels d'ordre au plus n. (L'ordre d'une fraction réduite a/b est égal à $\max(|a|, |b|)$.) Les arêtes frontières de P_n et extérieures sont retrouvées intactes dans $\tau(g^{-1})$ car $\text{Aut}(\tau_*) = \text{PSL}_2(\mathbb{Z})$. Par contre, les arêtes situées à l'intérieur de P_n peuvent être remplacées par de nouvelles arêtes. Grâce au lemme 4.4 de [Pe1] il existe $m(g) \in M$ tel que $m(g) \cdot \tau_* = \tau(g^{-1})$.

On définit $h'(g) = [m(g)]$. Si $g \in \text{PSL}_2(\mathbb{Z})$, on écrit $g = (g \circ h) \circ h^{-1}$, $g \circ h$ et h^{-1} possèdent alors quatre points de coupure si h en possède quatre, et on définit $h'(g) = [m(g \circ h)m(h^{-1})]$.

Soit maintenant $g \in \text{PPSL}_2(\mathbb{Z})$. Alors $h \circ h'(g) = f$, l'unique bijection de rigidité qui envoie l'arête orientée distinguée de $\tau(g^{-1})$ sur e_0 et les triangles de $\tau(g^{-1})^{(2)}$ sur ceux de $\tau_*^{(2)}$. Or par construction g possède ces propriétés donc $h \circ h' =$ id.

Dans la pratique, on peut calculer directement f_q à partir de q; on calcule ainsi que:

$$h([0]) \text{ vaut } \begin{cases} z - 1 & \text{sur } [0/1, 1/1] \\ \frac{z-1}{z} & \text{sur } [1/1, 1/0] \\ \frac{z}{z+1} & \text{sur } [1/0, -1/1] \\ \frac{-1}{z+1} & \text{sur } [-1/1, 0/1]. \end{cases}$$

$h([0])^2$ vaut $-1/z$.

$$h([2/1]) \text{ vaut } \begin{cases} \text{id} & \text{sur } [1/0, 0/1] \\ \frac{z}{z+1} & \text{sur } [0/1, 1/1] \\ \frac{-1}{z+3} & \text{sur } [1/1, 2/1] \\ z - 1 & \text{sur } [2/1, 1/0]. \end{cases}$$

En utilisant ce théorème et la présentation de $\text{PL}_2(S^1)$ déjà citée au §1.3, P. Lochak et L. Schneps ont trouvé une présentation remarquable de G en termes de deux générateurs et cinq relations (voir leur article dans ce volume). Ces deux générateurs sont $[0]$ et $[0][2/1][0]^3[2/1][0]$ qui correspond au générateur C de la présentation précitée.

Corollaire 2.4. *L'homéomorphisme entre $Homeo_+ \setminus \text{PSL}_2(\mathbb{R})$ et TESS est équivariant pour les actions respectives de $\text{PPSL}_2(\mathbb{Z})$ et G. De plus $\text{PPSL}_2(\mathbb{Z})$ (et donc G) agit sur l'espace universel de Bers $Homeo_{qs} \setminus \text{PSL}_2(\mathbb{R})$ où $Homeo_{qs}$ est le sous-groupe de $Homeo_+$ formé des éléments qui sont quasi-symétriques.*

Démonstration. On a les actions suivantes:

$$\text{PPSL}_2(\mathbb{Z}) \times Homeo_+ \setminus \text{PSL}_2(\mathbb{R}) \to Homeo_+ \setminus \text{PSL}_2(\mathbb{R})$$
$$(g, [h]) \mapsto [g \circ h^{-1}]$$

où [h] représente une classe à gauche; et

$$G \times \text{TESS} \to \text{TESS}$$
$$([m], [\tau]) \mapsto [m \cdot \tau].$$

Comme, d'après la preuve du théorème 2.3, on a $\tau(h \circ g^{-1}) = m(g) \cdot \tau(h)$, on a bien l'équivariance. La deuxième partie du corollaire résulte du fait suivant: les éléments de $\text{PPSL}_2(\mathbb{Z})$ sont de classe C^1 donc quasi-symétriques. ◇

On peut voir dans [Pe1], à la fin du §1, l'intérêt de la dernière action du corollaire.

Nous allons maintenant donner une deuxième démonstration du théorème 2.3, suggérée par V. Sergiescu, en démontrant que G est isomorphe au groupe DP formé de couples de polygônes.

Démonstration. Soit $q \in \mathbb{Q}$ strictement positif; on considère le q-mouvement sur τ_*. Alors on a déjà remarqué qu'il existe $n \geq 2$ tel qu'il n'y ait rien de changé à l'extérieur du polygône ouvert P_n. Soit I_n le polygône triangulé formé des géodésiques de τ_* reliant les rationnels positifs d'ordre inférieur ou égal à n; et soit S_n le polygône triangulé formé des géodésiques de τ_* reliant les rationnels négatifs d'ordre inférieur ou égal à n. On associe à q l'élément (R_n, Q_n, id) de DP, où R_n et Q_n sont construits ainsi : soit T un triangle de base, a et b deux de ses arêtes; on obtient R_n en collant T avec I_n (resp S_n) grâce à une instruction de collage de a avec l'arête $(0/1, 1/0)$ (resp. de b avec l'arête $(0/1, 1/0)$); enfin on obtient Q_n en remplaçant I_n par $q \cdot I_n$, le nouveau polygône triangulé obtenu après le q-mouvement. On procède de même pour q strictement négatif, en remplaçant S_n par $q \cdot S_n$. Enfin si $q = 0$, on lui associe l'élément $(R_2, R_2, (1234))$. On en déduit une application de M dans DP, et par définition de G et de son action sur les tesselations on a donc construit un homomorphisme injectif de groupes entre G et DP. La surjectivité de cet homomorphisme découle encore du résultat de transitivité finie cité à la fin du §1.4. Soit (P, Q, σ) un couple de polygônes triangulés ordonnés et une permutation de ses triangles externes; alors en injectant P dans un certain R_n en identifiant le triangle de base avec le triangle T, on sait que Q s'injecte dans le même polygône que P mais avec une triangulation différente, ce qui permet d'appliquer le lemme de transitivité finie, et en repassant à la tesselation de Farey de voir que (P, Q, σ) est l'image d'un élément de G. ◇

Remarque. Par définition l'arbre de Farey consiste en deux exemplaires de l'arbre infini dyadique défini au §1.3 reliés par une arête joignant les racines. On peut le construire également comme dual à la tesselation de Farey. Alors on démontre facilement que le groupe universel de Ptolémée G

est isomorphe au groupe des germes d'automorphismes à l'infini de l'arbre
de Farey. Cette action à l'infini sur l'arbre de Farey est utilisé dans [GrS]
pour construire une extension acyclique du groupe de tresses à partir du
groupe F mentionné au début du §1.1.

Références

[Br]   K.S Brown, Finiteness properties of groups, *J. Pure App. Alg.* **44**
         (1987), 45-75.

[CFP]   J.W. Cannon, W.J. Floyd, W.R. Parry, Notes on Richard Thomp-
         son's groups, Preprint (University of Minnesota, 1994), à paraître
         dans *Ens. Math.*

[Ch]   K. Chandrasekharan, Introduction to analytic number theory, Sprin-
         ger Verlag (1968).

[GhS]   E. Ghys, V. Sergiescu, Sur un groupe remarquable de difféomor-
         phismes du cercle, *Comm. Math. Helv.* **62** (1987), 185-239.

[GrS]   P. Greenberg, V. Sergiescu, An acyclic extension of the braid group,
         Comm. Math. Helv. **66** (1991), 109-138.

[Im]   M. Imbert, Constructions universelles dans la théorie des espaces
         de Teichmüller, Mémoire de DEA, Grenoble (1994).

[M,Th]   R. McKenzie, R.J. Thompson, An elementary construction of un-
         solvable word problems in group theory, *Word problems*, W. W.
         Boone, F.B Cannonito and R.C Lyndon, eds., Studies in Logic and
         the Foundation of Mathematics, Vol. **71**, Amsterdam (1973), 457-
         478.

[Pe1]   R. Penner, Universal constructions in Teichmüller theory, *Adv. in
         Math.* **98** (1993), 143-215.

[STT]   D. Sleator, R. Tarjan, W. Thurston, Rotation distance, triangula-
         tions, and hyperbolic geometry, *J. of the Amer. Math. Soc.* (1988),
         647-681.

The universal Ptolemy-Teichmüller groupoid

Pierre Lochak and Leila Schneps

Abstract

We define the *universal Ptolemy-Teichmüller groupoid*, a generalization of Penner's universal Ptolemy groupoid, on which the Grothendieck-Teichmüller group – and thus also the absolute Galois group – acts naturally as automorphism group. The essential new ingredient added to the definition of the universal Ptolemy groupoid is the *profinite local group* of pure ribbon braids of each tesselation.

§0. Introduction

The goal of this article is to give a completion (by braids) of Penner's *Ptolemy group* G such that there is a natural action of the Grothendieck-Teichmüller group (and a fortiori, the absolute Galois group $\mathrm{Gal}(\overline{\mathbb{Q}}/\mathbb{Q})$) on it. This work was motivated on the one hand by the deep relation of the Ptolemy group – shown to be isomorphic to Richard Thompson's group and to the group of piecewise $\mathrm{PSL}_2(\mathbb{Z})$-transformations of the circle – and mapping class groups and the geometry of moduli spaces in general, most visibly in genus zero, and on the other by the presence of the remarkable pentagonal relation, stimulating the natural reflex of the authors to associate every pentagon appearing in nature to that of the Grothendieck-Teichmüller group $\widehat{GT}$.

The difficulties in defining a $\widehat{GT}$-action on G were the following. Firstly, the profinite version of $\widehat{GT}$ which interests us (mainly by virtue of the fact that it contains the Galois group) acts on profinite groups, whereas via its isomorphism with Richard Thompson's group, G is known to be simple, and therefore its profinite completion is trivial. Furthermore, G contains no braids and $\widehat{GT}$ naturally seems to introduce them into every situation where it appears. The undertaking therefore framed itself as follows: instead of restricting attention to G, is it possible to extend G by braids, in such a way that it is possible to define a $\widehat{GT}$-action on a profinite version of the extension, in such a way that the pentagonal relation of G reflects that of $\widehat{GT}$? The answer turned out to be nearly yes, namely in order to succeed it was necessary to use not the group, but the groupoid interpretation of G, in which its elements are considered as morphisms between marked tesselations (this groupoid is known as the Ptolemy groupoid), and to *relax the condition of the Ptolemy groupoid stating that the group of morphisms from any marked tesselation to itself is trivial to a condition stating that the group of morphisms from any marked tesselation to itself is isomorphic to a certain ribbon braid group.* It is the profinite completion of the ensuing braid-groupoid which admits a $\widehat{GT}$-action.

In §1, we give the definition and presentation of the Ptolemy group, and its interpretation as a groupoid whose objects are marked tesselations of the Poincaré disk. In §2, we recall the definitions and important properties of braid and mapping class groups, and their generalizations to ribbon braid and mapping class groups, which will be the braid groups used to extend the Ptolemy groupoid to the Ptolemy-Teichmüller groupoid. §3 is devoted to the actual construction of the Ptolemy-Teichmüller groupoid $\mathcal{P}_\infty$, with a "physical" interpretation of the new, added groups of non-trivial morphisms from a tesselation to itself via braids of ribbons viewed as hanging from the intervals; at the end of the section we define the profinite completion of the groupoid $\hat{\mathcal{P}}_\infty$ by simply taking the profinite completions of each of the local groups, while preserving the set of objects (i.e. marked tesselations) and the basic (Ptolemy) morphisms from one to another. §4 contains the main theorem of the article (theorem 4) explicitly a $\widehat{GT}$-action on the universal profinite Ptolemy-Teichmüller groupoid $\hat{\mathcal{P}}_\infty$; the role of the two pentagons appears in lemma 5. Finally, in §5 we give a very brief discussion of the relation between the situation considered here and the geometry and arithmetic of genus zero moduli spaces.

This article was motivated by the idea of discovering a link between number theory and Penner's universal Ptolemy groupoid, an idea suggested to us by Bob Penner who had himself had conversation with Dennis Sullivan, and immediately seized upon by us because of the distant echo of the pentagonal defining relation of $\widehat{GT}$ which could be heard (upon listening carefully) when considering Penner's sequence of ten moves giving a fundamental defining relation of the Ptolemy group. We recall with great pleasure the enthusiasm of our early discussions with him about the possible links between those relations; later communal discussions with Dennis Sullivan and Vlad Sergiescu were both enlightening and stimulating. We warmly thank all three.

§1. The universal Ptolemy groupoid

A groupoid is a category all of whose morphisms are isomorphisms. We begin by giving some basic definitions leading to the definition of the universal Ptolemy groupoid from [P1, 4.1] (see also [P2]).

Identify the Poincaré upper half-plane with the Poincaré disk via the transformation $(z - i)/(z + i)$. Traditionally, points on the Poincaré disk are labeled by the corresponding upper half-plane, so that for instance the points -1, $-i$, 1 and i on the unit circle in $\mathbb{C}$ are labeled 0, 1, ∞ and -1, whereas the central point $0 \in \mathbb{C}$ is labeled i. In particular, $\mathbb{P}^1\mathbb{R}$ is wrapped once around the boundary of the disk, the rational numbers of course lie densely in it. We will be particularly concerned with these rational numbers.

Let a *marked tesselation* be a maximal (i.e. triangulating) tesselation of the Poincaré disk such that its vertices lie on the set of rational numbers on the boundary, equipped with a directed oriented edge. The *standard* marked tesselation is the dyadic tesselation T^* with the marked edge from 0 to ∞:

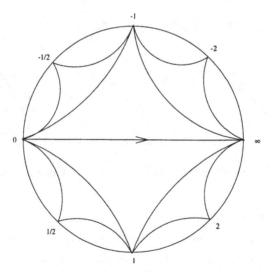

Figure 1. The standard dyadic tesselation T^* with its oriented edge

The *elementary move* on the oriented edge of a tesselation changes it from one diagonal of the unique quadrilateral containing it to the other by turning it counterclockwise; it is of order 4.

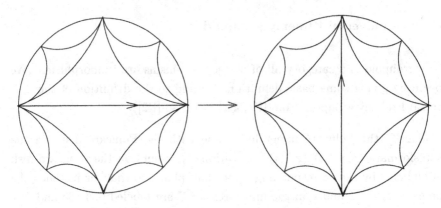

Figure 2. The elementary move on the oriented edge

An *arrow-moving* move on a marked tesselation is an operation on the
tesselation which moves the oriented edge to another edge without changing
the tesselation itself.

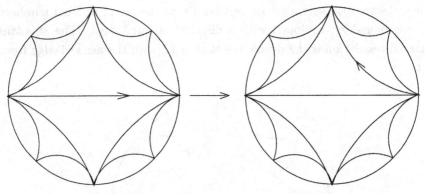

Figure 3. An arrow-moving move

Definition: The *universal Ptolemy groupoid* is the groupoid defined as
follows:

    Objects: The marked tesselations;

    Morphisms: Finite sequences (or chains) of elementary moves on the
                  oriented edge and arrow-moving moves;

    Relations: The only morphism from an object to itself is the trivial one.

Remark: The condition that the local groups of morphisms (i.e. groups of
morphisms from an object to itself) are trivial implies that if T_1 and T_2 are
two marked tesselations, then there is a unique morphism in the groupoid
from T_1 to T_2.

On any marked tesselation T, let α denote the elementary move on its oriented edge, as shown for the standard marked tesselation in Figure 2. Let the triangle to the "left" of the oriented edge of T denote the triangle lying to our left if we imagine ourselves to be lying face down on the tesselation along the oriented edge, with our head in the direction indicated by the arrow, and let β denote the move which sends the arrow counterclockwise to the next edge of the triangle to the left of the oriented edge, as in Figure 3. As noted by Penner, the universal Ptolemy groupoid can be given a group structure, simply because if we write a chain of α's and β's, it can be considered as a morphism on any given tesselation: at each point in the chain, the tesselation being acted on is the one resulting from application of all the previous moves. In particular, given a starting tesselation, a word in α and β uniquely determines a morphism in the groupoid, and conversely, every morphism in the groupoid can be given as a starting tesselation and a word in α and β. Moreover, clearly, if a word in α and β gives the trivial morphism from some tesselation to itself, then the same word will give the trivial morphisms from every tesselation in the Ptolemy groupoid to itself. We want to find exactly which words these are, namely to determine the relations in the group generated by α and β induced by the condition that the group of local morphisms of a tesselation is trivial. In other words, we need to determine the chains of α's and β's which bring a marked tesselation to itself.

Theorem 1. *All words in α and β taking a given tesselation to itself are generated by the following words (where square brackets denotes the commutator):*

$$\alpha^4, \quad \beta^3, \quad (\alpha\beta)^5, \quad [\beta\alpha\beta, \alpha^2\beta\alpha\beta\alpha^2], \quad [\beta\alpha\beta, \alpha^2\beta\alpha^2\beta\alpha\beta\alpha^2\beta^2\alpha^2].$$

Proof. In the contribution to this volume by M. Imbert, it is proved that Penner's group G is isomorphic to Richard Thompson's group. Therefore, it suffices to show that the group, let us call it G, defined by generators α and β and relations as in the statement of the theorem is isomorphic to Thompson's group. Let us give a presentation of Thompson's group which can be found on page 2 of Thompson's unpublished notes [T] (and under a different but recognizable notation, [CFP], lemma 5.2.). It is given by three generators, R, D and c_1, and six relations, namely: $[R^{-1}D, RDR^{-1}] = 1$, $[R^{-1}D, R^2DR^{-2}] = 1$, $c_1 = Dc_1R^{-1}D$, $RDR^{-1}Dc_1R^{-1} = D^2c_1R^{-2}D$, $Rc_1 = (Dc_1R^{-1})^2$ and $c_1^3 = 1$. We define a homomorphism ϕ from Thompson's group to G by setting $\phi(R) = \alpha^2\beta^2$, $\phi(D) = \alpha^3\beta$ and $\phi(c_1) = \beta$. We need to check first that ϕ is really a homomorphism, and second that it is invertible, so an isomorphism. To see that it is a homomorphism it suffices to compute the images of both sides of the six defining relations by ϕ. All are easily seen to hold in G. Indeed, the first two relations are the analogous commutator relations in G, and $c_1^3 = 1$ is sent by ϕ to $\beta^3 = 1$. The two sides of the relation $c_1 = Dc_1R^{-1}D$ are sent to β and $\alpha^3\beta\beta\beta\alpha^2\alpha^3\beta = \beta$ and the two sides of $Rc_1 = (Dc_1R^{-1})^2$ are sent to α^2 and $(\alpha^3\beta\beta\beta\alpha^2)^2 = \alpha^2$. Finally, rewriting the remaining relation as $D^2c_1R^{-2}DRc_1^{-1}D^{-1}RD^{-1}R^{-1} = 1$, we see that the left-hand side is sent by ϕ to

$$\alpha^3\beta\alpha^3\beta \cdot \beta \cdot \beta\alpha^2\beta\alpha^2 \cdot \alpha^3\beta \cdot \alpha^2\beta^2 \cdot \beta^2 \cdot \beta^2\alpha \cdot \alpha^2\beta^2 \cdot \beta^2\alpha \cdot \beta\alpha^2$$

$$= \alpha^2(\alpha\beta\alpha\beta\alpha\beta\alpha\beta\alpha\beta)\alpha^2 = \alpha^2(\alpha\beta)^5\alpha^2 = 1.$$

This shows that ϕ is a homomorphism; to show it is an isomorphism, it suffices to define $\phi^{-1}(\alpha) = c_1D^{-1}$ and $\phi^{-1}(\beta) = c_1$ (indeed, the relation $c_1 = Dc_1R^{-1}D$ shows that Thompson's group is generated by the two elements c_1 and D). $\diamond$

Remark. The generators of this group can be identified with the corresponding moves on marked tesselations in the Ptolemy groupoid. Indeed, the words α^4 and β^3 clearly bring a marked tesselation back to itself and are therefore trivial; similarly $(\alpha\beta)^5 = 1$ corresponds exactly to Penner's trivial sequence of 10 moves (cf. (c) on p. 179 of [P1]; note that $(\alpha\beta)$ simultaneously moves the two diagonals of a pentagon whereas Penner moves one at a time). Finally, the two commutation relations in G imply that that elementary moves on the diagonals of two neighboring quadrilaterals commute, and elementary moves on the diagonals of two quadrilaterals separated only by a triangle; it is a remarkable fact that the commutation of all pairs of elementary moves taking place in disjoint quadrilaterals (Penner's relation (d) on p. 179 of [P]) are consequences of the relations in G.

§2. Ribbon braids

Let us recall the definitions of the Artin ribbon braid groups and the ribbon mapping class groups. First recall the definitions of the usual Artin braid and mapping class groups. The Artin braid group B_n for $n \geq 1$ is generated by $\sigma_1, \ldots, \sigma_{n-1}$ with the relations

$$\sigma_i\sigma_{i+1}\sigma_i = \sigma_{i+1}\sigma_i\sigma_{i+1} \text{ for } 1 \leq i \leq n-2 \text{ and } \sigma_i\sigma_j = \sigma_j\sigma_i \text{ for } |i-j| \geq 2.$$

There is a natural surjection $\rho : B_n \to S_n$ for $n \geq 1$ which induces a natural surjection $\rho : M(0,n) \to S_n$ (for $n = 4$ we have the surjection $\rho :$

$B_3/Z \to S_3$). The kernel of ρ (denoted by K_n in B_n and $K(0,n)$ in $M(0,n)$) is known as the *pure* braid group resp. mapping class group. Both K_n and $K(0,n)$ are generated by the elements $x_{ij} := \sigma_{j-1} \cdots \sigma_{i+1}\sigma_i^2\sigma_{i+1}^{-1} \cdots \sigma_{j-1}^{-1}$ for $1 \le i < j \le n$. The center of B_n and of K_n is cyclic generated by the element $\omega_n = x_{12}x_{13}x_{23} \cdots x_{1n} \cdots x_{n-1,n}$. The mapping class group $M(0,n)$ (resp. the pure mapping class group $K(0,n)$) is the quotient of B_n (resp. K_n) by the following relations:

(i) $\omega_n = 1$;

(ii) $x_{i,i+1}x_{i,i+2} \cdots x_{i,n}x_{i,1} \cdots x_{i,i-1} = 1$; for $1 \le i \le n$, where the indices are considered in $\mathbb{Z}/n\mathbb{Z}$.

The *Artin ribbon braid group* B_n^* is a semi-direct product

$$B_n^* \simeq \mathbb{Z}^n \rtimes B_n;$$

it is generated by generators $\sigma_1, \ldots, \sigma_{n-1}$ of the B_n factor and $t_1, \ldots, t_n$ (all commuting) of the $\mathbb{Z}^n$ factor, with the "semi-direct" relations given by:

$$\begin{cases} \sigma_i t_i \sigma_i^{-1} = t_{i+1} \text{ for } 1 \le i \le n-1 \\ \sigma_i t_{i+1} \sigma_i^{-1} = t_i \text{ for } 1 \le i \le n-1 \\ (\sigma_i, t_j) = 1 \text{ for } 1 \le i \le n-1, \quad j \ne i, i+1 \end{cases}$$

This group is visualized like the usual braid groups, except that the strands are replaced by flat ribbons, so that a twist on any one of them is non-trivial. The subgroup of B_n^* consisting of "flat" braids of the ribbons (i.e. without twists on the ribbons, cf. Figure 4) is canonically isomorphic to B_n, and from now on we identify B_n and its generators σ_i with this subgroup of B_n^*.

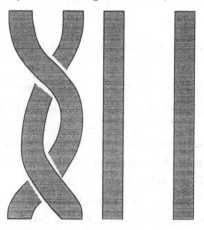

Figure 4. The flat braid x_{12}

We identify $\mathbb{Z}^n$ with the abelian subgroup of B_n^* generated by a full (2π) twist t_i on each of the n ribbons (cf. Figure 5).

Figure 5. A full twist on a ribbon

This visualization corresponds to the definition of B_n^* as a semi-direct product $\mathbb{Z}^n \rtimes B_n$ given above.

Since the twists t_i commute with pure braids, the pure ribbon braid subgroup K_n^* of B_n^* is just a direct product $\mathbb{Z}^n \times K_n$. Let us define the ribbon mapping class group $M^*(0, n)$ to be B_n^* modulo the following relations. Firstly, the center of B_n^* is generated by the element which also generates the center of K_n, namely $\omega_n = x_{12}x_{13}x_{23} \cdots x_{1n} \cdots x_{n-1,n}$, together with the subgroup $\mathbb{Z}^n$. The first relation we quotient out by is

(i') $\omega_n = \prod_i t_i$.

Now, instead of using the usual sphere relations by which we quotient B_n to obtain $M(0, n)$, we use the *ribbon sphere relations*:

(ii') $x_{i,i+1}x_{i,i+2} \cdots x_{i,n}x_{1,i} \cdots x_{i,i-1} = t_i^2$.

The quotient of B_n^* by the relations in (i') and (ii') is the *ribbon mapping class group $M^*(0, n)$*.

The surjection of B_n onto S_n extends to B_n^* by sending the subgroup $\mathbb{Z}^n$ to 1, and the kernel of this surjection is the *pure ribbon braid group K_n^**. The surjection passes to $M^*(0, n)$ and its kernel in this group is denoted by $K^*(0, n)$. Attention: although the group B_n^* is a semi-direct product of B_n with $\mathbb{Z}^n$ and the subgroup K_n^* is a direct product of K_n with $\mathbb{Z}^n$, analogous statements do not hold for $M^*(0, n)$ or $K^*(0, n)$; although these groups are extensions of $M(0, n)$ and $K(0, n)$ respectively by $\mathbb{Z}^n$, the extensions are not split. We refer to [MS], Appendix B for a detailed discussion of these groups in a similar but more geometric context.

Let us give some admirable properties of the ribbon groups.

(1) It is well-known that removing any strand gives a surjection from K_n onto K_{n-1}, which induces a surjection from $K(0,n)$ onto $K(0,n-1)$. There are analogous natural surjections from K_n^* into K_{n-1}^* and from $K^*(0,n)$ into $K^*(0,n-1)$ obtained by removing one ribbon. Removing several ribbons thus gives surjections from $K^*(0,n)$ onto $K^*(0,m)$ for $m < n$.

(2) The braid obtained by holding two adjacent ribbons i and $i+1$ firmly by their bottom ends and twisting them one full turn is equal to $t_i t_{i+1} x_{i,i+1}$. We denote this ribbon braid by $t_{i,i+1}$. It is the same as the full twist on the single "wide" ribbon obtained by sewing the two adjacent ribbons together. The expression for the simultaneous full twist of several adjacent ribbons can easily be deduced from this one by induction.

(3) There are natural injections K_m into K_n for $m < n$ given by dividing up the n strands into m adjacent packets (each of which can consist of one or more strands) called $A_1, \ldots, A_m$; the subgroup of K_n generated by the "flat" braids x_{A_i,A_j} (as in Figure 4, considering each packet as a ribbon) is isomorphic to K_m. Analogously, there are natural injections of K_m^* into K_n^* for any division of the n ribbons into m adjacent packets. Each packet, considered as adjacent ribbons sewn together, forms a "wide ribbon", and the group K_m^* of braids on these wide ribbons is naturally a subgroup of K_n^*.

Now, there is no such natural injection for the pure mapping class groups $K(0,n)$, because the twist on a packet of strands is non-trivial whereas the twist on a single strand is trivial. This problem is eliminated for the ribbon mapping groups where the ribbons and the wide ribbons behave in the same way with respect to twists. Therefore we have such injections for the pure ribbon mapping class groups $K^*(0,m) \hookrightarrow K^*(0,n)$; *this point represents the major advantage of the use of ribbon braid groups with respect to ordinary braid groups.*

§3. The universal Ptolemy-Teichmüller groupoid

The universal Ptolemy-Teichmüller groupoid $\mathcal{P}_\infty$ is a generalization of the universal Ptolemy groupoid in the sense that we add morphisms from a given marked tesselation to itself. The objects of $\mathcal{P}_\infty$ are those of the universal Ptolemy groupoid, namely marked braid tesselations; what we do here is to relax the condition stating that the local groups are all trivial to a condition defining the local groups as certain ribbon braid groups.

To explain exactly what is going on, we visualize a marked tesselation T a little differently; we assume that there is a ribbon hanging from the interval on the circle delimited by each edge of the tesselation. To be precise, each edge of the tesselation actually divides the circle into *two* intervals, and we want to choose only one of them; we choose to hang the ribbon from the interval lying entirely on one side of the oriented edge. This makes sense for every edge except the oriented one, to which we associate the interval lying to the left of it in the sense explained earlier. Note that since the ribbons are determined by the edges of T, assuming their presence does not add anything to T; *the point of adding them is that the non-trivial morphisms from T to itself which we are going to introduce correspond exactly to braiding them.* Before continuing, we note that if one considers the infinite trivalent tree *dual* to the tesselation, then it comes to the same thing to attach a strand to each of its "ends" (the rationals) and consider the set of strands in a given interval as forming a ribbon, and this in turn is equivalent to attaching a strand to each vertex of the tree (uniquely associated to a rational). This idea, due to Greenberg and Sergiescu (cf. [GS]) was one of the starting points of this article.

Consider thus from now on each marked tesselation T to come equipped with its ribbons. Each ribbon is automatically associated to an *interval* on the circle (delimited by an edge of T, a fortiori by two numbers in $\mathbb{P}^1\mathbb{Q}$). Two ribbons of T are said to be *disjoint* if their intervals are disjoint except for at most one common endpoint. They are said to be *neighbors* if their associated intervals are disjoint except for exactly one common endpoint. Two ribbons of T are said to be *adjacent* if their intervals are delimited by two sides of a triangle of the tesselation; thus, adjacent ribbons are of course neighbors and neighbors are disjoint, but the converses are not necessarily true.

Two neighboring ribbons of a given marked tesselation can always be made into adjacent ribbons of another tesselation which can be obtained from the first by a finite number of elementary moves, successively reducing to zero the (finite) number of edges coming out of the common endpoint of the two ribbons and lying inside the smallest polygon of the tesselation having as two of its edges those associated to the ribbons. On the left-hand side of Figure 6, we show two neighboring ribbons; the smallest polygon of the tesselation containing having their corresponding edges as edges is a

quadrilateral and there is just one edge coming out of the common endpoint
of the two ribbons and lying inside it (namely, its diagonal). Thus, after
the elementary move on that diagonal, the two ribbons become adjacent
(right-hand side), associated to two sides of the triangle A.

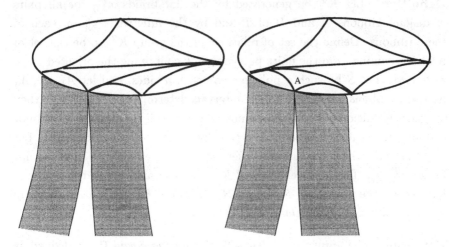

Figure 6. Adjacent and neighboring ribbons of a tesselation

Definition: Let A and B be disjoint ribbons of a given marked tesselation
T. Let t_A^T denote the full twist on A, oriented as in Figure 5, and let x_{AB}^T
denote the flat braid of A and B shown in Figure 7, where the ribbon on the
left-hand side passes in front of the right-hand one (whether the observer
stands inside or outside the tesselation).

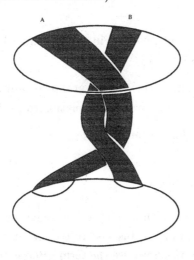

Figure 7. The braid x_{AB}^T

The *local group K^T at a marked tesselation T* is a group of morphisms from T to itself, essentially given by braiding the ribbons associated to intervals of T; it is defined as follows.

Definition. Let K^T be generated by the flat braids x_{AB}^T for all pairs of disjoint ribbons A and B of T and by the full twists t_A^T on each of these ribbons. Define the set of relations in the group K^T to be the set of relations coming from the finite polygons of T containing the oriented edge, as follows. Let S be such a polygon, say with n sides, and let $A_1, \ldots, A_n$ be the n ribbons associated to the intervals determined by the sides; they are pairwise disjoint (which would not be the case if the polygon S did not contain the oriented edge of T and thus lay entirely on one side of it). Let $K^T(S)$ denote the subgroup of K^T generated by the twists $t_{A_i}^T$ and the flat braids x_{A_i, A_j}^T. Then the relations of K^T are generated by all the relations induced by the assertion: *For every S, the group $K^T(S)$ is isomorphic to the pure ribbon mapping class group $K^*(0, n)$.*

Definition: The *universal Ptolemy-Teichmüller groupoid $\mathcal{P}_\infty$* is defined as follows:

Objects: Marked tesselations.

Morphisms: They are of two types: firstly, those of the universal Ptolemy groupoid, which act on marked tesselations as usual, and secondly, the groups $\mathrm{Hom}(T, T) \simeq K^T$ of morphisms from each marked tesselation to itself.

Relations: The full set of relations in the universal Ptolemy-Teichmüller groupoid $\mathcal{P}_\infty$ is given by:

(i)   those of the universal Ptolemy groupoid (any sequence of elementary moves leading from a tesselation to itself is equal to 1);

(ii) those of the local ribbon braid groups;

(iii) commutativity of these two types of morphisms as in equation (1) below.

The universal Ptolemy-Teichmüller groupoid contains the universal Ptolemy groupoid as a subgroupoid, because of (i). Let us explain (ii) and (iii) further. Firstly, from now on we use the term *interval* to denote an interval of the circle delimited by two rationals, and an *interval of T* or equivalently, a *ribbon of T* to denote the interval delimited by an edge of the tesselation T, as before (the one on the opposite side from the oriented edge of T). Not every interval of the circle is an interval of T, of course, but every interval is the union of a finite number of intervals of T; we call such an interval a *wide interval* or a *wide ribbon* of T (obtained by sewing together a finite number of neighboring ribbons of T). Thus every interval of the circle is associated to a ribbon or a wide ribbon of T.

Let A and B denote two disjoint intervals (recall that "disjoint" intervals may have one endpoint in common), equipped with ribbons. Let the braid x_{AB} denote the usual flat braid (as in Figure 7); this twist can be applied to ribbons corresponding to any two disjoint intervals of the circle, without needing to refer to a specific tesselation. However, fixing a tesselation T, we see that the ribbons associated to the intervals A and B are either ribbons or wide ribbons of T, which implies that the braid x_{AB} actually lies in K^T for all T. We write x_{AB}^T when we want to consider the braid x_{AB} *as an element of K^T*.

Recall that given marked tesselations T' and T, there is a unique morphism γ in the universal Ptolemy groupoid from T' to T. This gives rise to a canonical isomorphism between $K^{T'}$ and K^T via $K^{T'} = \gamma^{-1} K^T \gamma$. For all pairs of disjoint intervals A and B of the circle, this isomorphism has the property that

$$x_{AB}^{T'} = \gamma^{-1} x_{AB}^T \gamma \tag{1}$$

in the universal Ptolemy-Teichmüller groupoid. This is what is meant by the commutation of the morphisms of the universal Ptolemy groupoid with braids in (iii) above.

Proposition 2. *Let T be a marked tesselation.*

(i) A set of generators for the group K^T is given by the set of braids x_{AB}^T for all pairs of ribbons A and B corresponding to disjoint intervals of the circle, and the twists t_A^T on the wide ribbons of T corresponding to all intervals of the circle. The set of relations associated to this set of generators is independent of T.

(ii) A smaller set of generators for K^T is given by the twists t_A^T and x_{AB}^T where A and B are disjoint ribbons of T. A set of relations for K^T associated to this set of generators is given by the relations between the generators of each $K^T(S)$ for finite polygons S of T.

(iii) Another set of generators for K^T is given by the set of twists t_A^T on

all ribbons of T and x_{AB}^T where A and B are pairs of adjacent or neighboring wide ribbons of T. Indeed, if S a finite polygon of T containing the oriented edge and the ribbons of S are the ribbons of T associated to the edges of S, then each subgroup $K^T(S)$ is generated by the twists on ribbons of S and the x_{AB}^T where A and B are adjacent or neighboring wide ribbons of S.

Proof. (i) and (ii) are immediate consequences of the definition of K^T. For (iii), we start by showing the statement for $K^T(S) \subset K^T$. By definition, this subgroup is isomorphic to $K^*(0,n)$ where n is the number of edges of the polygon S.

A set of generators for the pure Artin braid group K_n is given by the elements $x_{i\cdots j,j+1\cdots k}$ for all $1 \leq i \leq j < k \leq n$; this braid denotes the flat braid of the "neighboring packets" of strands numbered $i \cdots j$ and $j+1 \cdots k$, which can be considered as ribbons; it looks like the one in Figure 4, with one or several strands in place of the ribbons. To see that this set really generates, it suffices to write each of the usual generators x_{ij} of K_n in terms of these, which can be done via the formula

$$x_{ij} = x_{i\cdots j-1,j}\, x_{i+1\cdots j-1,j}^{-1}$$

(draw the picture!) If $j = i+1$, the usual x_{ij} is itself a twist of neighboring packets, which consist of one strand each.

The flat braids of neighboring packets $x_{i\cdots j,j+1\cdots k}$ also generate the quotient $K(0,n)$ of K_n, so adding in the full twists on ribbons, we have a set of generators for $K^*(0,n)$. Since this group is isomorphic to $K^T(S)$, this proves the statement of (iii) for the groups $K^T(S)$. It follows immediately for K^T since the group K^T is generated by the subgroups $K^T(S)$ as S runs over all the finite polygons of T containing the oriented edge of T.   ◇

Let us now describe the "profinite completion" of the universal Ptolemy-Teichmüller groupoid $\mathcal{P}_\infty$. We need the following:

Lemma 3. *(i) If S and R are finite polygons containing the oriented edge of a given marked tesselation T, and S lies inside R, then the subgroup $K^T(S) \subset K^T$ is contained in $K^T(R)$.*

(ii) For $n \geq 4$, let S_n denote the 2^n-gon in the standard marked dyadic tesselation T^* obtained from S_4, the basic quadrilateral containing the oriented edge 0∞, by successively dividing every edge into two. Then

$$K^{T^*} = \bigcup_{n \geq 2} K^{T^*}(S_n),$$

where this union of groups is given by the natural inclusion of $K^{T^*}(S_n)$ into $K^{T^*}(S_{n+1})$ induced by the inclusion of S_n in S_{n+1}, as in (i).

(iii) Let $\hat{K}^{T^*}(S_n)$ denote the profinite completion of $K^{T^*}(S_n)$ for $n \geq 2$. Then the inclusion of S_n into S_{n+1} induces a natural inclusion of $\hat{K}^{T^*}(S_n)$ into $\hat{K}^{T^*}(S_{n+1})$.

Proof. Part (i) follows from the existence of injections $K^*(0, m) \to K^*(0, n)$ for $m < n$, sending ribbons in $K^*(0, m)$ to wide ribbons in $K^*(0, n)$, cf. property (3) in §2. Part (ii) is a corollary of this, since every polygon of T^* lies inside S_n for some sufficiently large n. Finally, (iii) is a consequence of the fact that like all braid and mapping class groups, the $K^{T^*}(S_n)$ inject into their profinite completions.   ◇

Let us define the *profinite local group* at T^* by

$$\hat{K}^{T^*} := \bigcup_{n \geq 2} \hat{K}^{T^*}(S_n).$$

For any pair of disjoint intervals A, B of the circle, there exists n such that $x_{AB}^{T^*}$ lies in $K^{T^*}(S_n)$, since every rational is a vertex of some S_n. Since $K^{T^*}(S_n)$ injects into its profinite completion, the braid $x_{AB}^{T^*}$ lies in $\hat{K}^{T^*}(S_n)$. The group $\hat{K}^{T^*}$ is topologically described by the same set of generators and relations as K^{T^*}. We define the *profinite local group* at any marked tesselation T to be the one obtained from $\hat{K}^{T^*}$ as in equation (1), i.e. by conjugating by the unique morphism γ in the universal Ptolemy groupoid which takes T^* to T.

Definition: Let the profinite completion $\hat{\mathcal{P}}_\infty$ of $\mathcal{P}_\infty$ be the groupoid defined as follows:

Objects: Marked tesselations T.

Morphisms: All the morphisms of the universal Ptolemy groupoid, together with the local groups $\mathrm{Hom}(T, T) = \hat{K}_T^*$ at each T.

Relations: (i) those of the universal Ptolemy groupoid;

(ii) those of the local profinite braid groups;

(iii) commutativity of these two types of morphisms as in equation (1).

§4. $\widehat{GT}$ and the automorphism group of $\hat{\mathcal{P}}_\infty$

Definition: The automorphism group $\text{Aut}^0(\hat{\mathcal{P}}_\infty)$ of the completed Ptolemy groupoid $\hat{\mathcal{P}}_\infty$ is defined to be the set of automorphisms of the groupoid $\hat{\mathcal{P}}_\infty$ which act trivially on the set of objects.

The goal of this article is to show that $\hat{\mathcal{P}}_\infty$, considered as a completion of the universal Ptolemy groupoid, has the property that the Grothendieck-Teichmüller groupoid lies in its automoprhism group $\text{Aut}^0(\hat{\mathcal{P}}_\infty)$ (cf. §0). To prove this, we begin by recalling the definition of $\widehat{GT}$ (cf. the survey [S] for references and details).

Definition: Let $\underline{\widehat{GT}}$ be the monoid of pairs $(\lambda, f) \in \hat{\mathbb{Z}}^* \times \hat{F}'_2$, satisfying the three following relations, the first two of which take place in the profinite completion $\hat{F}_2$ of the free group on two generators F_2, and the third in the profinite completion $\hat{K}(0,5)$ of the pure mapping class group:

(I) $f(y,x)f(x,y) = 1$;

(II) $f(z,x)z^m f(y,z)y^m f(x,y)x^m = 1$, where $m = \frac{1}{2}(\lambda - 1)$ and $z = (xy)^{-1}$;

(III) $f(x_{34},x_{45})f(x_{51},x_{12})f(x_{23},x_{34})f(x_{45},x_{51})f(x_{12},x_{23}) = 1$.

Under a suitable multiplication law, this set forms a monoid. The group $\widehat{GT}$ is defined to be the group of invertible elements of the monoid $\underline{\widehat{GT}}$. Drinfel'd and Ihara showed that the group $\widehat{GT}$ contains the absolute Galois group $\text{Gal}(\overline{\mathbb{Q}}/\mathbb{Q})$ as a subgroup in a natural way (again, cf. [S] for all relevant references). In particular, whenever $\widehat{GT}$ acts on an object, this object becomes equipped with a Galois action, indicating a – sometimes quite unexpected – link with number theory, which was one of the main motivations behind this article.

Theorem 4. *There is an injection* $\widehat{GT} \hookrightarrow \text{Aut}^0(\hat{\mathcal{P}}_\infty)$.

Proof. Let $F = (\lambda, f) \in \widehat{GT}$. Then we let F act trivially on the set of objects of $\hat{\mathcal{P}}_\infty$. The proof is outlined as follows: first we define the action of F to be trivial on arrow-moving morphisms, next we give its definition on the morphisms of the universal Ptolemy groupoid (considered as a subgroupoid of $\hat{\mathcal{P}}_\infty$) and prove in proposition 5 that the relations of the universal Ptolemy groupoid are respected, and finally we define the action on a set of generators of the profinite local group $\hat{K}^T$ – using lemma 5 to show that the action on the generators is well-defined – and prove in lemma 6 that the action extends to an automorphism of $\hat{K}^T$. This takes care of two of the three types of morphisms in $\hat{\mathcal{P}}_\infty$; to conclude, we show that the commutation relations of equation (1) are respected.

Let us proceed to the definition of the action of F on the morphisms of $\hat{\mathcal{P}}_\infty$.

Arrow-moving morphisms. F acts trivially on these.

Elementary morphisms. The oriented edge of a marked tesselation T determines two pairs of adjacent edges forming a quadrilateral called the *basic quadrilateral of* T; we call the ribbons hanging from the intervals delimited by these four edges X_T, Y_T, Z_T and W_T respectively, going around counterclockwise from the point of the arrow.

Let α_T denote the elementary move on the oriented edge of T. We set

$$F(\alpha_T) = \alpha_T \cdot f(x^T_{X_T Y_T}, x^T_{Y_T Z_T}). \tag{2}$$

The profinite braid $f(x^T_{X_T Y_T}, x^T_{Y_T Z_T})$ is a morphism from T to itself, lying in $\hat{K}^T$, so $F(\alpha_T)$ is a morphism of $\hat{\mathcal{P}}_\infty$.

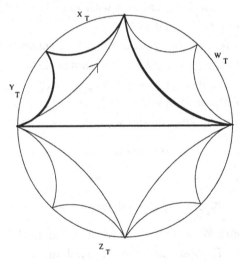

Figure 8. The basic quadrilateral of a marked tesselation

Lemma 5. *This action of* $\widehat{GT}$ *respects all the relations of the universal Ptolemy groupoid.*

Remark. The lemma can be restated by saying that every element of $\widehat{GT}$ is a groupoid-isorphism from the universal Ptolemy subgroupoid of $\hat{\mathcal{P}}_\infty$ to its image. This shows that $\widehat{GT}$ respects the first of the three types of relation in $\hat{\mathcal{P}}_\infty$.

Proof. The point is to check that if $F = (\lambda, f)$ is an element of $\widehat{GT}$, then its action on any two finite sequences of elementary moves leading from T to T' is the same, since in the universal Ptolemy groupoid two such sequences give the same morphism. It is equivalent to check that F respects the relations in the groupoid, and these relations were given explicitly in theorem 1. So we simply need to compute the action of F on the five relations in α and β given in the statement of theorem 1. Here we index the moves according to the tesselation they are acting on, so that the group can be recovered from the groupoid by dropping all indices.

The relation $\beta^3 = 1$ is trivially satisfied since F fixes β. It is also clear that F respects the two commutation relations since they involve braids on disjoint ribbons (or at worst, one braid is made of ribbons all of which are contained in a single ribbon of the other braid) and such braids commute in the local braid groups. As for the pentagon relation $(\alpha\beta)^5 = 1$, or equivalently $(\beta\alpha)^5 = 1$, we have $F(\beta\alpha) = \beta\alpha f(x_{XY}, x_{YZ})$, and five repeated applications of this map, together with the use of equation (1) to push all the factors of $\beta\alpha$ to the left, leave us with exactly the famous pentagon relation (III) defining $\widehat{GT}$, equal to 1.

Let us check the remaining relation, $\alpha^4 = 1$. To start with, fix a marked tesselation T_0, so that α_{T_0} and β_{T_0} are the moves shown in figures 2 and 3. Then by equation (2), $F(\alpha_{T_0}) = \alpha_{T_0} f(x_{X_{T_0} Y_{T_0}}^{T_0}, x_{Y_{T_0} Z_{T_0}}^{T_0})$. Let T_1, T_2 and T_3 denote the tesselations obtained from T_0 via α_{T_0}, $\alpha_{T_1}\alpha_{T_0}$ and $\alpha_{T_2}\alpha_{T_1}\alpha_{T_0}$ (i.e. α, α^2 and α^3 in the group), and write $\alpha_i = \alpha_{T_i}$, so that

$$\alpha_3\alpha_2\alpha_1\alpha_0 = 1.$$

Note that for the four tesselations T_i, we have

$$x_{X_i Y_i}^{T_i} = x_{Z_i W_i}^{T_i} \text{ and } x_{Y_i Z_i}^{T_i} = x_{W_i X_i}^{T_i},$$

where X_i, Y_i, Z_i and W_i are the ribbons attached to the four intervals of the each tesselation T_i shown in figure 8. Applying F to this relation, we obtain

$$F(\alpha^4) = F(\alpha_3\alpha_2\alpha_1\alpha_0) =$$

$$\alpha_3 f(x_{W_3 X_3}^{T_3}, x_{X_3 Y_3}^{T_3}) \alpha_2 f(x_{Z_2 W_2}^{T_2}, x_{W_2 X_2}^{T_2}) \alpha_1 f(x_{Y_1 Z_1}^{T_1}, x_{Z_1 W_1}^{T_1}) \alpha_0 f(x_{X_0 Y_0}^{T_0}, x_{Y_0 Z_0}^{T_0})$$

$$= \alpha_3\alpha_2\alpha_1\alpha_0 f(x_{Y_0 Z_0}^{T_0}, x_{X_0 Y_0}^{T_0}) f(x_{X_0 Y_0}^{T_0}, x_{Y_0 Z_0}^{T_0}) f(x_{Y_0 Z_0}^{T_0}, x_{X_0 Y_0}^{T_0}) f(x_{X_0 Y_0}^{T_0}, x_{Y_0 Z_0}^{T_0})$$

$$= \alpha_3\alpha_2\alpha_1\alpha_0 = \alpha^4 = 1,$$

since f satisfies $f(x, y) = f(y, x)^{-1}$ by relation (I) of $\widehat{GT}$. This takes care of the relation $\alpha^4 = 1$.

$\diamond$

Braids: Let us define the action of $\widehat{GT}$ on the local groups $\hat{K}^T$. We begin by defining an action on a set of generators of $\hat{K}^T$. By (iii) of proposition 2, a set of (topological) generators of $\hat{K}^T$ is given by the braids x_{AB}^T where A and B are neighboring wide ribbons of T, i.e. (finite) unions of neighboring ribbons corresponding to neighboring edges of a finite polygon of T containing the oriented edge. An example is shown in figure 9.

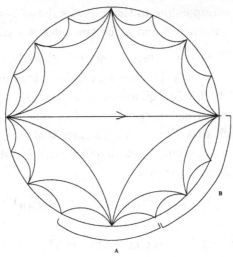

Figure 9. Neighboring intervals corresponding to wide ribbons of T

To define the action of F on all the x_{AB}^T for neighboring wide ribbons A and B of T, it suffices to define it only on the x_{AB}^T for *adjacent* (possibly wide) ribbons A and B (recall that adjacent ribbons are ribbons attached to intervals corresponding to two edges of a triangle of T) for all tesselations T; we can then extend it to pairs of neighboring wide ribbons by equation (1) and the fact that we know the action of F on morphisms of the universal Ptolemy groupoid. This works as follows. Firstly, if A and B are adjacent (possibly wide) ribbons of T we set

$$F(x_{AB}^T) = (x_{AB}^T)^{\lambda}. \tag{3}$$

If A and B are neighboring (possibly wide) ribbons of T, i.e. (unions of) intervals corresponding to edges of a finite polygon S of T, then we change T into another tesselation T' such that the intervals A and B are two edges of a triangle of T', via a *finite number of elementary moves on T, all taking place inside S.* By writing them down explicitly and using (2) and the action of $\widehat{GT}$ on elementary moves, we can compute the explicit expression for $F(x_{AB}^T)$ as follows. Choose a finite sequence of marked tesselations $T^* = T_0, \ldots, T_r$ such that

(1) for $i > 0$, T_i is obtained from T_{i-1} by one elementary morphism g_i on some edge lying inside S (not an edge of S);

(2) each T_i contains the polygon S and is identical to T^* outside of S;

(3) the ribbons corresponding to the intervals A and B are adjacent ribbons of T_r, i.e. the intervals A and B are two sides of a triangle of T_r.

Let $\gamma = g_r \circ g_{r-1} \circ \cdots g_1$; then no matter what choices we make for $T_1, \ldots, T_r$ and $g_1, \ldots, g_r$, γ is the *unique* morphism of the universal Ptolemy groupoid taking T^* to T_r. By equation (1), we have $\gamma^{-1} x_{AB}^{T_r} \gamma = X_{AB}^{T^*}$. By repeated applications of equations (1) and (2), we find an element $\eta \in K^*(0, s) \subset K_{T^*}^*$ such that $F(\gamma) = \gamma\eta$; the fact that η is well-defined is a consequence of lemma 5. Thus, the action of F on $X_{AB}^{T^*}$ when A and B are neighboring wide ribbons is given by

$$F(x_{AB}^T) = \eta^{-1} \gamma^{-1} (x_{AB}^{T_r})^\lambda \gamma \eta = \eta^{-1} (x_{AB}^T)^\lambda \eta. \qquad (4)$$

Step 2: This action of $\widehat{GT}$ on the generators of $\hat{\mathcal{P}}_\infty$ extends to a groupoid automorphism of $\hat{\mathcal{P}}_\infty$. We must check that all the relations of the groupoid are respected.

Lemma 6. *The action defined above of $\widehat{GT}$ on the generators $x_{AB}^{T^*}$ of $\hat{K}_{T^*}^*$ for all pairs of adjacent or neighboring clumps A and B of T^* determines an automorphism of $\hat{K}_{T^*}^*$.*

Proof. Let $F \in \widehat{GT}$. Then the action of F on the generators of each of the groups $\hat{K}^*(0, S_n)$ for $n \geq 2$ extends to an automorphism. Indeed, it is known (cf. [PS], chapter II) that $\widehat{GT}$ is an automorphism group of the pure mapping class group $\hat{K}(0, 2^n)$ in many ways, corresponding to the trivalent trees with 2^n edges; the action we consider here corresponds to the trivalent tree dual to the polygon S_n. Now, we have the exact sequence

$$1 \to \mathbb{Z}^{2^n} \to \hat{K}^*(0, S_n) \to \hat{K}(0, 2^n) \to 1,$$

and it is easily seen that the $\widehat{GT}$-action on $\hat{K}(0, 2^n)$ extends to an automorphism of $\hat{K}^*(0, S_n)$ simply by letting $F = (\lambda, f) \in \widehat{GT}$ act on each twist t_A^T by sending it to $(t_A^T)^\lambda$.

Now let us show that the automorphisms of each $\hat{K}^*(0, S_n)$ given by $F \in \widehat{GT}$ respect the natural inclusions $\hat{K}^*(0, S_n) \hookrightarrow \hat{K}^*(0, S_{n+1})$. Recall that S_{n+1} is obtained from S_n by subdividing each interval of S_n into two. If A and B are neighboring wide ribbons of T, then x_{AB}^T lies in $\hat{K}^*(0, S_n)$ if and only if A and B are actually supported on S_n, i.e. correspond to intervals delimited by a finite number of neighboring edges of S_n. Supposing this is the case, then of course x_{AB}^T is also supported on S_{n+1}, so x_{AB}^T also lies in $\hat{K}^*(0, S_{n+1})$, as it should since $K^*(0, S_n)$ injects into $\hat{K}^*(0, S_{n+1})$. Furthermore, in order to compute $F(x_{AB}^T)$, we need to use a finite series of elementary morphisms as explained in the definition of the $\widehat{GT}$-action on braids, and they consist of moves on edges lying in S_n, and are therefore the same whether x_{AB}^T is considered as lying in $\hat{K}^*(0, S_n)$ or $\hat{K}^*(0, S_{n+1})$, so that the expression of $F(x_{AB}^T)$ is not dependent on n, i.e. F respects the injection $\hat{K}^*(0, S_n) \hookrightarrow \hat{K}^*(0, S_{n+1})$. $\diamond$

Lemma 7. *For A and B disjoint intervals of the circle and T and T' different marked tesselations, the commutativity relations $x_{AB}^{T'} = \gamma^{-1} x_{AB}^T \gamma$ of equation (1) are respected by the action of $\widehat{GT}$.*

Proof. Recall that γ is a finite chain of morphisms in the universal Ptolemy groupoid taking T' to T. In the case where A and B are actually adjacent ribbons for T, the lemma follows immediately from the definition of the action of F on the braids x_{AB}^T. If A and B are not adjacent ribbons for T, it suffices to take a third tesselation T'' such that they are adjacent ribbons for T'' and then again use the definition of F on the braids x_{AB}^T and $x_{AB}^{T'}$, by commuting them to T'' via an element of the universal Ptolemy groupoid. $\diamond$

Lemmas 5, 6 and 7 show that the action of $\widehat{GT}$ respects all defining relations of the universal Ptolemy-Teichmüller groupoid, and this concludes the proof of theorem 4. $\diamond$

§5. Relations with the ordered Teichmüller groupoids

Let us very briefly sketch the relationship between the universal Ptolemy-Teichmüller groupoid and the fundamental Teichmüller groupoids of genus zero moduli space. Let $\mathcal{M}_{0,n}$ denote the moduli space of Riemann spheres with n ordered marked points. The Teichmüller groupoids are the fundamental groupoids $\pi_1(\mathcal{M}_{0,n}; \mathcal{B}_n)$ of the moduli spaces $\mathcal{M}_{0,n}$ for $n \geq 4$ of genus zero Riemann surfaces with n ordered marked points, on the set $\mathcal{B}_n$ of base points *near infinity* of maximal degeneration. The use of these groupoids was suggested in [D], page 847, and their structure was investigated in detail in [PS], chapters I.2 and II. The set $\mathcal{B}_n$ is essentially described by isotopy classes of numbered trivalent trees with n leaves. Let us define *ordered* Teichmüller groupoids to be the fundamental groupoids of the moduli spaces $\mathcal{M}_{0,n}$ on a certain subset $\mathcal{C}_n$ of the base point set $\mathcal{B}_n$, so that the ordered Teichmüller groupoids are subgroupoids of the full Teichmüller groupoids. We define the base point sets $\mathcal{C}_n$ of the ordered Teichmüller groupoids to be the set of base points near infinity in $\mathcal{M}_{0,n}$ corresponding to trivalent trees whose n leaves are numbered in cyclic order $1, \ldots, n$. The set of associativity moves acts transitively on such trees, so that the paths of the ordered Teichmüller groupoids are of two types: the braids (local groups), i.e. the fundamental groups of $\mathcal{M}_{0,n}$ based at each base point, which are all isomorphic to $K(0,n)$, and associativity moves going from one base point to another.

The universal Ptolemy-Teichmüller groupoid covers all the ordered Teichmüller groupoids for $n \geq 4$ in the sense that these groupoids naturally occur as quotients of subgroupoids in many ways. Indeed, for $n \geq 4$, choose a n-sided polygon S in any given tesselation T, and consider the set of elementary paths on T which act only on edges inside the polygon. In other words, consider the finite set of tesselations T' differing from T only inside the polygon S. Now consider the set of marked tesselations obtained from these by marking any chosen edge of T not in the interior of S, and the same edge on the other tesselations differing from T only inside S. This gives a subgroupoid of $\mathcal{P}_\infty$ on a finite number of base points. Now we quotient the local group at each tesselation by suppressing all the ribbons except those attached to intervals delimited by edges of the polygon S. The quotient of K_T^* obtained in this way is exactly the pure mapping class group $K^T(S)$ on these ribbons, isomorphic to $K^*(0,n)$; thus we obtain the ordered Teichmüller groupoid as a quotient of $\mathcal{P}_\infty$. It is shown in [S1] that there is a $\widehat{GT}$-action on the profinite completion of the fundamental groupoid $\pi_1(\mathcal{M}_{0,n}; \mathcal{B}_n)$; this action fixes the objects of the groupoid, i.e. the elements of $\mathcal{B}_n$, so it restricts to an action of $\hat{\pi}_1(\mathcal{M}_{0,n}; \mathcal{C}_n)$. The relation with the main theorem of this article is that our $\widehat{GT}$-action on $\hat{\mathcal{P}}_\infty$ passes to the quotient described here, and gives exactly the usual one on $\hat{\pi}_1(\mathcal{M}_{0,n}; \mathcal{C}_n)$.

References.

[CFP] J.W. Cannon, W.J. Floyd and W.R. Parry, Notes on Richard Thompson's groups, preprint (University of Minnesota, 1994), to appear in *Enseignement Math.*

[D] V.G. Drinfel'd, On quasitriangular quasi-Hopf algebras and a group closely connected with Gal($\overline{\mathbb{Q}}/\mathbb{Q}$), *Leningrad Math. J.* **2** (1991), 829-860.

[GS] P. Greenberg and V. Sergiescu, An acyclic extension of the braid group, *Comm. Math. Helv.* **66** (1991), 109-138.

[IS] M. Imbert, Sur l'isomorphisme du groupe de Richard Thompson avec le groupe de Ptolémée, this volume.

[MS] G. Moore and N. Seiberg, Classical and Quantum Conformal Field Theory, *Commun. Math. Phys.* **123** (1989), 177-254.

[P1] R.C. Penner, Universal Constructions in Teichmüller Theory, *Adv. in Math.* **98** (1993), 143-215.

[P2] R.C. Penner, The universal Ptolemy group and its completions, this volume.

[PS] *Triangulations, courbes arithmétiques et théories des champs*, ed. L. Schneps, issue to appear of *Panoramas et Synthèses*, Publ. SMF, 1997.

[S] L. Schneps, On $\widehat{GT}$, a survey, *Geometric Galois Actions*, volume I.

[S1] ———, On the genus zero Teichmüller tower, preprint.

[T] R. Thompson, unpublished handwritten notes.

URA 762 du CNRS, Ecole Normale Supérieure, 45 rue d'Ulm, 75005 Paris

UMR 741 du CNRS, Laboratoire de Mathématiques, Faculté des Sciences de Besançon, 25030 Besançon Cedex

Errata for

Tame and stratified objects*

Bernard Teissier

I wish to thank Selma Kuhlmann for bringing to my attention some errors and omissions in the text.

Page 234, line 15: a parenthesis is missing after $(f_j)_{j\in J}$.

Page 234, line -11: ....consists only of *finite* unions of intervals.

Page 235, line 12: language

Page 235, line -18: $G_1(x), \ldots, G_i(x)$ (and not $G_p(x)$).

Page 235, line -17: $k[x_1, \ldots, x_{m+i}]$

Page 235, line -13: ....proved the model completeness of the structure...

Page 235, line -12: ....adding *the* exponential ...

---

* This article appeared in volume I of *Geometric Galois Actions*.

Printed in the United States
By Bookmasters